The Language of Rubber

An introduction to the specification and
testing of elastomers

The Language of Rubber

An introduction to the specification and testing of elastomers

L P Smith PhD, DIC

Published in association with Du Pont de Nemours International SA

Butterworth-Heinemann Ltd
Linacre House, Jordan Hill, Oxford OX2 8DP

⟨R⟩ A member of the Reed Elsevier group

OXFORD LONDON BOSTON
MUNICH NEW DELHI SINGAPORE SYDNEY
TOKYO TORONTO WELLINGTON

First published 1993

© Du Pont de Nemours International SA 1993

All rights reserved. No part of this publication
may be reproduced in any material form (including
photocopying or storing in any medium by electronic
means and whether or not transiently or incidentally
to some other use of this publication) without the
written permission of the copyright holder except in
accordance with the provisions of the Copyright,
Designs and Patents Act 1988 or under the terms of a
licence issued by the Copyright Licensing Agency Ltd,
90 Tottenham Court Road, London, England W1P 9HE.
Application for the copyright holder's written permission
to reproduce any part of this publication should be addressed
to the publishers

British Library Cataloguing in Publication Data
Smith, Len
    Language of Rubber, The
    I. Title
    678

ISBN 0 7506 1413 7

Composition by Genesis Typesetting, Laser Quay, Rochester, Kent ME2 4HU
Printed and bound in Great Britain on recycled paper

Contents

Foreword by Sir Harry Melville KCB, FRS vii
Preface ix
Author's Profile xi
Acknowledgements xiii
Notice xiv

Part 1 Standard laboratory tests 1

1 Introduction 3
  1.1 What is rubber? 3
  1.2 Natural and synthetic 3
  1.3 Compounding 3
  1.4 Vulcanisation 4
  1.5 Processing 4
  1.6 Rubber is not a simple material 5
  1.7 Laboratory tests 5

2 Standards 6

3 Du Pont elastomers 7
  3.1 Comparison of properties 10

4 Hardness 11
  4.1 Measuring hardness by durometers 11
  4.2 Dead-load hardness testers 11
  4.3 Interpretation 12
  4.4 Hardness of Du Pont elastomers 13

5 Tensile stress–strain 14
  5.1 Test procedure 14
  5.2 Interpretation 14
  5.3 Design calculations 16

6 Permanent set 18
  6.1 Tension set 18
  6.2 Compression set 18
  6.3 Interpretation 21

7 The relationship between permanent set, creep and stress relaxation 23

8 Creep 24
  8.1 Creep in compression 24
  8.2 Creep in shear 25
  8.3 Interpretation of creep results 25

9 Stress relaxation 27
  9.1 Test equipment 27
  9.2 Standards 31
  9.3 Interpretation 32

10 Short-term stress–strain 34
  10.1 Relaxed modulus 34
  10.2 Compression modulus 35
  10.3 Shape factor 35
  10.4 Shear modulus 36
  10.5 Tear strength 37
  10.6 Cutting resistance 38

11 Dynamic stress-strain 39
  11.1 Combinations of dynamic and static deformation 42
  11.2 The measurement of dynamic properties 43
  11.3 Rebound resilience 43
  11.4 Effect of compounding 45
  11.5 Torsion pendulum 45
  11.6 Yerzley oscillograph 46
  11.7 Forced vibration 47
  11.8 Resonance 48
  11.9 Dynamic mechanical analysis 48
  11.10 The Dynaliser™ 50
  11.11 Interpretation 51

12 Transmissibility 54
  12.1 Interpretation 54

13 Fatigue 56
  13.1 Flex cracking and cut growth 57
  13.2 Heat buildup 58
  13.3 Fracture mechanics 60
  13.4 Fatigue tests in tension 61
  13.5 Interpretation 63

14 Abrasion 64
  14.1 The ISO abrader 64
  14.2 Other abraders 66
  14.3 Interpretation 67

15 Electrical properties — 68
- 15.1 Resistivity — 68
- 15.2 Insulation resistance — 71
- 15.3 Conductive and antistatic elastomers — 71
- 15.4 Electric strength — 72
- 15.5 Tracking — 73
- 15.6 Permittivity — 74
- 15.7 Power factor — 74
- 15.8 Wires and cables — 75

16 Thermal properties — 77
- 16.1 Thermal analysis — 77
- 16.2 Thermal conductivity — 78
- 16.3 Thermal coefficients of expansion — 78

17 The Gough-Joule effect — 80
- 17.1 Practical implications — 81

18 Low temperature properties — 84
- 18.1 Brittle point — 84
- 18.2 Crystallization — 84
- 18.3 Low temperature tests — 86
- 18.4 Interpretation — 90
- 18.5 Crystallization tests — 91

19 Heat ageing — 93
- 19.1 Interpretation — 94
- 19.2 Stress relaxation — 95
- 19.3 Interpretation — 95

20 Adhesion — 97
- 20.1 Adhesion to metals — 97
- 20.2 Adhesion to fabrics — 100
- 20.3 Adhesion to cord — 102
- 20.4 Interpretation — 103

21 Permeability — 104
- 21.1 Permeability to gases — 104
- 21.2 Vapour permeability — 106
- 21.3 Volatile liquids — 106
- 21.4 Interpretation — 108

22 Resistance to weathering — 109
- 22.1 Assessment of deterioration — 109
- 22.2 Natural weathering — 110
- 22.3 Artificial weathering — 111
- 22.4 Interpretation — 112

23 Resistance to gases — 113
- 23.1 Resistance to ozone — 113
- 23.2 Interpretation — 117

24 Resistance to liquids — 118
- 24.1 Oil and chemical resistance — 120
- 24.2 Water resistance — 120
- 24.3 Interpretation — 120

25 Storage — 121

26 Which elastomer? — 122
- 26.1 ASTM D2000/SAE J200 — 122

Part 2 Simulated service tests — 125

27 Introduction — 127
- 27.1 Purchase specifications — 127

28 Automobile coolant hose — 129
- 28.1 Simulated service test conditions — 129
- 28.2 Overall specifications — 130

29 Elastomeric seals — 131
- 29.1 Valve stem seals — 133
- 29.2 Bellows — 134

30 Flue duct expansion joints — 136
- 30.1 Simulated service test rig — 137
- 30.2 Test conditions — 138
- 30.3 Results — 139

31 V-belts — 140
- 31.1 Industrial V-belts — 141
- 31.2 Automotive V-belts — 142

Part 3 Vocabulary of the rubber industry — 145

Indexes to vocabulary — 218
- French index — 218
- German index — 224
- Italian index — 230

Part 4 SI units and conversion factors — 237

SI units — 239
Conversion factors — 243

Index — 248

Foreword

Natural and synthetic rubbers have been used for many decades as engineering materials in tyres, belts and hose, where the manufacturer has to meet close specifications. However, the specifications have generally been selected and controlled by technologists skilled in the art of designing rubber compounds. The introduction and acceptance of the high performance synthetic elastomers, particularly by the automotive industry, have made it necessary for engineers and designers to have a better understanding of the intricacies of rubber technology. Without this knowledge it is impossible for them to prepare product specifications that will meet their exacting requirements at reasonable cost.

Rubbers are materials that are often used in dynamic conditions and the understanding of the relationship between stress and strain and of the mechanisms controlling stress relaxation, fatigue and fracture have considerably improved. It is no longer sufficient to consider only the simple properties, such as hardness, elongation at break and ultimate tensile strength, when preparing specifications. The behaviour of rubbers under the conditions in which they will be used must also be taken into account. This requirement has led to the more sophisticated end-users installing dynamic testing equipment in their own laboratories rather than relying wholly on the product manufacturers.

This volume is especially valuable in that it explains the differences in terminology used by the rubber technologist and the engineer. It also outlines the internationally accepted test methods that are used by the rubber industry for the selection and quality control of rubber compounds. More importantly, the book provides information about the limitations of these tests and the difficulties that can arise when attempting to predict actual performance from laboratory tests. Service testing is all important.

The inclusion of the eight language vocabulary and the section on SI units add to the value of the book as a work of reference for rubber technologists as well as for engineers and designers.

Sir Harry Melville KCB, FRS

Preface

Science and technology do not stand still, particularly in the field of high performance polymers. This second edition of *The Language of Rubber* has, therefore, been completely revised and enlarged and has many more illustrations. The original references to ASTM (American Society for Testing and Materials) have been replaced by those of ISO (International Organization for Standardization) which are now accepted throughout Europe and other parts of the world. The International System of Units (SI) has been used throughout the work and additional emphasis has been placed on the effects of fatigue on rubber products and the increased importance of dynamic testing.

The first edition of *The Language of Rubber* was published in 1957 by the Elastomer Chemicals Department of E I Du Pont de Nemours & Co Inc of Wilmington, USA. The principal market for both natural and synthetic rubbers was at that time, and still is, for automotive tyres. However, the advantages of the new high performance engineering elastomers were beginning to be appreciated and it became necessary to explain the differences in terminology used by rubber technologists and engineers. The names are often the same but the meanings are different.

The articles which formed the basis for the first edition originally appeared in the Du Pont house magazine 'Neoprene Notebook' which later, as the number of high performance elastomers increased, became the 'Elastomers Notebook'. Such was the popularity of the book that reprinting was required in 1963, 1980 and 1986 and French, German, Italian and even Japanese translations were made. However, apart from the introduction of metric in addition to the foot-pound-second units, the majority of the 1957 text remained the same.

In 1986 Du Pont de Nemours International SA in Geneva, Switzerland decided to prepare a complete revision. As before, the new edition was published in parts in successive issues of 'Elastomers Notebook' (Nos. 147 to 160). 'Elastomers Notebook' ceased publication at the end of 1991 with more than half of the text of *The Language of Rubber* remaining unpublished. Du Pont therefore arranged with Butterworth-Heinemann for the publication of the present book.

A possible criticism of the book is that, for the most part, only Du Pont elastomers have been used for its illustrations. Although this is true, it is also true that most of the high performance engineering elastomers have been discovered and/or introduced commercially by Du Pont. Neoprene polychloroprene, the first commercial synthetic rubber, was invented by the legendary W H Carothers, whilst he was working for Du Pont. HYPALON chlorosulphonated polyethylene, VITON fluoroelastomer, ELVAX ethylene-vinyl acetate copolymer, NORDEL hydrocarbon rubber, KALREZ perfluoroelastomer, HYTREL thermoplastic elastomer, VAMAC ethylene-acrylic elastomer, and ALCRYN melt-processable rubber followed to provide Du Pont with an unequalled range of high performance elastomers. Full credit has also been given in the text to the Malaysian Rubber Producers Research Association (MRPRA) and RAPRA Technology whose work with natural rubber has made a significant contribution to the understanding of the engineering properties of elastomers.

The importance of simulated service testing before preparing purchasing specifications has been emphasized in the second part of the new work, as it was in the first edition. Many of the tests used by rubber technologists are only valuable for production control and often do not give any guide to performance. Testing related to the application is essential if specifications are to be meaningful.

A valuable feature of the first edition of *The Language of Rubber*, which did not appear in subsequent reprints, was the eight-language vocabulary. This has been included in the present edition and has been enlarged and verified by the numerous local offices maintained by Du Pont throughout Europe. Alphabetical indices to the vocabulary in French, German and Italian have been added.

The first edition was published before the introduction of the International System of Units (SI) and Part 4 now contains a guide to the use of the system as it applies to the rubber industry. In addition, the universal availability of calculators has made it possible to replace detailed conversion tables by simple multiplying factors.

Finally, by the very nature of this work it has been necessary to reproduce certain parts of the ISO standards almost verbatim and many of their illustrations have had to be used. This book, however, is far from being an alternative to the full standards, which should always be consulted when preparing specifications.

Note on trade marks

Throughout this book, Du Pont trade marks are given in capital letters. Those belonging to other companies are followed by the ™ symbol.

L P Smith PhD, DIC
Spring 1993

Author's Profile

Dr L P Smith qualified as a chemical engineer at Imperial College, London, where he obtained his PhD under Professor D M Newitt. After industrial experience, which included the development of the first company in the United Kingdom to manufacture and market plastics rainwater goods, he became Technical Editor of *Rubber and Plastics Age*. He was also a Director of Rubber and Technical Press Ltd, organizers of a series of international symposia on synthetic rubber.

After a period as Manager of *The Engineer* he went into private practice as a consulting engineer. He was European Editor of *Rubber Age* (later *Elastomerics*), where he contributed a monthly EuroNews column, for eleven years. In 1976 he was invited to give the prestigious Keynote Address to the Akron Rubber Group, choosing as his subject 'Europe – The Uncommon Market'.

He has been a consultant to Du Pont de Nemours International SA in Geneva since 1985 and has written the *Language of Rubber* at their behest.

He is on the Court of the Worshipful Company of Horners, an ancient guild that now embraces the plastics industry as well as its original craft of working with horn. He is also Chairman of the Committee that makes the annual Horners Award for imaginative and commercially viable uses of plastics.

Acknowledgements

The basic technical authority for *The Language of Rubber* has been provided by the Du Pont organization and, in particular, the technical departments at Hemel Hempstead, UK and Meyrin, Geneva. Many people have been involved but special mention should be made of V N Chatterton, B Gottleib, J G Pillow and J E A Williams, who between them have read and commented upon the whole manuscript.

This edition of the book was the conception of G J Swaelens, one time member of the Marketing Communications Department (MCD) of Du Pont de Nemours International SA in Geneva and Editor of 'Elastomers Notebook' until its demise in 1991. Apart from being responsible for the production of the individual sections that appeared in 'Elastomers Notebook', it was he who undertook the revision of the Vocabulary and instigated the three language indexes.

Within the MCD, acknowledgement should also be made to C A Koster and F R Gaag who successively were responsible for overseeing the programme and to W Schmidlin who has brought the project to completion.

Grateful acknowledgement is also due to R Brown, Technical Manager, RAPRA Technology, UK who has made many helpful suggestions. His own book *Physical Testing of Rubber*, published by Elsevier Applied Science Publishers, is a treatise highly regarded by rubber technologists and it has been the source of much useful background information for the present work. Figures 9.1, 10.4, 11.4 to 11.7, 15.1, 15.2, 15.4, 18.4, 20.6, and 23.1 have been redrawn but are based on illustrations used in his book. Figure 2.6 has been redrawn from *Engineering Design with Rubber* by A R Payne and J R Scott, with the permission of RAPRA Technology.

Valuable advice on fatigue testing has been provided by Dr A Stevenson, Materials Engineering Research Laboratories (MERL), who also supplied the information for Figure 13.5 and the photograph for Figure 13.6.

Table 15.2 and Figures 6.1, 8.1 to 8.3, 11.3, 11.10, 11.13, 13.1 to 13.4, 14.1, 15.3, 18.2, 18.3, 18.5, 20.1 to 20.4, 21.2, 23.2, 24.1, 26.2, and 31.4 have been redrawn but are based on illustrations in various standards published by the International Organization for Standardization, Geneva. A complete list of the standards mentioned appears in the index.

The illustration used in Figure 13.7 is reproduced by permission of the Malaysian Rubber Research Association (MRPRA) and thanks are due to the Association's Information Department for its help in providing reprints and technical information. Figures 12.1 and 12.2 are based on information contained in a paper presented by A R Payne to the 'Use of rubber in engineering' Conference organized by the MRPRA in London in 1966.

The photograph and line drawing used in Figure 5.1 were supplied by Monsanto Rubber Instruments Group, UK, who also supplied technical data about its range of instruments. H W Wallace and Company, Croydon, UK, supplied the photographs used in Figures 9.6 and 21.1 and the company's technical data sheets, covering its extensive range of rubber testing equipment, were a valuable source of information. Wykeham Farrance Engineering, UK, supplied the line drawing used in Figure 9.7. TA Instruments Inc, USA, provided the illustrations for Figures 11.14, 11.15 and 16.1 and Barco Industries, Belgium, supplied the photographs for Figure 11.16. The help received from these companies is gratefully acknowledged.

The photographs used in Figures 9.2, 9.4, 9.5, 14.2 and for the front cover were commissioned by Du Pont and were taken by John Lawrence of London, UK.

Thanks are also due to Volvo Car Corporation, Sweden, Carl Freudenberg, Germany and Gates Europe, Belgium, for providing the information about their simulated service tests and for the illustrations used in Part 2.

The interest shown by Sir Harry Melville KCB, FRS, who contributed the Foreword, is also greatly appreciated.

Finally, acknowledgement is due to those members of the Du Pont Elastomer Chemicals Department in Wilmington, USA, in the mid-1950s, whose names have long been lost in the archives but who had the original idea for *The Language of Rubber*.

Notice

The author, copyright holder and the publisher have used their best efforts to prepare this book, and while every due care has been taken in compiling this material, the author, copyright holder and publisher make no warranty, implicit or explicit, about the information. Neither will they be liable under any circumstances for any direct or indirect damages arising from any use, direct or indirect, of the information contained in this book.

Part 1

Standard laboratory tests

1
Introduction

Rubber has been used in engineering applications for well over one hundred years. Yet engineers and designers still have difficulty in correlating the terms and expressions used by the rubber technologist with those they use themselves. Tensile strength, hardness, elongation, and creep, for example, are terms familiar to engineers but their meaning in rubber technology can often be quite different. *The Language of Rubber* is an attempt to explain these differences and to provide basic information which will help in the design of satisfactory components made from rubber. It will also indicate how to specify rubber in a manner which will enable the components to be made economically and to give the performance in use that is required.

1.1 What is rubber?

Rubbers are loosely described as materials which show 'elastic' properties. Such materials are generally long chain molecules known as 'polymers' and the combination of elastic and polymer has led to the alternative name of 'elastomers'. Rubbers and elastomers will be considered to be synonymous in this work.

One easily understood definition of a rubber or elastomer is a material which at room temperature can be stretched repeatedly to at least twice its original length and, upon immediate release of the stress, will return with force to approximately its original length.

1.2 Natural and synthetic

Natural rubber (NR) is generated in the *Hevea brasiliensis* tree as an emulsion of *cis*-polyisoprene and water, known as latex. The milky liquid is exuded from the tree when it is cut and is collected in small cups. Latex is also obtained in small quantities from the Guayule shrub. The latex is coagulated and then dried to produce a clear crepe rubber. If it is dried in the presence of smoke it becomes a light brown colour and is called smoked sheet. Natural rubber was the only rubber available for more than a century but the growth in the demand for tyres has outstripped the available supply and today NR represents less than 33% of the total usage of rubber.

Synthetic rubber is prepared by reacting suitable monomers to form polymers and can be obtained as a water emulsion or as a suspension in water or solvents. Small quantities of methyl rubbers were made during the First World War but the first commercially successful synthetic rubber was Du Pont's Neoprene, polychloroprene rubber, introduced in 1931. Since that time, Du Pont alone has introduced nine distinct groups of synthetic elastomers and there are no less than eight general classes of synthetic rubber, with 44 sub-classes, listed by the American Society for Testing and Materials (ASTM).

1.3 Compounding

The raw or base polymers vary from soft plastic materials to tough gristly substances and, generally, they are not suitable for use in the form in which they are supplied. Their elastomeric properties have to be developed by further compounding and the possible permutations and combinations are infinite. Fillers, such as carbon black and finely ground silica, can be used to provide reinforcement; oils, waxes and fatty acids can be used to improve processability and colours can be obtained by incorporating suitable pigments. Other additives are used to improve chemical resistance and to assist in the curing process (see Section 1.4). Rubber compounds usually contain less than 50% of the raw polymer and in some cases, e.g. flooring, the rubber content can be less than 25%.

1.4 Vulcanisation

Most rubbers when they have been compounded need to be vulcanised or cured. Chemically the process produces crosslinks in the molecular structure, which provide the physical properties required and which give the finished rubber, chemical and thermal stability. However, there is one group of rubbers, the thermoplastic elastomers (TPE), which do not require vulcanising. Examples include Du Pont's HYTREL engineering thermoplastic elastomer and ALCRYN melt-processable, halogenated polyolefin rubber.

1.5 Processing

Rubber compounding is generally carried out on open rubber mills or large internal mixers. Open mills consist of two rollers (typically 2 m across and 0.6 m in diameter) which rotate in opposite directions. The rolls can be heated or cooled as necessary. The rubber is placed on the rolls and mixing is achieved by the shearing action induced at the 'nip' between the rolls. Additives are added in carefully weighed quantities during the mixing process. After the mixing operation is complete, the compound is removed from the mill in the form of sheet.

Internal mixers have an enclosed chamber in which two rotors with helical blades turn in opposite directions at slightly different speeds. The blades impart a shearing action to the rubber which quickly produces a homogeneous mix. Cooling water or steam is circulated through the rotors and various parts of the casing to maintain the optimum temperature conditions. Mixing is by batch and the compound is dropped from the bottom of the internal mixer onto an open mill and then removed as sheet.

The sheet generally requires pre-shaping into blanks of suitable dimensions for moulding into finished parts. The three main methods of moulding are by compression, and by transfer and direct injection. A blank is placed between two halves of a mould in compression moulding and is then squeezed under temperature and pressure into the shape of the cavity. Typical temperatures and pressures are between 150° and 180°C and 7 to 15 MPa of projected mould cavity area. During the moulding process the compound is vulcanised and the finished moulding generally only requires trimming before it is ready for use.

For transfer moulding the rubber blank is placed in a pot outside the mould cavity and the rubber is forced under pressure through a series of feed gates into the cavity itself.

For direct injection moulding, the rubber compound is fed into a screw extruder, where it is preplasticised, and is then forced under pressure through a system of runners and feed gates into the mould cavity.

Sheeting is made on calenders which have three or four rolls. Products, such as tyres and hose can be built up from calendered sheet by both hand and automatic processes. Rubbers can also be bonded to steel and other metals to form construction mountings, shock absorbers and similar springs.

Products, such as tubing and glazing gaskets, which are required in continuous lengths are made by plasticising the rubber in a screw extruder and forcing it through a shaped die. Vulcanisation is then achieved by heating the extrudate by UHF (ultra-high frequency) or LCM (liquid curing medium) in continuous tunnels. Alternatively, the extrudate may be collected in batches and cured in steam-heated autoclaves.

The thermoplastic elastomers can be shaped more easily, using the techniques used by the plastics industry, including injection and blow moulding, extrusion, etc.

1.6 Rubber is not a simple material

It is hoped that the above brief summary of rubber processing will be sufficient to prove that rubber is not a simple material. In addition, it is a material whose properties change with time and which is affected by the environment in which it is to be used. Metals are subject to corrosion but rubber can deteriorate in contact with heat, oils, chemicals, sunlight, and the weather. A rubber compound which is apparently suitable from laboratory tests may be quite unsatisfactory after a few months' exposure to service conditions. For example, standard abrasion tests are a fairly reliable index of the wearing qualities of rubber heels and soles – if the shoes are to be worn under normal conditions. But the same tests would give entirely different results after the shoes had been worn for some time in a machine shop, gasoline station or other oily location.

Another problem is that the physical properties of a rubber composition vary greatly with the test conditions; much more than do the properties of steel or concrete. For example, consider the effect of temperature. The tensile strength of a rubber composition at 100°C may be as little as 10 to 20% of the same composition tested at 28°C. The rate of application of a load also has an effect. The load required to produce a given elongation in a rubber compound under conditions of rapid loading may be as much as three times that required when tested at a low rate of loading. Still another factor is previous stress history. Repeated loading cycles may result in substantial changes in measured property values.

The interrelation between the properties of rubber is also sometimes confusing and can be the reverse of the familiar relationships in, say, steel. For example, an increase in the hardness of steel is accompanied by an increase in tensile strength. The correlation between the two is quite close and well known. However, this relationship does not hold at all for rubbers. As hardness increases, the tensile strength of a rubber may increase to a maximum and then decrease, or it may decrease from the outset, depending on the formulation of the compound.

1.7 Laboratory tests

Laboratory tests in the rubber industry are generally used to establish satisfactory and reproducible products rather than to provide design information. For example, the tensile properties of rubber as measured by laboratory tests have little, if any, direct bearing on serviceability and hardly ever affect a design calculation. Similarly the standard laboratory test for abrasion, in which a sample of the rubber is subjected to a moving abrasive surface (sandpaper or emery wheel), gives completely false results for the use of a rubber as a hose for sandblasting. The sandblasting hose is subjected to an abrasive stream rather that to a rigid abrasive surface and those compounds that have proved to be best in service are decidedly inferior by the standard laboratory test.

These difficulties do not mean that the laboratory tests are superfluous. With understanding, the pitfalls can be avoided and reliable and useful information can be obtained. Tests can be used to decide which compounds warrant the expense of field testing and to eliminate those which are grossly inadequate. Also, once a compound has proved to be serviceable, physical test data are required to ensure uniformity in the commercial product.

The Language of Rubber will attempt to provide the basic knowledge required. Each commonly used physical property will be defined in rubber industry terms and the test methods used to measure it will be described. The interpretation of the results will be discussed and emphasis will be placed on their significance, or lack of it, as an indication of serviceability. However, serviceability of rubber products can only be accurately predicted by actual service or by good simulated service tests.

2

Standards

The principal national authorities issuing standards for rubber materials and products are:–

France	Association Française de Normalisation (NF)
Germany	Deutsches Institut für Normung (DIN)
UK	British Standards Institution (BSI)
USA	American National Standards Institute (ANSI), American Society for Testing and Materials (ASTM)
Former Soviet Union	State Committee for Standards (GOST)

The standards issued by each authority are generally similar in principle but they may differ in detail. In Europe, The International Organization for Standardization (ISO) issues internationally agreed standards which are either derived from existing national standards or are recommended to form the basis for future national standards. As far as possible, the information in *The Language of Rubber* is based on the ISO standards. However, when product specifications are prepared there may be instances where it is more appropriate to refer to the national standards.

3
Du Pont elastomers

As mentioned previously, Neoprene chloroprene rubber (CR) was the first of Du Pont's high performance engineering elastomers. Introduced in 1931, it was originally developed as an oil resistant substitute for natural rubber (NR) and it is still widely used for this purpose. Neoprene compounds have a practical temperature range for continuous service of −40° to 100°C. They have very good resistance to waxes, fats, oils, greases and many other petroleum products; good resistance to alkalis, dilute mineral acids and inorganic salt solutions and excellent ozone, weather, abrasion and impact resistance when properly compounded. The chlorine content in Neoprene compounds makes them inherently more resistant to burning than exclusively hydrocarbon polymers.

HYPALON chlorosulphonated polyethylene (CSM) was introduced in 1951. General purpose compounds can operate between −20° and 135°C. Special compounds can be formulated which will withstand intermittent use at temperatures up to 175°C. Formulations can also be made which can be used intermittently at temperatures as low as −40°C. HYPALON has superb resistance to ozone and weathering, even in light coloured products; excellent resistance to microbiological attack; good electrical insulating properties up to 600 volts; resistance to a wide range of oxidizing agents and corrosive chemicals; excellent resistance to abrasion, wear and other forms of mechanical abuse and, like Neoprene, its chlorine content makes it inherently more resistant to burning than exclusively hydrocarbon polymers.

VITON fluoroelastomer (FKM) was introduced in 1957 and is one of the most heat- and fluid-resistant rubbers ever developed. It withstands hundreds of standard and difficult fluids over a temperature range of −20° to 200°C and provides reliable leak-free service long after other rubbers have failed. Special compounds can be used for limited periods outside this range. In addition, VITON has exceptionally good resistance to compression set, atmospheric oxidation, sun, weather and biological attack. It also has extremely low permeability to a broad range of substances, excellent electrical properties and is flame retardant.

ELVAX ethylene-vinyl acetate copolymers (EVA), introduced in 1960, are thermoplastic resins with elastomer-like properties. They are inherently tough, resilient and flexible over the temperature range −30° to 70°C without the addition of any plasticisers and they have excellent resistance to stress-cracking. Their clarity varies from translucent to transparent depending upon the vinyl acetate content of the resin. They have very high impact strength and good chemical resistance combined with good weathering and ozone resistance. Many grades of ELVAX are approved for use in contact with food. EVA can be processed by conventional thermoplastic or rubber equipment and can be foamed and/or crosslinked if required. ELVAX copolymers and terpolymers with organic acids are used in hot melt adhesives and as processing aids for rubber compounding.

NORDEL, a terpolymer of ethylene propylene and a non-conjugated diene (EPDM), was launched in 1963 and is a versatile rubber with excellent heat-, weather- and ozone-resistance. General-purpose compounds can be used between −50° and 145°C and special formulations can be made which can operate intermittently at −70° and up to 185°C. Abrasion and tear resistance are good, even at elevated temperatures.

KALREZ, a perfluoroelastomer (FFKM), was developed by Du Pont in 1968 and is unique in that it is the only elastomer the company produces as finished parts. KALREZ parts have outstanding chemical and heat resistance, exceeding those of any other elastomer. Parts retain their elastic properties in long-term service at temperatures as high as 290°C and in intermittent service up to 315°C, even in contact with corrosive chemicals.

HYTREL engineering thermoplastic elastomer, introduced in 1972, is a high performance alternative to thermoset rubber, which does not require vulcanisation. It can be processed by a variety of thermoplastic techniques including injection moulding, extrusion, rotational moulding, melt casting and blow moulding. HYTREL has exceptional resistance to temperatures from −40° to 120°C

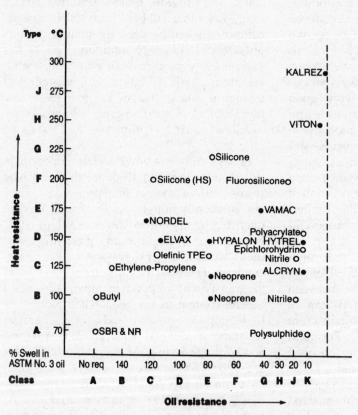

Notes
1) The SAE J200/ASTM D2000 classification only covers crosslinked elastomers. HYTREL, ALCRYN and the other thermoplastics are included for comparison.
2) The temperature of testing in ASTM No. 3 oil varies with the heat resistance classification (Type) as follows:

Type	A	B	C	D-J
Test temperature, °C	70	100	125	150

3) Specific points are shown for clarity. In practice, there will be a 'spread' depending upon compounding and grades of polymers used.

Figure 3.1 *Heat and oil resistance of elastomers*

Table 3.1 *Comparative properties of Du Pont elastomers*

Property	Natural Rubber (NR)	Neoprene (CR)	HYPALON (CSM)	VITON (FKM)	ELVAX (EVA)	NORDEL (EPDM)	KALREZ[5]	HYTREL (YBPO)	VAMAC (AEM)	ALCRYN
Service temperature, °C	70	100	135	200	70	120	290	120	170	90
Service temperature (excursion), °C	85	120	150	300	–	150	316	150	200	120
Hardness range, durometer A or D	30–90 A	40–95 A	50–95 A	55–95 A	40–95 A	40–90 A	65–95 A	35–72 D	40–95 A	60–80 A
Tensile strength (pure gum), MPa	22	20	22	15	4.5–25	17	14	13–44.1	15	13.1
Tensile strength (black-loaded stock), MPa	–	–	–	–	–	–	–	–	–	–
Specific gravity (base material)	0.93	1.23	1.12–1.28	1.85	0.93–0.97	0.86	2.01	1.17–1.25	1.08–1.12	1.21–1.25
Vulcanising characteristics	E	E	E	G/E	NA	E	F	NA	G	NA
Adhesion to metals	E	E	E	G/E	E	G	G	G	G	G/E
Adhesion to fabrics	E	E	G	G/E	E	F	E	G	G	G/E
Tear resistance	G	G	F	F/G	G	G	F	E/O	G	E
Abrasion resistance	G	E	E	G/E	F	VG	G/E	F/P	G	E
Rebound, cold	E	G	F	F	–	G	–	F/VG	P	F/G
Rebound, hot	E	VG	G	E	–	VG	–	G/E	F	P

Electrical properties

Dielectric strength	E	G	E	G	G	E	E	F/G	G	–
Electrical insulation	G/E	F/G	G	F/G	G	E	E	F/G	F/G	F

Fluid resistance

Permeability to gases[6]	F	G	VG	VG	P	F	E	G	F	–
Acid resistance, dilute	F/G	E	E	E	VG	E	E	G	G	E
Acid resistance, conc.	F/G	G	VG	E	G	E	E	P	P	–
Solvent resistance:										
aliphatic hydrocarbons	P	G	G	E	F	P	E	E	G	G
aromatic hydrocarbons	P	F	F	E	F	P	E	F/G	F	G
oxygenated (ketones, etc.)	F/G	G	P	P	F	G	E	F/G	P	F
paint solvents	P	P	P	P	F	P	E	F/G	P	P
Resistance to:										
swelling in lubrication oil	P	G	G/E	E	O	P	E	G/E	G	G
oil and gasoline	P	F	G	E	O	O	E	VG/E	P	E
animal and vegetable oils	P/G	G	G	G/E	F	G	E	VG/E	G	E
water absorption (hydrolytic stability)	E	G	VG	VG	G	E	VG	G	G[1]	G

Resistance to environmental factors

Oxidation	F	E	E	O	O	E	O	E	E	E
Ozone	F	E	O	O	O	O	O	E	O	O[7]
Sunlight ageing	P	VG	O	O	VG[2]	VG	O	VG[2]	O	O[7]
Flame[3]	P	VG	VG	E	P	P	O	G[4]	P	P
Heat	G	VG	E	O	F	E	O	E	E	VG
Cold	E	VG	G	F	VG	E	P/F	E	G	G

0 = outstanding
E = excellent
VG = very good
G = good
F = fair
P = poor

[1] Up to 60°C; up to 100°C with 10 MS additive
[2] With additives
[3] These evaluations are qualitative and comparative only. They should not be construed as recommendations. Specific compounding is required to optimise performance. Elastomer choice should be based upon a practical consideration of the potential fire hazards involved in each individual case and, if applicable, the results of appropriate flame tests
[4] Will melt
[5] Available only as finished parts made by Du Pont.
[6] Good is equivalent to low permeability
[7] Non black types need additional protection

and has an outstanding resistance to stiffening at low temperatures. It has excellent flex fatigue performance, abrasion resistance and high resistance to deformation under moderate strain conditions. In addition, the load bearing capacity and resistance to fatigue in cyclic load-bearing applications are outstanding and it retains its properties over long periods because it contains no plasticiser. Resistance to solvents and hydrocarbon fluids (including gasoline and oil) is also outstanding.

VAMAC, a terpolymer of ethylene, methyl acrylate and a cure site monomer (AEM), was introduced in 1975. It is a high quality multi-functional, hot-oil-resistant rubber with unique damping and fire-resistant properties. In dry heat VAMAC outlasts most other oil-resistant rubbers and retains its elasticity after continuous exposure to air at 170°C. It can even withstand exposure at temperatures of 204°C for 7 days. Outstanding performance in hot water, superb resistance to ozone, UV and weather, good tear resistance, tensile strength and flex life, exceptionally low compression set and good vibration-damping characteristics are additional features. VAMAC when properly compounded is non-corrosive, flame retardant and has a low smoke emission when exposed to fire.

ALCRYN halogenated polyolefin was introduced in 1985 and is a mid-performance industrial rubber which can be processed on either rubber or plastics equipment. It has a service temperature range of −40° to 90°C, outstanding oil resistance, excellent ozone and weather resistance and good tear strength and abrasion resistance. It feels like a vulcanised rubber but as it is melt-processable all scrap is reusable.

3.1 Comparison of properties

The physical properties of the nine elastomers are qualitatively compared with natural rubber in Table 3.1 and their relative heat and oil resistances are shown in Figure 3.1.

4

Hardness

In Europe, hardness tests on vulcanised rubber generally follow ISO 48. This is based on measuring the depth of indentation by a rigid ball under a dead load, the indentation being converted to International Rubber Hardness Degrees (IRHD), the scale of which ranges from 0 (infinitely soft) to 100 (infinitely hard). The ISO standard allows for a 'normal' test and a 'micro' test. The former requires test pieces at least 4 mm but preferably 8 to 10 mm thick. The latter is an approximately 1:6 scaled-down version of the 'normal' test using thinner sections, preferably 1.5 to 2.5 mm thick, which can often be taken from finished products. The two forms of test give similar results if the ratio of normal to micro test piece thickness is about 5:1.

It is important to emphasize that the results of any indentation test depend on the thickness of the rubber, unless this is much greater than the depth of indentation. Hence the specified standard thickness should be used wherever possible and the thickness used should always be quoted.

4.1 Measuring hardness by durometers

Hardness should always be specified by a dead-load test method (see Section 4.2) but simpler spring-loaded 'pocket' hardness gauges, known as durometers, are widely used for rapid checks. In these devices the indentor point projects below the flat bottom of the case and is held in the zero position by a spring. When pressed against a sample, the indentor point is pushed back into the case against the spring and this motion is translated through a rack and pinion into movement of the pointer on the dial. The harder the sample, the farther it will push back the indentor point and the higher will be the numerical reading on the dial.

The Shore durometer is typical of these hand instruments and has a frustoconical (type A) or pointed conical (type D) indentor with a spring force that decreases with increasing indentation, the readings being in arbitrary degrees. Shore A degrees approximate to IRHD but are lower in the range below 30. Shore D has a more open scale intended for use on very hard materials.

4.2 Dead-load hardness testers

Dead-load hardness testers are less convenient to use than durometers but the measurements are more reproducible. As the name implies, a dead load is applied to the indentor for a specified time and the hardness is obtained from the depth of the indentation.

Typical table top instruments are the Monsanto Durolab™ and the Wallace microhardness tester. The latter permits the measurement of hardness on very small areas (down to 1 mm) and on curved surfaces. Using standard flat samples, results can be correlated directly with the IRHD tester. On non-standard samples and, in particular, on curved surfaces such as O-rings, results will not correlate with IRHD but will provide meaningful relative hardness comparisons for samples of the same size and shape.

The hardnesses of some typical rubber products are given below:

	IRHD
Tap washer, flooring, typewriter platen	90 ± 5
Shoe sole	80 ± 5
Solid tyre, shoe heel	70 ± 5
Tyre tread, hose cover, conveyor belt cover	60 ± 5
Inner tube, bathing cap	50 ± 5
Stationer's rubber band	40 ± 5

Hardness values below 30 are sometimes found in such products as erasers and printing rolls.

4.3 Interpretation

Hardness is one of the most useful and often quoted properties of rubber but, in fact, the figures can be quite misleading. Firstly, the measured values, especially by durometer, are often unreliable because of the mechanical limitations of the instruments and because of operator error. Hardness degrees should therefore never be quoted to better than 5°. Secondly, the characteristic that is measured, surface indentation, rarely bears any relation to the ability of the rubber product to function properly.

The lack of significance can best be understood by considering three products: a hose cover, a gasket to be used between rough flanges and an automobile mounting. It is fairly obvious that the ease or difficulty of indenting the surface of the hose cover has nothing to do with utility. The important properties in a hose cover are abrasion resistance and resistance to oil, weather and other conditions relating to its service. The case of the gasket is unusual because surface indentation *does* have some significance. Indentation by the point of the test instrument is similar, to some extent, to the indentation the gasket will receive from protrusions on the sealing surface.

There is not much danger of attributing false significance to hardness in the above cases but the danger is very real in the case of the motor mounting. Motor mounts are typical of many rubber products which are required to carry load and in which the relationship between load and deformation, called stiffness, is a critical design factor. Hardness cannot be assumed to be a close

Table 4.1 *Hardness ranges of Du Pont elastomers*

	Hardness
Natural rubber (NR)	30–90 A
Neoprene (CR)	40–95 A
HYPALON (CSM)	40–95 A
VITON (FKM)	55–95 A
ELVAX (EVA)	40–95 A
NORDEL (EPDM)	40–90 A
KALREZ (FFKM)	65–95 A
HYTREL (YBPO)	40–72 D
VAMAC (AEM)	40–95 A
ALCRYN	60–80 A

Table 4.2 *National and international standards for hardness testing*

International Standard Hardness Test
ISO 48
ASTM D1415
BS 903:Part A26
DIN 53519
NF T46-003

Durometer A
ISO 7619
ASTM D2240
DIN 53505
NF T46-052

Note: The International Standard Hardness Test is recommended for both specification and control purposes. One International Rubber Hardness Degree (IRHD) represents approximately the same proportionate difference in Young's modulus. The Durometer A test is empirical. For substantially elastic or isotropic rubbers the two scales are comparable but for markedly plastic or anisotropic materials the relationship is less precise.

measure of stiffness. It is true that hardness and stiffness are both stress–strain relationships but the relationships are established for two entirely different kinds of deformation. Hardness measurements derive from small deformations at the surface. Stiffness measurements derive from gross deformations of the entire mass. Because of this difference, hardness is not a reliable measure of stiffness. Even if hardness and stiffness did have a better correlation, the irreducible five-point variation in durometer readings would be equivalent to a 15 to 20% variation in stiffness as measured by a compression-deflection test. Hardness measurements would not, therefore, be sufficiently accurate for design purposes.

The misuse of hardness to measure stiffness is very common and causes much confusion. Wherever possible, simulated service tests should be used rather than hardness testers.

4.4 Hardness of Du Pont elastomers

It can be seen from Table 4.1 that most of the Du Pont elastomers can be compounded to give hardnesses between 40 and 95 IRHD. HYTREL is the exception and gives compounds with hardnesses between 40 and 72 Shore D. Hardnesses below 40 can be obtained by the incorporation of plasticisers and extenders such as oils, resins or waxes but generally some other physical characteristics have to be sacrificed.

5

Tensile stress–strain

The definitions of tensile strength, elongation and modulus and the test procedures to be used are standardized in ISO 37.

Tensile strength is the maximum tensile stress reached in stretching a test piece, usually a flat dumb-bell shape, to its breaking point. By convention, the force required is expressed as force per unit area of the original cross section of the test length.

Elongation, or strain, is the extension between bench marks produced by a tensile force applied to the test piece and is expressed as a percentage of the original distance between the marks. Elongation at break, or ultimate elongation, is the elongation at the moment of rupture.

Tensile stress, or modulus, is the stress required to produce a certain elongation. Thus if a stress of 7 MPa produces an elongation of 200%, the rubber is said to have a 200% modulus of 7 MPa.

Table 5.1 *National and international standards for tensile stress–strain*

Tensile stress – strain
ISO 37
ASTM D412
BS 903: Part A2
DIN 53504
NF T46-002

Unlike metals, the stress in a rubber is not directly proportional to strain and modulus is therefore the stress at a certain strain. It is neither a ratio nor a constant but merely the coordinates of a point on the stress–strain curve.

For steel and other metals, modulus is stress divided by strain (Young's modulus) and is both a ratio and a constant.

5.1 Test procedure

The tensile stress–strain properties of rubbers are measured with a tensile testing machine, illustrated in Figure 5.1. The early heavy pendulum dynamometers have largely been replaced by inertia-less transducers which convert force into an electrical signal. Measurements of stress and strain are taken continuously from zero strain to breaking point and are recorded graphically.

Dumb-bell shaped test pieces 10 or 13 cm long are die-cut from flat sheet and marked in the narrow section with benchmarks 2.5 or 5 cm apart. The ends of the test piece are placed in the grips of the testing machine and the lower grip is power driven at 50 cm per minute so that the test piece is stretched until it breaks. As the distance between the benchmarks widens, measurement is made between their centres to determine the elongation. The standard temperature for conditioning and testing the samples is $23 \pm 2°C$.

5.2 Interpretation

Tensile strength and elongation are useful to the rubber technologist for compound development, manufacturing control and for determining a compound's resistance to attack by various chemicals.

Tensile tests are universally used as a means of determining the effect of various compounding ingredients and are particularly useful when such ingredients affect the rate and state of vulcanisation of the rubber.

Similarly, tensile tests are excellent for controlling product quality once the compound has been selected. The tests are sensitive to changes in manufacturing conditions

Tensile stress—strain 15

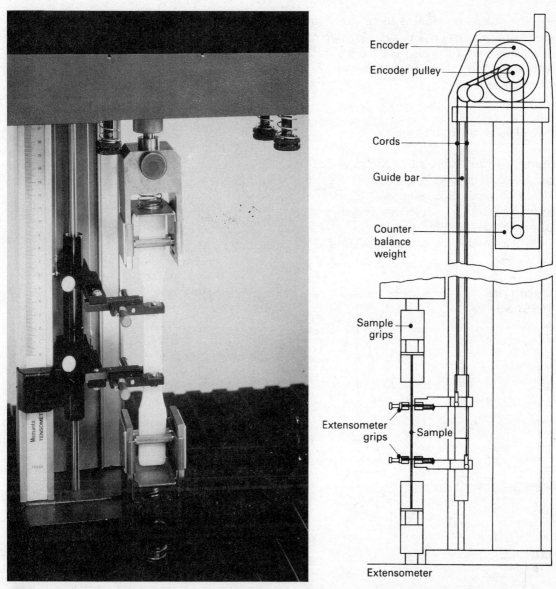

Figure 5.1 *Modern testing machines measure elongation or strain with an extensometer. Two small clamps are attached to the dumb-bell test piece at a selected standard distance apart (usually 10, 20, 25 or 50 mm). Cords attached to each clamp transmit the movement to two encoder units mounted at the top of the extensometer column. The encoders produce pulses which drive a digital strain display and a recorder chart, if fitted*

and can be used to identify under- or over-vulcanisation, bad blending and the presence of foreign matter.

Tensile tests can also be made before and after an exposure test to determine the relative resistance of a group of compounds to deterioration by heat, oil, ozone, weathering, chemicals, etc. Even a small amount of deterioration causes appreciable changes in tensile properties. However, it should be noted that the *retention* of tensile properties is much more significant than the absolute values before and after the exposure test. For example, if the tensile strength of one rubber falls from 21 to 14 MPa after exposure to oil and the fall for a second rubber is from 12 to 11 MPa, it is the second one that is more likely to be successful in practice. The superior retention of tensile strength is also a good indication that the second rubber will retain most of its other properties.

5.3 Design calculations

Despite their usefulness to a rubber technologist, tensile properties are of limited use to the design or applications engineer. They cannot be used in design calculations and they bear little relation to performance in practice. Rubber components are seldom loaded in

Table 5.2 Tensile strength of Du Pont elastomers

	Tensile strength, MPa	
	Pure Gum	Black-loaded stock
Natural rubber (NR)	20.7	20.7
Neoprene (CR)	20.7	20.7
HYPALON (CSM)	17.2	20.7
VITON (FKM)	12.4	13.8
ELVAX (EVA)	4.5–25.0	–
NORDEL (EPDM)	–	20.7
KALREZ (FFKM)	–	13.8
HYTREL (YBPO)	25.5–44.1	–
VAMAC (AEM)	–	17.2
ALCRYN	–	13.1

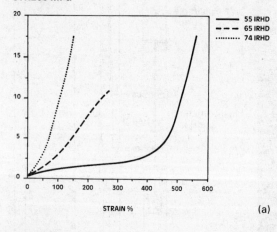

(a)

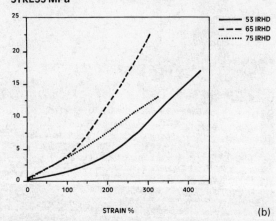

(b)

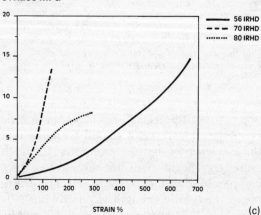

(c)

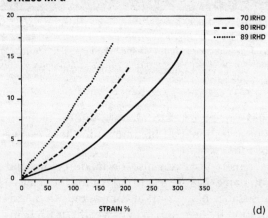

(d)

Figure 5.2 *Stress-strain curves in tension*

tension and never to a degree even approaching their ultimate strength or elongation. Belts, hose, O-rings, mountings, packings, etc. are rarely, if ever, subjected to tension stresses above 1 MPa, so whether their ultimate tensile strengths are 10 or 20 MPa can hardly affect their ability to perform their functions.

Neither are tensile properties a reliable indication of the quality of the rubber. It is true that compounds with tensile strengths below 7 MPa are usually rather poor in most mechanical properties and those where the tensile strength is greater than 21 MPa have good mechanical properties. However, in the middle range, say from 10 to 20 MPa, which covers the great majority of rubber products, the correlation between tensile strength and such properties as resilience, abrasion resistance, compression set and flex life is, at best, haphazard.

In the middle range, it is possible to compound two stocks with identical tensile strength and elongation and yet find no similarity in service life. This is especially true if different rubbers are being compared. Tyre treads based on natural rubber, for example, often have tensile strengths between 31 and 34.5 MPa, whereas treads with an equivalent service life made from synthetic rubber may have tensile strengths of no more than 17 MPa.

Purchasers of rubber products who over-specify tensile properties may find that the price they pay is unnecessarily high, or that they are sacrificing other properties which may be more important. With established products, purchasers often will not consider a new compound, which is actually better for their application, if it happens to have lower tensile strength and elongation than the original specification. Once again, it is necessary to emphasize that rubber specifications should be based on performance rather than on arbitrary physical tests.

The tensile properties of elastomers can be varied over wide limits by suitable compounding and can be adjusted to meet any particular requirement. Typical values are given in Table 5.2 and stress–strain curves for HYPALON, Neoprene, NORDEL and VITON are given in Figure 5.2 (a) to (d) for typical compounds with three different hardnesses. These curves are instructive to the rubber technologist but should not be used in design calculations or for selecting a compound for a specific application.

6

Permanent set

Rubbers deform under load and rarely return completely to their original dimensions when the load is removed. The difference between the original and final dimensions, expressed in various ways, is known as 'permanent set' and can be measured in tension, compression or shear. However, permanent set in shear is not often required and there are no recommended standards for its measurement.

In practice, the measurement of permanent set depends upon carefully defining the conditions of the test, the time for which the test is conducted and the time that is allowed for the test piece to recover. Although standardized, these conditions are arbitrary and it is sometimes difficult to obtain reproducible results and to relate them to how the rubber will perform under service conditions.

6.1 Tension set

The methods for measuring tension set in Europe generally follow ISO 2285 and consist of stretching a standard strip or dumb-bell test piece (such as that shown in Figure 6.1) to a constant strain, holding the elongation for a standard time, removing the load and allowing the test piece to recover for 30 minutes. The increase in length between the reference marks or narrow section of the test piece, expressed as a percentage of the original length of the narrow section, is the permanent set in tension.

$$\text{Tension set} = \frac{L_1 - L_o}{L_s - L_o} \times 100$$

where,

L_o is the original unstrained reference length
L_s is the strained reference length and
L_1 is the reference length after recovery

The percentage strain value should be selected in accordance with the final application of the vulcanisate and with reference to its breaking elongation and the test temperature. A value of 100% strain is preferred but 25, 50, 200 and 300% are acceptable alternatives. The test can be made at a temperature of 23°, 70°, 85°, 100°, 125° or 150°C and the duration of the test can be 24, 72 or 168 hours, the test period commencing 30 minutes after the specified strain has been reached.

The measurement of tension set under constant stress and the use of ring-type test pieces are no longer recommended in some countries in Europe although two sizes of rings are standardized in ISO 2285.

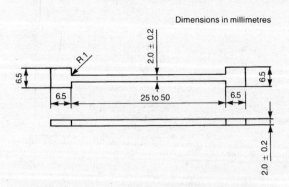

Figure 6.1 *Test piece with enlarged ends for tension set measurements. Test pieces of the shape shown are usually cut with a sharp die from a flat sheet, 2 mm thick, of the material under test. The length of the narrow section (reference length) must be between 25 and 50 mm. The sheets may be prepared by moulding or from finished articles by cutting and buffing*

6.2 Compression set

The practice in Europe is to measure compression set after constant strain at ambient or high temperatures following ISO 815. ISO

Table 6.1 *National and international standards for permanent set*

Tension set
ISO 2285
ASTM D412
BS 903: Part A5
DIN 53518
NF T46-009

Compression set at ambient and elevated temperatures
ISO 815
ASTM D395
BS 903: Part A6
DIN 53517
NF T46-011

Compression set at low temperatures
ISO 1653
ASTM D1229
BS 903: Part A39

1653 is used for measurements at low temperatures. The measurement of compression set after constant stress is no longer recommended.

Small cylindrical disks of known dimensions are compressed to a fixed height in a simple jig. The jig consists of two or more flat parallel metal plates which are sufficiently rigid to withstand the stress without bending and are of adequate size to hold the test piece or pieces within the area of the plates. The plates are clamped together with nuts and bolts. Steel spacers of the appropriate thickness, in the form of rings around each bolt, are placed between the plates to control the thickness of the test pieces whilst compressed.

The recommended sizes for the disks are either 13 mm diameter by 6.3 mm thick or 29 mm by 12.5 mm, as shown in Figure 6.2. The two sizes do not necessarily give the same values for compression set and comparisons should always be made with similar test pieces. The use of a lubricant on the contact surfaces of the plates is optional and may, in some cases, give more reproducible results. However, the lubricant may also affect the compression set values obtained and, again, comparison should only be made when the test conditions are similar.

It is recommended that three test pieces be used, either separately or as a set, for each determination and the results averaged. The bolts are tightened so that the plates are drawn together uniformly until they are in contact with the appropriate spacers,

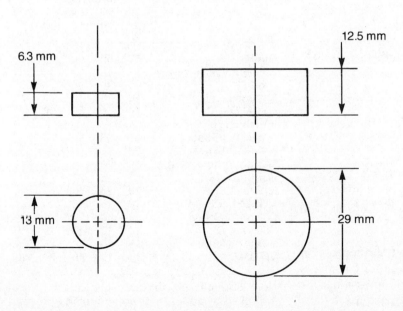

Figure 6.2 *Compression set test pieces – actual size*

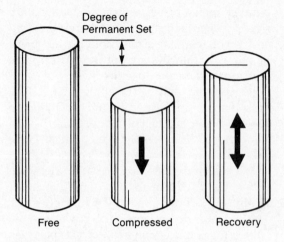

Figure 6.3 *Diagram illustrating compression set*

generally sized to give a compression of 25%. The apparatus containing the test pieces is introduced without delay into the central part of an oven which is maintained at the test temperature. Recommended temperatures are 23°C or one of nine temperatures between 70° and 250°C. The duration of the test is 24 hours for tests at elevated temperatures or 72 hours at 23°C.

At the end of the specified time, the test pieces are removed from the jig and allowed to recover at 23°C for 30 minutes before the thickness is re-measured. The compression set is the difference between the original thickness of the test piece and that after recovery, as shown in Figure 6.3, expressed as a percentage of the initially applied compression.

$$\frac{\text{Compression set at}}{\text{constant strain}} = \frac{t_o - t_r}{t_o - t_s} \times 100$$

where,

t_o is the original thickness of the test piece,
t_r is the thickness of the test piece after recovery, and
t_s is the thickness of the spacer

Thus, if there is no recovery the compression set is 100% and if the test piece fully recovers to its original thickness the compression set is 0%. Typical values for compression set for a variety of elastomers are provided in Table 6.2.

The measurement of compression set at low temperatures requires a slightly different procedure. The residual deformation at low temperatures is not truly permanent and is usually reversed as the temperature rises. The final measurement of thickness must therefore be made at the temperature of the test in order to obtain meaningful results.

ISO 1653 calls for exposure times of either 24 or 72 hours and the test temperatures can be 0°, −10°, −25°, −40°, −55° or −75°C. The jig and procedure are similar to those already described for elevated temperatures, with the additional requirement that the apparatus for measuring the final thickness must be introduced into the low-temperature cabinet at least 30 minutes before it is used. At the end of the test period, the test pieces are released from the jig as quickly as possible and a stopwatch is started simultaneously. Final thickness measurements are made within the cabinet at regular intervals for up to 2 hours. The final thickness is plotted against time on semi-logarithmic graph paper, as shown in

Table 6.2 *Typical values for compression set*

	Temperature °C				
	20 (%)	70 (%)	100 (%)	150 (%)	200 (%)
ALCRYN	18	50	60	–	–
Efficient vulcanisation nitrile	5	10	15	–	–
Fluorosilicone (FMQ)	10	10	10	20	70
HYPALON (CSM)	20	50	80	90	–
HYTREL (YBPO) constant load	10	28	50	–	–
KALREZ (FFKM)	20	–	–	–	70
Natural rubber (NR)	5	20	45	–	–
Neoprene (CR)	5	15	25	–	–
Nitrile (NBR)	10	20	40	–	–
NORDEL (EPDM)	10	15	35	60	–
Polynorbornene	15	50	85	–	–
Silicone (Q)	10	10	10	15	40
VAMAC (AEM)	15	15	20	25	–
VITON (FKM)	10	10	15	20	25

Note: Compression set can usually be improved by appropriate compounding.

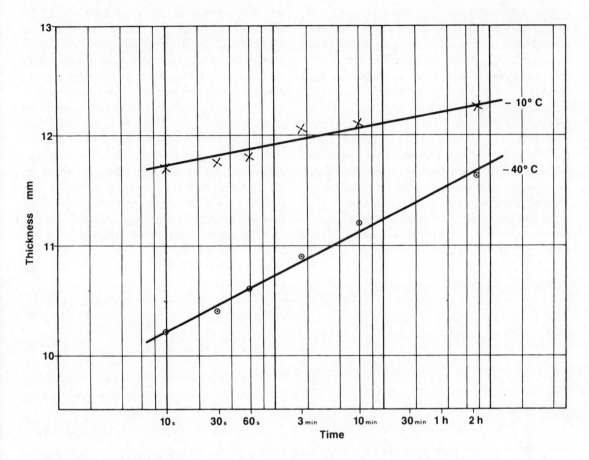

Figure 6.4 *Graph of final thickness vs time for measurement of compression set of a Neoprene WRT vulcanisate at low temperatures. Final thickness measurements are made within the low temperature cabinet at regular intervals for up to 2 hours. The final thickness is plotted against time on semi-logarithmic graph paper and should give an approximately straight line. The final thickness is then read from the graph at recovery times of 10 seconds and 30 minutes using the formula given in the text*

Figure 6.4, and should give an approximately straight line. The final thickness is read from the graph at recovery times of 10 seconds and 30 minutes and the compression set is calculated for these two times using the formula given previously.

Large fluctuations in the results may occur when measurements are made at temperatures at which changes take place in the physical characteristics of the polymer (known as the transition temperature) or when the compression set values are below 15%.

6.3 Interpretation

Permanent set measurements can be useful for production control because they provide an indication of the degree of vulcanisation that has taken place. They can also be helpful when selecting a compound from a number of alternatives. Unsuitable compounds can often be identified if the permanent set falls in the high range. On the other hand, if the permanent set is in the low range, differences in values will probably not give a true indication of performance in service.

The main difficulty with the interpretation of permanent set measurements is that the testing time is so short in relation to the time in service. Gaskets for water mains, for example, must retain their sealing properties for decades, whereas the tests are only made for hours. Also, the standard tests call for a recovery time of only 30 minutes after removal of the load. Experiments have shown that in many cases considerable recovery occurs after this time and, in some instances, the compression set after 24 hours can be as little as one-sixth of that after 30 minutes.

Although, many specifications include a requirement for permanent set, it is often really stress relaxation or strain relaxation (usually known as 'creep') which should be specified. This subject will be considered in more detail in the following section but it should be noted that there are a number of misconceptions relating permanent set to other physical properties. For example, it is sometimes claimed that low compression set is accompanied by high resilience and low creep but this is not necessarily true. Neither is it any longer true to say that synthetic elastomers always have higher permanent set than natural rubber. The wide variety of base polymers and the improvements in formulation will usually enable a compounder to meet most requirements for permanent set.

7

The relationship between permanent set, creep and stress relaxation

As discussed in Section 6, the deformation that remains in an elastomer after the removal of an applied stress or strain, i.e. its permanent set, does not necessarily provide a good indication of how the elastomer will perform in practice. Of equal or greater importance is how the elastomer behaves whilst it is held under conditions of constant stress or strain for long periods.

When a constant load (stress) is applied to an elastomer the deformation is not constant but increases gradually with time; this behaviour is known as 'strain relaxation' or more generally as 'creep'. Conversely, when an elastomer is subjected to a constant strain, a decrease of stress takes place with time; this behaviour is called 'stress relaxation'. Both phenomena have great significance. For example, creep is important in engine mountings, since it influences the space relationship between the various parts of the equipment, and stress relaxation is the factor that controls the effectiveness of static seals, such as pipe gaskets held between flanges.

The phenomena of permanent set, creep and stress relaxation are the result of physical and chemical changes in the elastomer, both of which can occur simultaneously. The physical changes are due to the viscoelastic nature of elastomers and the chemical effects to changes in the molecular structure by ageing, oxidation or similar processes. In general, physical effects are most important at normal or low temperatures and/or short times and chemical effects at high temperatures and/or long times.

Unfortunately, creep and stress relaxation are very time- and temperature-dependent and they are seriously affected by the media with which the elastomer is in contact. A number of different methods of measuring creep and stress relaxation have been developed but the results obtained from one method may be difficult to compare directly with those from another. In practice, correlation between creep and stress relaxation is rarely attempted. The choice of test usually depends upon the application, some methods being more relevant to certain conditions than others. Simulated service tests are again to be recommended whenever practicable.

In addition, extreme caution should be exercised before attempting to extrapolate results obtained by short time tests to considerably longer periods or using results obtained at higher temperatures as accelerated tests to provide information for lower temperatures.

8

Creep

Creep is usually defined as the increase in deformation which has occurred after a specified time interval, during application of a constant force, expressed as a percentage of the test piece deformation at the commencement of that time interval. It should be noted that this definition differs from that commonly used for metals and plastics. In their case, creep is the percentage deformation relative to the original thickness of the test piece, i.e. the percentage strain at any given time.

At normal or low temperatures and/or short times the physical process is dominant and creep is found to vary directly with time plotted on a logarithmic scale. At high temperatures and/or long times chemical processes are dominant and chemical creep is often found to vary directly with time.

In the absence of chemical effects, the longer the relaxation processes continue the slower they become. For example, the amount of creep occurring in a decade of time, from say one to ten minutes after loading, can be the same as the amount in the much longer decade from one to ten weeks after loading. In such cases, it is usual to give test results as 'creep rate', which is defined as the ratio of creep to the common logarithm of the time interval over which it is measured, expressed as per cent per decade. In practice it is customary to take the creep one minute after the deformation is imposed as the base, hence,

$$\text{Creep (\%)} = \frac{(D_t - D_1)}{D_1} \times 100$$

and

$$\text{Creep rate (\% per decade)} = \text{slope} \times \frac{100}{D_1}$$

where,

D_1 is the deformation of the test piece after 1 minute

D_t is the deformation of the test piece after t minutes, and

the slope is obtained from the straight line graph of D_t vs $\log_{10}$ time.

ISO 8013 covers creep but it is restricted to tests in which the elastomer is deformed in either compression or in shear. In practice, rubbers are seldom used where they are continuously deformed in tension.

Table 8.1 *National and international standards for creep*

Creep (strain relaxation)
ISO 8013
ASTM D412
BS 903: Part A15
DIN ISO 8013

8.1 Creep in compression

The test pieces used for measuring creep in compression are similar to those described in Section 6.2 for the measurement of compression set but the plane surfaces of the disks are bonded to rigid metal end pieces. The test piece is rigidly held between a fixed and a movable steel plate, as shown in Figure 8.1, the latter being equipped with a mechanism for applying a constant force within six seconds with negligible overshoot. The deformation is measured by any means which will give an accuracy of ±1% of the test piece thickness.

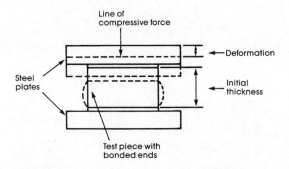

Figure 8.1 *Test piece for measuring creep in compression*

Creep rates depend on whether the elastomer is new and unworked or has been pre-loaded to the working deformation at some time before the test (known as 'scragging'). The standard therefore calls for the test piece to be compressed by about 25% at 23°C and returned to zero deflection five times before testing. Sixteen to 48 hours must be left between the mechanical conditioning and the creep test.

After assembly, the test rig is kept in a suitable temperature controlled chamber maintained at the specified test temperature. The force is then applied so that a strain of 20 +2% is produced in the elastomer when measured after one minute from the moment that the full force is reached. The deformation of the test piece is measured at 1, 10, 100 and 1,000 minutes for short duration testing or between one and 28 days for longer duration tests. Creep and creep rate (if applicable) are calculated as shown above.

8.2 Creep in shear

When measuring creep in shear the test piece consists of two rubber blocks bonded to three steel plates as shown in Figure 8.2. The test pieces may be either circular or square in cross-section and no absolute dimensions are specified. The diameter (or side, in the case of a square cross-section) of the rubber blocks must be at least four times the thickness to ensure that the deformation is a simple shear.

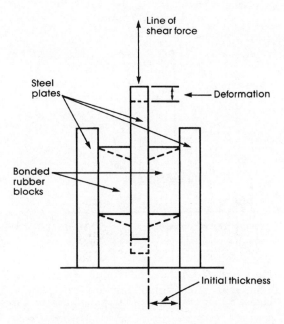

Figure 8.2 *Test piece for measuring creep in shear*

A typical test piece will contain rubber blocks with a circular cross-section of 25 mm and a thickness of 5 mm. A suitable testing fixture is illustrated in Figure 8.3. In this case, the force is applied to the central plate and the outer plates are held rigidly but alternatively, the force can be applied to the outer plates whilst the central plate is held rigidly.

The method of testing is in principle the same as that for creep in compression described above.

8.3 Interpretation of creep results

Creep varies considerably with the composition and type of elastomer but, in general, it has been established that the higher the initial strain, the higher the creep and that creep increases with temperature. Creep is also dependent upon the type of strain. It is greater under tension strain than under an equal shear strain and it is greater under shear strain than under an equal compression strain. It is also greater under dynamic loading than under static loading. Creep also increases with the amount of filler present.

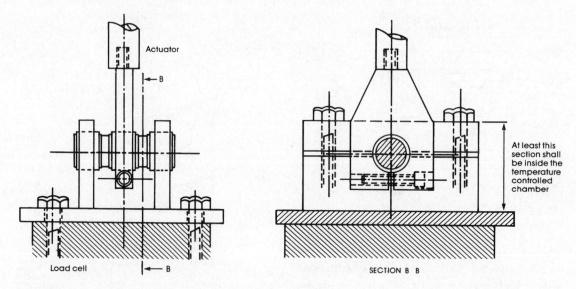

Figure 8.3 *Typical testing fixture for double bonded shear test piece*

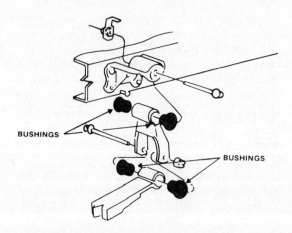

Figure 8.4 *Moulded bushings of HYTREL have replaced a rubber-metal assembly consisting of a metal-inner and -outer ring separated by a layer of rubber in this truck leaf spring mounting. The ability of HYTREL to replace metal-rubber assemblies illustrates the combined elasticity and creep resistance of the material*

Chemical effects can cause creep to occur more rapidly than short term tests predict. High humidity for example can double the creep rate. Oxidation also has an effect but this can be reduced by the use of suitable antioxidants. However, large rubber components, such as bridge bearings, are usually protected from oxidation simply because their bulk prevents easy diffusion of oxygen to the interior.

In engineering applications creep is most rapid during the first few weeks under load but should not exceed 20% (for 70 IRHD) of the initial deformation in this period. Allowance can be made for this in the design. Thereafter, in correctly compounded rubbers, only a further 5 to 10% increase in deformation should occur over a period of many years.

9

Stress relaxation

As mentioned in Section 7, stress relaxation is the change of stress with time when the rubber is held under constant strain. It is the converse of creep, which is the increase of strain with time under constant force.

Stress relaxation measurements can be made in compression, shear or tension; the first being the most important for testing materials for use as seals and gaskets. Stress relaxation tests can also be used to provide a general guide to ageing, particularly when chemicals are involved, and the tests are then normally made in tension. These will be described in Section 19 which will cover ageing in detail.

ISO has standardized three methods for the measurement of stress relaxation in compression using cylindrical test pieces or rings. Rings are important because they allow the maximum surface area of the test piece to be exposed to a test liquid and they simulate the actual working conditions of an O-ring. The same principle is used in most methods of measuring stress relaxation in compression and is illustrated in Figure 9.1. The test piece is compressed between platens to a constant strain and the force exerted by the test piece is measured at intervals by applying a very small additional strain. The force-measuring device records the force at the moment the top platen separates from the body of the jig. The principle is simple but two major practical difficulties exist. They are to identify the exact moment when the platen separates and to prevent the platens tilting without introducing excessive friction. Alternatively, the force-measuring device can continuously balance the force exerted by the test piece but this requires a head for every test piece, which can become expensive.

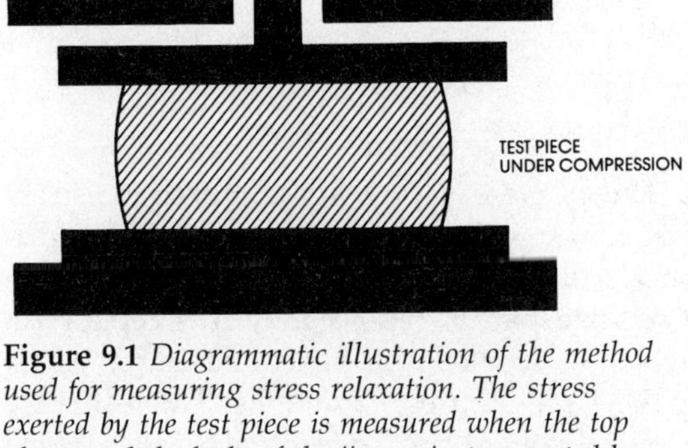

Table 9.1 National and international standards for stress relaxation

Stress relaxation
ISO 3384
ASTM D1390
BS 903: Part A42
DIN 53537
NF T46-044

9.1 Test equipment

It is possible to use a simple arbor press fitted with a load cell for measuring stress relaxation in compression. The arm of the press is lowered manually and the compressive force measured by breaking an electric circuit. However, the operation is extremely delicate and it is preferable to use more sophisticated equipment.

Figure 9.1 Diagrammatic illustration of the method used for measuring stress relaxation. The stress exerted by the test piece is measured when the top platen and the body of the jig are just separated by application of a small additional force

28   *The Language of Rubber*

Figure 9.3 *Exploded view of a Lucas test-jig. In use, the base plate is located over the setting tool and the test ring is placed centrally on the base plate. The compression plate is lowered on to the test ring and the fixed top plate is bolted into position using the three pillars and nuts. The three compression control spacers are inserted between the base and compression plates and are used to centralize the test ring. The adjustment screw is then lowered by means of a hexagon socket wrench until the test ring is compressed to the diameter of the small rods on the spacers, which are usually machined to provide a standard compression of between 15 and 30%. After compression the spacers are removed. Finally, the jig is removed from the setting tool and is ready for the required test procedure*

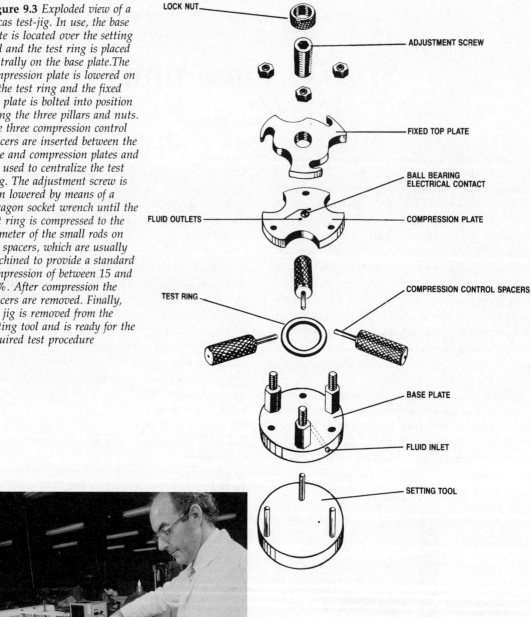

Figure 9.2 *Lucas compression stress relaxometer*

One of these is the Lucas compression stress relaxometer shown in Figure 9.2. The force-measuring device is a beam balance in which the scale pan is replaced by a flat plate on which is located the jig containing the test piece – see Figures 9.3 and 9.4. The test piece is usually an O-ring or toroidal washer which is located between the base plate and the

Stress relaxation 29

Figure 9.4 *Parts of the Lucas test-jig*

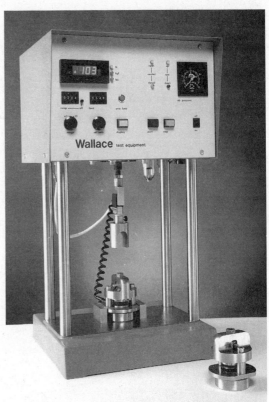

Figure 9.6 *Shawbury-Wallace compression stress relaxometer Mk II*

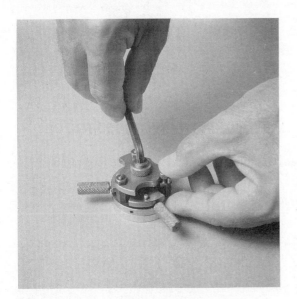

Figure 9.5 *Pressure is applied to the compression plate of a Lucas test-jig by means of a hexagon socket wrench. The amount of the compression is controlled by the spacers which are subsequently removed*

compression plate of the jig. It is compressed to the required percentage of its cross section by means of an adjustment screw fitted in the top plate of the jig as shown in Figure 9.5. The assembly sits on the platform of the modified balance. When the loading screw on the balance makes contact with the loading platform it completes an electrical circuit through the jig. When a force is applied to the compression plate of the jig which is slightly greater that that exerted by the test piece, the contact is broken and an indicator light in the electrical circuit dims and indicates that the force should be measured. The jig is then removed from the instrument and subjected to the selected ageing procedure, which may include circulating fluids through the centre of the jig. After each period of ageing the force exerted by the test piece is measured, thus enabling the decay in the stress exerted by the rubber under a given set of ageing conditions to be determined.

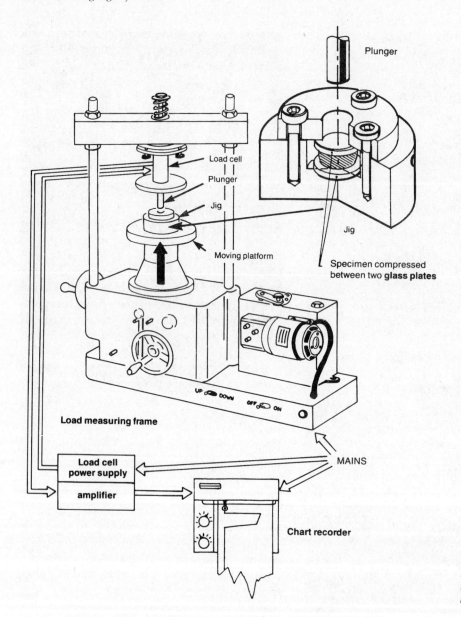

Figure 9.7 *Wykeham Farrance compression stress relaxation apparatus*

The Lucas equipment is used extensively but requires considerable manipulative ability. In addition, it cannot be used on solid disk test pieces or flat samples taken from finished articles. The Rubber and Plastics Research Association (now RAPRA Technology Ltd) in the UK therefore developed the apparatus shown in Figure 9.6 to overcome these problems. A load cell is driven onto the jig by a pneumatic ram; at the moment when a very small additional compression of the test piece is detected electrically the ram is automatically stopped, the force reading digitally recorded and the ram reversed. This measuring head can be used with a variety of jig designs including those recommended for the Lucas and other test methods.

Another device to measure stress relaxation is the Wykeham Farrance unit, shown diagrammatically in Figure 9.7, which was developed by the Institute of Polymer Technology at Loughborough University together with other research establishments in the UK. The test piece is compressed between two glass plates and a plunger applies the load to the top plate. At the point that the load cell force exceeds that of the specimen the recorder shows a significant change in the force/time response. The gradient of this response provides a measure of the instantaneous modulus, in addition to the time dependent force at the discontinuity. This method allows the value of the initial measurement to be obtained at the same time as the test piece attains the required strain level.

Finally, it is possible to use a universal tensile testing machine as described in Chapter 5 for measuring stress relaxation. Although this offers some reduction in experimental difficulty, it must be operated at a very slow speed in order to avoid overshoot because the increase in compression is very small, less than 0.05 mm in tests conducted to meet ISO 3384.

9.2 Standards

The methods of measuring stress relaxation are covered by ISO 3384. Cylindrical or ring test pieces may be used depending upon the circumstances. The cylindrical test pieces are the same as those used for measuring compression set and creep and consist of cylindrical disks, either 13 mm diameter by 6.3 mm thickness or 29 mm diameter and 12.5 mm thickness. The smaller test piece is preferred. The test piece is not bonded to end pieces and, if necessary, the ends may be lubricated with a silicone or fluorosilicone fluid or molybdenum disulphide.

The reproducibility of stress relaxation tests for some elastomers, particularly those containing substantial proportions of fillers, may be improved by mechanically conditioning the test piece. In such cases the test piece is conditioned by compressing it to the same strain that will be used during the rest of the test and then immediately returning it to zero deflection. This procedure is repeated five times. Mechanical conditioning must be followed by thermal conditioning at ambient temperature for between 16 and 48 hours immediately before testing.

Thermoplastic elastomers may contain moulding stresses and thermal conditioning for 3 hours at 70°C may improve the reproducibility of the results.

There are three test procedures. A compression of 25% is recommended but lower values may be used if necessary. In method A, the test piece is compressed at the test temperature and all force measurements are made at that temperature. In method B the compression and force measurements are made at ambient temperature and the test piece is subjected to the test temperature for various periods. In method C the compression is applied at ambient temperature and all counterforce measurements are made at the test temperature. The three methods do not give the same values of stress relaxation and comparisons between them should be avoided. The duration of the test is normally 168 hours after the initial compression but measurement of the counterforce may also be made at intermediate times, in which case, 24 and 72 hours are preferred. In method B, the time required each time for the jig to cool to the ambient temperature is not included in the total duration of the test.

It is frequently necessary to measure the effect of stress relaxation over much longer periods than 168 hours and in such cases it is convenient to use a logarithmic time scale.

The compression stress relaxation, $R(t)$, after a specified duration of test, t, expressed as a percentage of the initial counterforce, is given by,

$$R(t) = \frac{F_o - F_1}{F_o} \times 100$$

where,

F_o is the initial counterforce measured 30 minutes after completing the compression in methods A and B and 2 hours after completing the compression in method C, and

F_1 is the counterforce measured after the specified duration of the test.

For some applications it may be more useful to calculate compression stress ratios, i.e. F_1/F_o, after different times of exposure rather than stress relaxation values. The compression stress ratio values may be presented graphically as a function of time, using a logarithmic scale if necessary.

The specification for ring test pieces in ISO 3384 is either 2 mm square section and 15 mm internal diameter or O-rings with a cord diameter of 2.65 mm and an internal diameter of 14 mm. The apparatus used is similar to that used for cylindrical test pieces but with the addition of a method for circulating fluids within the volume enclosed by the ring. This arrangement provides a relatively large surface area to volume ratio so equilibrium swelling is reached reasonably quickly.

The standard does not cover testing at temperatures below ambient. The methods specified have been used for low-temperature testing but their reliability under these conditions has not been proven. One difficulty is that rubbers stiffen at low temperatures and require relatively large forces to obtain the necessary over-compression.

9.3 Interpretation

Most of the comments about the interpretation of the results of creep tests made in Section 8.3 also apply to stress relaxation. Nevertheless the results obtained from stress relaxation tests give a far better idea of how a seal will perform in practice than those obtained by measuring compression set, volume swell or similar property. In addition, even though it is not possible to extrapolate the stress relaxation results with any degree of certainty, the tests can be a very useful way of comparing the performance of different rubbers.

For example, the graph in Figure 9.8 shows the percentage retained sealing force in O-rings tested in the Lucas apparatus after contact with air at 150 and 200°C. After 100 hours in air at 150°C the seal of VITON fluoroelastomer (FKM) retained more than 90% of its original sealing force, whilst the other materials only retained 62, 61, 51 and 32% respectively. When the test was concluded after 8,000 hours (333 days), VITON still retained over 70% of its sealing force. None of the other materials survived. In addition, the broken line of the graph shows that the VITON seal provided a better sealing effect at 200°C than any of the others at 150°C. Investigations limited to the comparison of compression set under these conditions would not have led to the same conclusion.

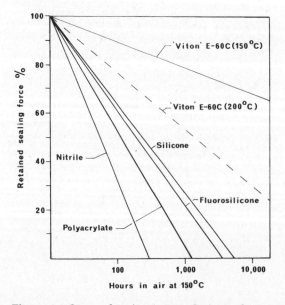

Figure 9.8 *Stress relaxation tests on O-ring seals*

It is also important to appreciate that elastomeric seals are mostly in contact with more than one fluid. A pipe seal, for example, will be in contact with the fluid being retained on one side and with air on the other. There may also be a temperature differential within the seal. Such conditions can be fairly reasonably simulated when measuring stress relaxation.

Problems may also arise with fluids that take a very long time to permeate the test piece. In these cases, the effect of other parameters, such as swelling, can become important. Another possible hazard is that tests undertaken at high temperatures may give misleading results because of the degradation of the fluid in contact with the elastomer. Fuels, in particular, can form gums which will stick to the seals or to their metal housings. These gums may improve rather than lessen the effectiveness of the seals.

10

Short-term stress–strain

Stress–strain tests on elastomers are made in two ways. Short-term or static stress–strain tests, such as those already described in previous sections, are made at somewhat arbitrary chosen strain rates and the effects of long times and cycling are ignored. Dynamic tests, on the other hand, subject the elastomer to a deformation pattern from which the cyclic stress–strain behaviour can be calculated. However, cyclic tests in which the main object is to fatigue the elastomer will be treated as a separate subject in Section 13.

When a vulcanised rubber containing fillers is deformed, physical breakdown occurs and its stress–strain curve changes. Usually the most obvious result is a reduction in stiffness, which is defined as,

$$\text{Stiffness} = \frac{\text{Load}}{\text{Deformation}} = \frac{\text{Modulus} \times \text{Area}}{\text{Thickness}}$$

Each subsequent deformation has a lesser effect, which indicates that eventually an equilibrium stress–strain curve should be reached. The effect of the deformations is not permanent but the recovery to the original stress–strain curve may be very slow, even at elevated temperatures.

The importance of pre-stressing or mechanical conditioning (scragging) a test piece to be used in a static test will depend on the service required from the finished product. If the part is to be subjected to repeated deformations it will be advisable to pre-stress the test piece. On the other hand, if the part has to behave as an elastomer after being held in a state of rest for some time, e.g. a constant velocity bellows for an automobile, pre-stressing will not provide a true indication of the performance in practice. In many dynamic tests, e.g. where the test piece is subjected to continuous oscillation, pre-stressing to the strain level of the test is automatic.

A requirement for mechanical conditioning is incorporated into some test procedures but there is no general rule. It is, therefore, important to check whether mechanical conditioning is called for and to decide whether it has any significance for the application under consideration.

10.1 Relaxed modulus

Tensile stress–strain tests, as described in Chapter 5, measure the stress at a given strain and can be considered to give an indication of the stiffness of elastomers. Another indication of stiffness can be obtained by measuring the 'relaxed modulus', which is the tensile stress at a given elongation after a fixed time of relaxation. This test is effectively a measurement of short-term stress relaxation.

There is no international standard for relaxed modulus and individual laboratories use their own methods. An instrument developed by the Malaysian Rubber Producers' Research Association (MRPRA) is commercially available which measures relaxed modulus at either 50 or 100% elongation. It is particularly suitable for evaluating the vulcanisation characteristics of elastomers because it measures modulus at ambient rather than at curing temperatures.

Similar data are provided using ASTM D1456, which recommends applying a fixed stress and measuring the elongation after one minute.

Table 10.1 *National and international standards for short-term stress–strain. See also Tables 4.2 and 5.1*

Compression modulus
ISO 7743
ASTM D575
BS 903: Part A4
GOST 265

Relaxed modulus
ASTM D1456

Shear modulus
ISO 1827
BS 903: Part A14
NF T46-023

Tear strength
ISO 34
ISO 816
ASTM D624
BS 903: Part A3
DIN 53506
DIN 53507
DIN 53515
GOST 262
GOST 12014
NF T46-007
NF T46-033

10.2 Compression modulus

Elastomers for engineering applications are more often used in compression or shear than in tension and ISO 7743 provides a method for determining the compression stress–strain relationship. The test piece is a cylindrical disk of 29 ±0.5 mm diameter and 12.5 ±0.5 mm thickness, which is the same as the larger size test piece recommended for measuring compression set (see Section 6.2).

The test piece is compressed axially in a universal tensile testing machine used in its compression mode, with autographic recording of force and deflection i.e. stress and strain. The test piece is placed between two sheets of fine glass-paper, with the abrasive side against the elastomer, in order to resist lateral slip. Two conditioning cycles are made in which the test piece is compressed to 5% greater compression than is required for the standard. The results obtained from the third compression are expressed either as the compression stress at a specified strain or series of strains or as the compression strain at a specified stress or series of stresses. When practical, the standard recommends that the results be expressed as a graph of compression stress against compression strain. The standard test temperature is 23 ± 2°C.

10.3 Shape factor

The stiffness of rubber in compression, when the loaded surfaces are prevented from slipping by bonding or by mechanical location, is affected by its shape. For this purpose rubber may be considered to be incompressible and there will be no reduction in volume when the piece is compressed. The reduction in height must, therefore, be accommodated by bulging at the sides and this will depend upon the free area of the unloaded surfaces. It follows that a circular cylinder, which has less free area on its vertical sides, will bulge less easily under a given load than a rectangular block. Allowance is made for this effect by calculating the 'shape factor', which is defined as the ratio of the area of one of the loaded surfaces to the area of all of the unloaded surfaces which are free to bulge – see Figure 10.1. Thus, for cubes

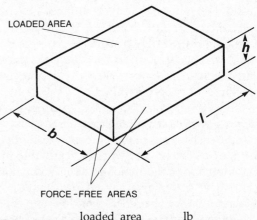

$$\text{Shape factor} = \frac{\text{loaded area}}{\text{force-free area}} = \frac{lb}{2h(l + b)}$$

Figure 10.1 *Shape factor for a rectangular block*

the shape factor is 0.25 and for cylinders of radius, r, and height, h, it is,

$$\text{Shape Factor} = \frac{\pi r^2}{2\pi rh} = \frac{r}{2h}$$

It has been shown empirically that blocks of rubber having the same shape factor and composition, will require approximately the same compressive stress to produce a given deflection. The shape factor is often required in calculations for rubber springs and damping devices. It follows that if the edges of a rubber device are constricted, the stiffness will increase and this is a feature of some designs of rubber compression springs.

ISO 7743 makes no allowance for shape factor so the results obtained will not necessarily be comparable with those obtained with test pieces of a different shape.

Bridge bearings require a high vertical stiffness to prevent appreciable changes in height of the bridge deck under changing load. At the same time they require a relatively low horizontal stiffness to prevent excessive loading of the supporting piers as a result of the thermal expansion and contraction of the bridge deck. The cross-sectional area of the bearing will depend upon the allowable pressure on the support. Given this area, the thickness of rubber necessary to limit the horizontal stiffness can be determined. The required vertical stiffness can then be obtained by inserting a sufficient number of metal spacer plates, which will reduce the effective shape factor and hence the freedom of the rubber to bulge. The metal plates do not affect the shear stiffness but their thickness must not be included when calculating the thickness of rubber required.

Although ISO and other authorities have always considered that static tests provide data which are suitable for the design of bridge bearings, recent work indicates that the pads are also under dynamic stress because of the small movements caused by changes in temperature. Experience has shown that Neoprene-bearing pads have a proven life of over 30 years, yet their performance as predicted by low temperature laboratory tests indicates that they would be inadequate. Again, it is essential to ensure that the conditions used for laboratory tests accurately reflect those encountered in practice.

10.4 Shear modulus

Elastomers are used in shear for many engineering applications and at least one authority, the MRPRA, claims that shear modulus is the most useful single property in design calculations. However, there appears to be some uncertainty about the best method of making the measurements and there is no agreed standard that is used universally.

A simple shear test is described in ISO 1827. The test piece comprises four rubber blocks 4 mm thick, 20 mm wide and 25 mm long, bonded on each of their two largest opposite faces to the mating faces of four rigid plates of the same width as shown in Figure 10.2. The rubber is bonded to the metal supports during vulcanisation and with a thickness to length ratio of 0.16 the error due to bending is negligible. The test piece assembly is strained in a tensile testing machine at 25 ± 5 mm per minute, at least five conditioning cycles being applied before the measuring cycle. The force at a given deflection or the deflection at a given force is then recorded. The strain is measured with a dial gauge or by the movement of the crosshead. The standard test temperature is 23 ± 2°C.

The shear stress, expressed in pascals (N/m^2) is calculated by dividing the applied force by twice the bonded area, i.e. twice 20 × 25 × 10^{-6} m^2. The shear strain is calculated by

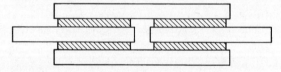

Figure 10.2 *Quadruple shear test piece. The shear strain is half the measured deformation divided by the thickness of one rubber block. The shear stress is the applied force divided by twice the area of a bonded face of one block. The mean apparent shear modulus, expressed in pascals, at any value of the shear strain is calculated as the ratio of the corresponding shear stress to the shear strain*

dividing one-half of the actual deformation of the test piece by the sheared thickness (both expressed in the same units of length). The mean apparent shear modulus, expressed in pascals, at any value of the shear strain is calculated as the ratio of the corresponding shear stress to the shear strain.

10.5 Tear strength

An important factor in the failure of rubber products is the ease with which a tear can be initiated and propagated. However, the conditions under which tearing can occur vary so widely that use has to be made of tests which are only really suitable for laboratory comparisons. When these tests are included in purchase specifications they only provide a measure of quality control for production. In addition, the tearing is generally induced by the test piece having a cut. These results, therefore, only measure the force required to propagate the tear and give no indication of the force required to initiate it.

There are three basic configurations for testing tear strength as shown in Figure 10.3. In the first, the force is applied in the plane of the test piece and parallel to its length and in the second, the force is normal to the plane. In both cases the applied stresses are essentially tensile. In the third, known as the trouser test piece, an element of shear stress is also included.

The standards for determining tear strength in Europe are ISO 34 and 816. The former recommends four procedures using the shapes of test piece 1 to 3 shown in Figure 10.4 and the trouser test piece. The last has the advantage of being relatively easy to prepare accurately and is not sensitive to the length of cut. The angle shape can be used with or

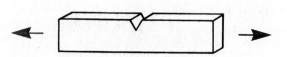

1 Tension force applied in the plane of the test piece

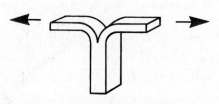

2 Tension force applied at right angles to the plane of the test piece

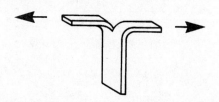

3 Force applied to produce a shearing effect (trouser test)

Figure 10.3 *The three basic configurations for testing tear strength*

1 Crescent with enlarged ends for gripping

2 Simple crescent with multiple cuts

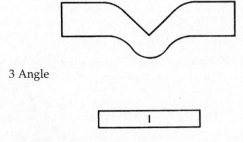

3 Angle

4 'Delft'

Figure 10.4 *Test pieces used when measuring tear strength*

without a nick and the crescent shape provides a re-entry angle without the difficulty of providing or measuring the size of the nick. ISO 816 covers the use of the so-called 'Delft' test piece which has an internal slit and is of such a small size that it can be cut from finished products. There are also many other variations in shapes of test pieces that are used for specific applications. The results from one test method are not in general comparable with those measured by another, although the relative rankings may be similar. At least one authority considers that the same ranking would be obtained from tensile testing and that tear tests are therefore unnecessary.

Traditional tensile testing machines, as described in Chapter 5, are used for the tests. As relatively low forces are required for tearing tests a sensitive load-measuring device and a machine with low inertia are essential. The test pieces are gripped in the usual manner and a stretching rate of 100 ± 10 mm/min is specified for the trouser test and 500 ± 50 mm/min for the others. The tearing energy can vary with the speed of stretching and these rates must be strictly observed. It can also vary with the thickness of the test pieces and ISO 34 calls for a variation in thickness of less than ±7.5%.

The results for tests made under ISO 34 are expressed as the force in kilonewtons per metre of thickness, obtained by dividing the force required to produce the tearing in newtons, by the thickness of the test piece in mm. When the trouser method is used the force varies during the test and a median force has to be calculated in accordance with ISO 6133. The maximum force experienced is used with the other test pieces.

The results for ISO 816 are expressed differently and give the force to tear a test piece of standard width and thickness using the formula,

$$F_o = \frac{8F}{b_3 \, d}$$

where, F is the force in newtons required to tear the test piece, b_3 is the width in mm of rubber torn in the test piece and d is the thickness of the test piece in mm. The tear strength of all elastomers can be varied by compounding and by the degree of vulcanisation.

10.6 Cutting resistance

Resistance to cutting is another important factor for which there is no standard test. It is lessened if the rubber is under stress because the stress reduces the friction. As most of the ad hoc tests which have been devised operate under unknown and arbitrary friction conditions they do not measure an intrinsic strength property.

11

Dynamic stress–strain

To a rubber technologist 'modulus', as measured by the short-term methods described previously, means the stress in an elastomer at a certain strain. That is, it is a point on a stress–strain curve. Engineers, who are used to working with materials that obey Hooke's Law, where stress is directly proportional to strain, define modulus of elasticity as the ratio of stress to the strain produced by that stress. That is, the slope of a linear relationship between stress and strain. When materials are in tension or compression the engineering definition is usually known as Young's modulus.

The stress–strain (i.e. load-deformation) curves for an elastomer in tension and compression are approximately linear for strains of the order of a few percent and values of the engineering Young's modulus, E_o, can be obtained from these linear regions. Fortunately many engineering applications in which the relationship between stress and strain is important, e.g. engine mounting and vibration damping, only require the elastomer to be submitted to low strains. As the curves are continuous through the origin, the values of Young's modulus in tension and compression are approximately equal. Shear modulus, G, can be obtained in a similar manner but the values are about one quarter to one third of those for Young's modulus. As strain increases the departure from linearity must be taken into account. In addition, when elastomers are in compression and the loaded surfaces are prevented from slipping (by bonding or by mechanical location), the effect of the shape factor (see Section 10.3) must also be considered.

In a cyclic or dynamic situation, the fact that elastomers are viscoelastic materials and therefore contain both elastic and viscous elements is very important. The energy required to deform a perfectly elastic material is completely recovered when the force is removed but the viscous element, which is caused by internal molecular friction, retards elastic deformation and energy is lost. This lost energy is dissipated in the form of heat and the consequent temperature rise in the elastomer is called the 'heat buildup'. The percentage energy loss per cycle of deformation is known as the 'hysteresis'. If force is plotted against deflection for one cycle of deformation a hysteresis loop, such as is shown in Figure 11.1, is obtained. The area under the loading (top) curve is proportional to the energy input and the area under the unloading curve is proportional to the energy

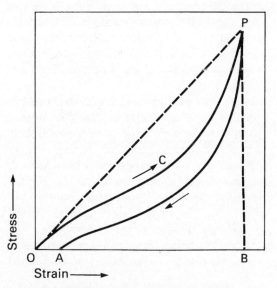

Figure 11.1 OCP is the stress–strain curve of extension for an elastomer and PA that for retraction. The energy absorption is proportional to the area OCPB. The energy returned on releasing the extension is proportional to the area APB and this area is also a measure of the resilience. The loss of energy, known as the hysteresis, is proportional to the area OCPA and is dissipated by the generation of heat in the elastomer

Table 11.1 National and international standards for dynamic stress–strain

General
ISO 2856
ASTM D2231
NF T46-026

Rebound resilience
ISO 4662
ASTM D1054
ASTM D2632 (falling weight)
BS 903: Part A8
DIN 53512
NF T46-036

Free vibration
ISO 4663
ASTM D945
BS 903: Part A31

Forced vibration
ISO 4664
BS 903: Part A24
DIN 53513

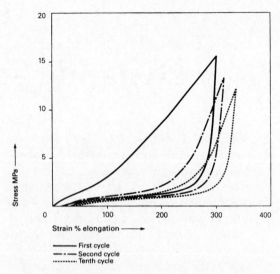

Figure 11.2 *The first, second and tenth stress–strain loops of a Neoprene W compound containing carbon black filler. Most of the structure breakdown and set occurs on the first cycle*

returned. The difference is the 'hysteresis energy loss'. If the relationship between stress and strain were perfectly linear the loop would be elliptical and if the sample were perfectly elastic the ellipse would collapse to a straight line.

When the loading and unloading cycle is repeated the shape and position of the hysteresis curves change. Figure 11.2 shows the first, second and tenth cycles for a Neoprene W compound containing carbon black filler. It can be seen that most of the breakdown of the structure occurs in the first cycle with the shift away from the origin being caused by set. Incidentally, it should be remembered that conventionally the stress and strain are calculated on the dimensions at the start of the first cycle.

Resilience is a term often used in rubber technology and is defined as the ratio of the energy returned on recovery from deformation to the energy required to produce deformation and is usually expressed as a percentage. It follows that hysteresis is 100% minus the resilience. Rebound resilience is used as a test of dynamic properties and is discussed in Section 11.3.

The dynamic properties of elastomers can be visualized in terms of a specimen undergoing uniform sinusoidal deformation as shown in Figure 11.3. The in-phase stress is due to the elastic component in the elastomer and the out-of-phase stress is due to the viscous component. The names arise because the former is in-phase and the latter out-of-phase with the strain curve. Because of the hysteresis losses, the strain lags behind the resultant of the two stresses by an amount which is usually known as the phase or loss angle, δ. The more viscous the material, the greater the loss angle.

The tangent of the loss angle, $\tan \delta$, in the simplest terms is the viscous modulus divided by the elastic modulus. For mathematical convenience, the viscous modulus is sometimes considered to be imaginary, although there is nothing imaginary about it in practice! The resultant or complex modulus is then defined as,

$$G^* = G' + iG'' \quad \text{for shear modulus}$$

or $\quad E^* = E' + iE'' \quad$ for tension or compression moduli

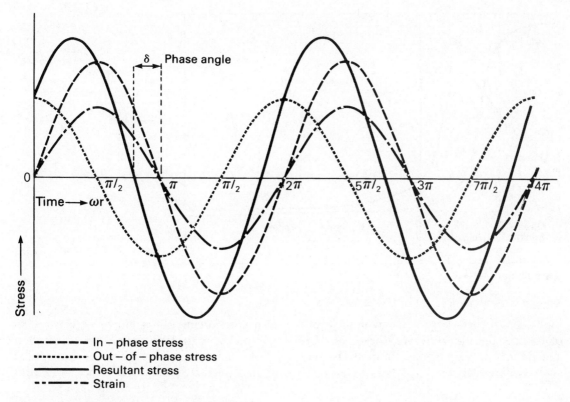

Figure 11.3 *Sinusoidal stress–strain time cycles*

The moduli measured by the short-term static tests described in previous Sections are assumed to be G' and E', and G'' and E'' are the out-of-phase or loss moduli.

The results of dynamic tests are dependent on the test conditions, test piece shape, mode of deformation, strain amplitude, strain history, frequency and temperature. It is usually recommended that dynamic tests should be made in shear because stress and strain are linearly related over a greater strain range than in tension or compression. In addition, with shear measurements the shape of the test piece can be chosen so that the strain is homogeneous and a shape factor correction need not be applied. Measurements at two strain amplitudes will show up any strain dependence.

The repetition of strain cycles, as is the normal procedure in a forced oscillation test, may cause a progressive change in the dynamic properties. At the beginning of the test, there may be stress softening as a result of mechanical conditioning (see Section 10) and the dynamic properties will, for practical purposes, reach a steady equilibrium level after a few cycles. Repetition of strain cycles also generates heat which can raise the temperature of the test piece and change its properties. This is most likely to occur when elastomers having a large loss factor are tested at high strain amplitudes. In one extreme case, during a test made at 15 Hz, a temperature rise of 2°C was observed after only 30 seconds.

Dynamic properties are dependent on both frequency and temperature and the general form of the effect of temperature on complex modulus and tan δ is shown in Figure 11.4. The effect of increasing or decreasing frequency is to shift the curves along the temperature axis to the right or left respectively. At room temperature the order of magnitude of the effect of temperature on modulus is 1%

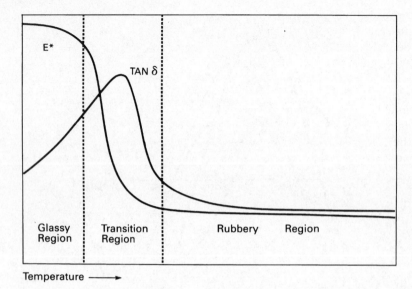

Figure 11.4 *Effect of temperature on dynamic properties*

per °C and the effect of frequency in the order of 10% per decade. However, these are average figures and particular compounds may show wide variations.

11.1 Combinations of dynamic and static deformation

It is commonly found, at least with elastomers containing reinforcing fillers such as carbon black, that if a dynamic cycle is superimposed on a static deformation the effective modulus over the dynamic cycle is greater than the slope of the static stress–strain curve at this point. This is illustrated in Figure 11.5, where a dynamic cycle represented by the loop DE is centred about the static loading point A. It is customary to describe the slope of the tangent FG to the static curve as the static modulus and that of DE as the dynamic modulus. There is an additional modulus corresponding to the slope of OA which is known as the chord or secant modulus. This modulus is also generally less than the dynamic modulus because the curve OAB is usually concave to the stress axis.

The relatively simple relationships between modulus, frequency and temperature apply to pure gum rubbers without fillers and fairly well to relatively soft rubbers containing little or no reinforcing fillers. However, elastomers reinforced with large quantities of carbon black do not have a unique value of modulus at any given frequency and temperature and the slope of DE will depend on the amplitude used in the test. That is, the dynamic modulus is a function of the extent of the deformation.

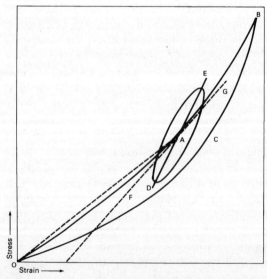

Figure 11.5 *Dynamic loop superimposed on a static stress–strain curve. Static or tangent modulus is the slope of FG, dynamic modulus is the slope of DE and chord modulus is the slope of OA*

In practice, if the reinforced elastomer is to be used under conditions involving a static load but with superimposed deformations over large cycles, its effective modulus for the static deformation will be lower than that measured by a test on the elastomer in its initial state. The static modulus used for calculations must then be taken from curve OCB in Figure 11.5 after the test piece has been mechanically conditioned several times.

11.2 The measurement of dynamic properties

It has been shown earlier that the dynamic properties of an elastomer are dependent upon the method of measurement, especially if fillers are present. It is, therefore, always important to define the test parameters used when testing.

The general requirements for dynamic testing are outlined in ISO 2856 but the standard does not give any details of the machines themselves. These may be classified according to their method of operation as follows:–

(a) Free vibration
(b) Forced vibration
   (i) away from resonance
   (ii) at or near to resonance
(c) Propagation
   (i) continuous waves
   (ii) pulses

Machines using free vibration set the test piece in oscillation and the amplitude is then allowed to decay due to the damping in the system. With forced vibration, oscillation of the test piece is maintained by external means and the frequency is adjusted either to be non-resonant or resonant. Methods based on propagation of ultrasound operate at much higher frequencies (kHz to MHz range) and as they are not covered by ISO 2856 will not be considered further in this work. The various forms used for dynamic testing are shown in Figure 11.6.

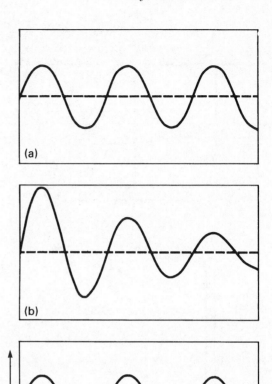

Figure 11.6 *Forms of strain or stress cycles used in dynamic tests (a) continuous constant amplitude (b) continuous decaying amplitude (c) successive half waves*

11.3 Rebound resilience

A very basic form of free vibration dynamic test consists of allowing a free-falling weight or an indentor attached to a pendulum to impact on the test piece and to measure the rebound. The rubber is thus subjected to only one half-cycle of deformation as shown in Figure 11.6(c). The rebound resilience is defined as the ratio of the energy of the indentor after impact to its energy before impact expressed as a percentage.

Thus, if the indentor falls under gravity the rebound resilience is equal to the ratio of the rebound height to the drop height. Resilience

44 The Language of Rubber

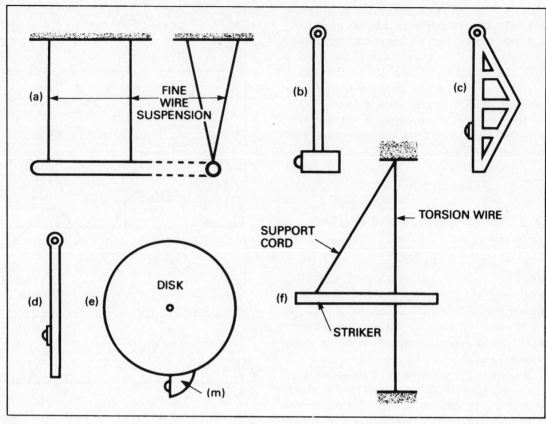

Figure 11.7 *Principles of some of the pendulums used for measuring rebound resilience (a) Lüpke (b) Schob (c) Dunlop (d) Goodyear-Healey (f) Zerbini. The principle of the Dunlop tripsometer with off-centre mass, m, is shown at (e)*

is not an arbitrary parameter and is approximately related to tan δ.

The rebound test has the attraction of simplicity and many types of apparatuses are used. However, rebound resilience varies with temperature, strain distribution, strain rate, strain energy and strain history and tests should be made under conditions applicable to service unless they are only required for quality assurance.

The method of testing using a pendulum, for elastomers with hardnesses ranging from 30 to 85 IRHD, is covered in ISO 4662. No particular design of apparatus is recommended but the Lüpke, Schob and Zerbini are mentioned as machines which, when suitably designed, will give results which will conform to the standard. The basic principles of six of the most widely used apparatuses are illustrated in Figure 11.7.

The characteristics for a pendulum testing machine as recommended in ISO 4662 are,

Indentor diameter (D) : 12.45 to 15.05 mm

Test piece thickness (d) : 12.5 ± 0.5 mm

Impacting mass (m) : $0.35 \genfrac{}{}{0pt}{}{0}{-0.1}$ kg

Impact velocity (v) : $1.4 \genfrac{}{}{0pt}{}{+0.6}{-0}$ m/s

Apparent strain energy density (mv²/Dd²) : $351 \genfrac{}{}{0pt}{}{+112}{-27}$ kJ/m³

The specified test piece is a disk with thickness 12.5 ± 0.5 mm and 29.0 ± 0.5 mm diameter but a number of non-standard alternatives may be used for comparative measurements. ISO 4662 calls for not less than three and not more than seven pre-conditioning impacts, followed by three impacts at the same velocity. The recommended test temperature is 23°C but as rebound resilience varies with temperature other test temperatures between −75° and +100°C are quoted. If the temperature is to be other than ambient it is necessary to use prescribed thermal conditioning procedures.

The method of calculating the rebound resilience varies with the type of equipment used. However, most modern machines provide scales reading directly in, or which are easily converted into, rebound resilience percentages or have data computers which provide the results in digital form with statistical evaluation and printout available if required.

An alternative to a pendulum is to use a falling weight apparatus in which a shaped plunger or a steel ball falls under gravity on to the test piece. An early falling weight instrument was the Shore sclerometer and another is covered by ASTM D2632. However, the falling weight method is not generally popular in Europe and there is no ISO standard.

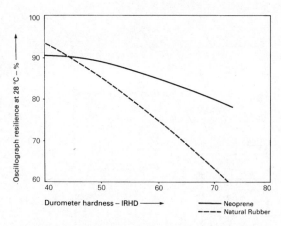

Figure 11.8 *Resilience of Neoprene and natural rubber at various hardnesses*

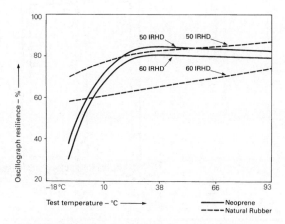

Figure 11.9 *Resilience of Neoprene and natural rubber at various temperatures*

11.4 Effect of compounding

The resilience of all elastomers can be varied through a wide range by appropriate compounding. Figure 11.8 shows how the resilience of a Neoprene (CR) and a natural rubber (NR) varies over a range of hardnesses. The curves show that NR 'gum' rubber compositions have slightly higher resilience than similar CR compounds. However, as the stiffness is increased by adding fillers, the resilience of the NR decreases much more rapidly than does that of CR. At hardnesses greater than IRHD 45, which is the range of greatest practical interest, CR's resilience is higher than that of NR.

Similarly Figure 11.9 shows how the resilience of CR and NR varies with temperature.

At temperatures above 10°C the CR 60 IRHD material has a significantly higher resilience than that of the NR.

11.5 Torsion pendulum

A widely used form of free vibration test is the torsion pendulum. The method is covered by ISO 4663 and consists of suspending a strip test piece of uniform cross-section to a fixed clamp at one end and fastening an appropriate mass which is free to rotate at the other. If the strip is twisted and released it will describe a series of decaying torsional oscillations. Several types of suitable apparatus are

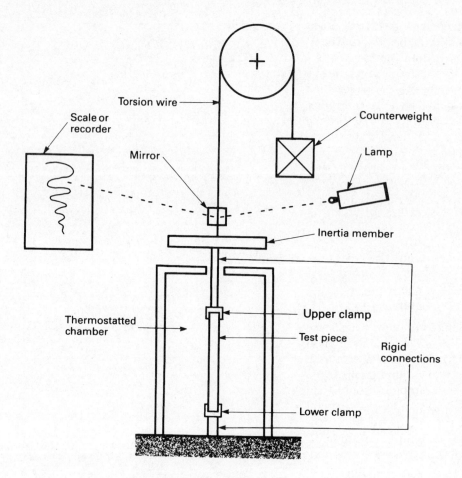

Figure 11.10 *Uncompensated free oscillation apparatus with counterweighted inertia member suspended above the test piece*

available but the basic principle is illustrated in Figure 11.10. It is also possible to add devices which can supply energy to the system and enable measurements to be made under conditions of forced vibration (see Section 11.7).

The dynamic properties of the rubber are determined from the frequency of oscillation, the logarithmic decrement of the compound pendulum and the geometry of the test piece. When necessary, the calculations must allow for the effect of the wire support and reference should be made to the standards for the full details.

The method is only suitable for frequencies from 0.1 to 10 Hz and is not particularly accurate for the determination of absolute values of dynamic modulus. It is mostly used for the determination of the temperature at which a rubber shows transitions in visco-elastic properties by plotting observed values of modulus and damping as a function of temperature.

11.6 Yerzley oscillograph

The Yerzley oscillograph is used in the USA as a free vibration test but is not much used in Europe. It is specified in ASTM D945 and is shown schematically in Figure 11.11. A horizontal beam is pivoted so as to oscillate

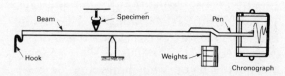

Figure 11.11 *Diagram of the Yerzley oscillograph*

vertically and in so doing to deform a test piece mounted between the beam and a fixed support. A pen attached to one end of the beam records the decaying train of oscillations on a revolving drum chart. Again, the possible frequency range is limited to a few hertz.

11.7 Forced vibration

Forced vibration methods of measuring dynamic properties are generally preferred for design purposes. However, as the relationship between dynamic stiffness and the basic moduli may be complex and only approximate, it is often preferable to work in stiffness at conditions relevant to the service. If, in a dynamic test with forced sinusoidal oscillations, force is plotted against deflection an hysteresis loop is obtained as shown in Figure 11.12 and,

Dynamic stiffness $S^* = \dfrac{\text{force amplitude}}{\text{deformation amplitude}} = \dfrac{f_o}{x_o}$

from which it can be shown that,

$\tan \delta = \dfrac{f_2}{f_1}$

where

f_o and x_o are the maximum force and deflection amplitudes respectively,

f_1 is the force at a deflection of x_o and

f_2 is the force at zero deflection.

Tests in shear are preferred but even in this case in practice the hysteresis loop for a filled rubber may be deformed as described in Section 11.1.

The general principles covering the determination of dynamic properties by forced sinusoidal shear strain are given in ISO 4664. The test piece recommended is as shown in Figure 11.13. The rubber elements should be firmly attached to the metal end plates either by direct bonding during vulcanisation or by using a suitable adhesive such as a cyanoacry-

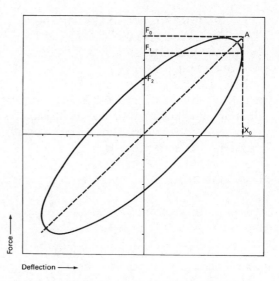

Figure 11.12 *Hysteresis loop in dynamic stress–strain cycle*

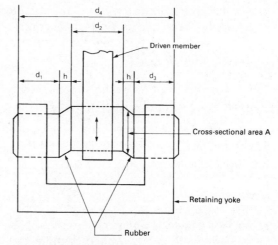

Figure 11.13 *Schematic arrangement of double shear test piece and mounting*

late. The actual dimensions will depend upon the forces and amplitudes that can be generated by the particular testing machine being used.

ISO recommends that at least six preconditioning cycles should be made before recording the measurements of force amplitude, displacement and damping. As materials with high loss angles may generate sufficient heat to cause a significant increase in the temperature of the test piece, it is also recommended that all measurements should be made within 60 seconds of starting the test. Preferably, measurements of force and displacement amplitudes should be made before those of damping.

There are many machines available for measuring dynamic properties by forced vibration and they may be classified according to their drive mechanism as mechanical, hydraulic or electromagnetic.

Mechanical

At least three mechanical methods are available:

a) A screw machine which operates in a similar way to a universal testing machine can be made to apply force or displacement cycles but it is limited to low frequencies, usually below 2 Hz.
b) A rotating eccentric weight can be used to apply force cycles and an eccentric cam can be used to apply displacement cycles (Roelig).
c) An electric motor can be used to impose a sinusoidal motion of the test piece via a sliding block yoke (Wallace-RAPRA).

The frequencies obtainable with mechanical drives are limited to less than 50 Hz and their use is decreasing in Europe.

Hydraulic

Servo valves which admit oil from a compressor-reservoir to a hydraulic cylinder can be used to provide sinusoidal or any other wave pattern such as step function, square or random vibrations. A wide variety of servo-hydraulic machines are available and in general can be used up to frequencies of 100 Hz. Power limitations make it difficult to achieve high force, amplitude and frequency at the same time but it is the most versatile method available, albeit at high capital cost.

Electromagnetic

Electromagnetic vibrators can cover a very wide frequency range, even up to 10,000 Hz. Quite high power is obtainable but the machines are usually relatively small and operate at low strains. It is the favoured method of drive when the force is applied at resonant frequencies. One of the best known examples of an electromagnetic machine is the Rheovibron™ developed in Japan, which operates between 3.5 and 110 Hz and over a range of temperatures. It uses small test pieces in tension but can also be used in compression and shear and at high elongations.

A number of small but versatile machines have been developed that are known as dynamic mechanical analysers (DMA). These machines are very sensitive and can provide dynamic data quickly and with small test pieces. They are generally computer controlled and can provide printouts of the data as well as graphical interpretations. It has been claimed that these machines can provide as much useful design information as simulated service tests but the absolute values of the results are often limited by the geometry of the test rig.

11.8 Resonance

Much of the pioneering work on dynamic properties was undertaken with forced vibration resonant testers in which the tests were made by tuning to the frequency which gives maximum amplitude of oscillation. By this means moderate amplitudes and high frequencies can be obtained from low power electromagnetic vibrators. The dynamic properties can be calculated with the aid of a set of equations based on linear vibration theory. Operating at resonance can detect subtle transitions in the properties of an elastomer but the lack of precision with high loss materials, for which tuning is difficult, and the distortion of the resonance peak due to non-linearity have led to a preference for the use of forced non-resonant methods.

11.9 Dynamic mechanical analysis

Typical of the modern methods of generating dynamic data is the TA Instruments 983

Dynamic stress–strain 49

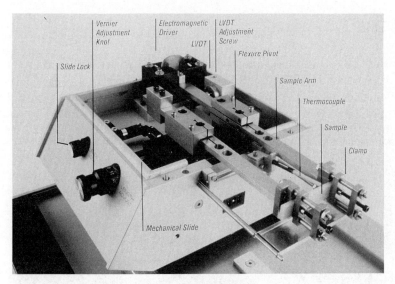

Figure 11.14 *Mechanical components of the TA Instruments 983 dynamic mechanical analyzer with the clamping mechanism for holding samples in a vertical configuration. Easily interchanged clamp faces are available to accommodate rectangular shapes and cylinders or tubing. A horizontal clamping system with smooth clamp faces is available to hold soft test pieces and supported liquids*

dynamic mechanical analysis system. The general arrangement of the mechanical components of the system is shown in Figure 11.14. The test piece is clamped between the ends of two parallel arms, which are mounted on low-force flexure pivots which only allow motion in the horizontal plane. The distance between the arms is adjustable by means of a precision mechanical slide to accommodate a wide range of test piece lengths. An electromagnetic motor attached to one arm drives the arm/test piece to a strain (amplitude) selected by the operator. As the arm/test piece system is displaced, the test piece undergoes a flexural deformation as depicted schematically in Figure 11.15. A linear variable differential transformer (LVDT) mounted on the driven arm measures the test piece's response (strain and frequency) to the applied stress and provides feedback control to the motor. The test piece is positioned in a temperature-controlled chamber which contains a radiant heater and a coolant distribution system.

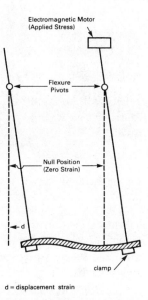

Figure 11.15 (a) *Schematic of method of operation of the TA Instruments 983 dynamic mechanical analyzer.*

The unit can be operated in a number of modes. When operating in the fixed frequency mode the test piece is subjected to a sinusoidal oscillation at a frequency and amplitude selected by the operator. The unit can be programmed to measure the viscoelastic characteristics of a test piece at up to 57 frequencies during a single test. The test piece is allowed to reach mechanical and thermal equilibrium at each frequency before the data is collected. After all the frequencies have been scanned, the test piece is automatically stepped to the next temperature and the frequency scan is repeated.

In the resonance mode, the unit operates on the mechanical principle of forced resonant

Figure 11.16 (a) *Barco Industries Dynaliser*™

vibratory motion at a fixed amplitude (strain), which is selected by the operator. The arms and test piece are displaced by the electromagnetic driver, subjecting the test piece to a fixed deformation and setting the system into resonant oscillation.

In free vibration, a test piece will oscillate at its resonant frequency with a decreasing amplitude of oscillation. The resonance mode differs from the free vibration mode in that the electromechanical driver puts energy into the system to maintain a fixed amplitude. The make-up energy, oscillation frequency and test piece geometry are used by the DMA software to calculate the desired viscoelastic properties.

The other two operating modes are used to measure stress relaxation and creep as functions of time and/or temperature.

11.10 The Dynaliser™

The Dynaliser™, supplied by Barco Industries in Belgium and illustrated in Figure 11.16, can

Figure 11.16 (b)

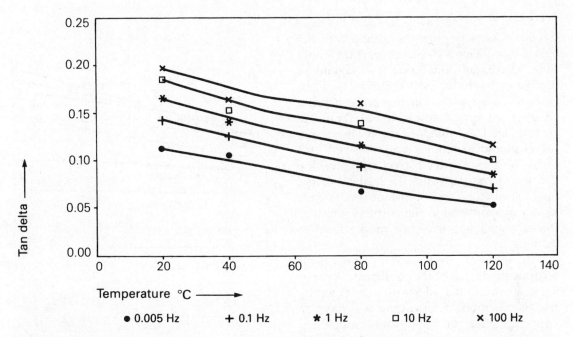

Figure 11.17 *The graph shows the variation of tan δ with temperature and frequency for a typical HYPALON elastomer, as measured with a Dynaliser*™

be used to determine the dynamic properties of an elastomer over the frequency range 0 to 100 Hz. It is different from most dynamic test machines in that it performs a strain controlled relaxation test by imposing a known deformation on the sample and measuring the variation of the reaction force as a function of time.

The test piece, which must be at least 7 mm thick but otherwise can be of any shape or size, is deformed by the fast penetration of a spherical indentor. The indentation is kept constant for the minute or less required for the test and an integrally mounted transducer produces the data for a force relaxation curve. A computer then calculates the elastic modulus, E', the viscous modulus, E", and tan δ and provides a printout of the results. The correlation between the results obtained with this instrument and other dynamic test methods is extremely good and the method has the advantages that it is non-destructive, can be made on finished parts and is performed very quickly.

A typical graph obtained with the Dynaliser™ is provided in Figure 11.17, which shows the variation of tan δ with temperature and frequency for a typical HYPALON elastomer.

11.11 Interpretation

The results of dynamic tests on elastomers are dependent on the test piece shape, mode of deformation, strain amplitude, strain history, frequency and temperature but they are of direct use to the engineer. This is because the measurements are usually made in shear or compression and at low strains, which is the way most elastomers are stressed in service. In addition, the test pieces can be subjected to mechanical and/or thermal conditioning related to actual practice and the data supplied can be applicable to a range of conditions rather than to a single point. Finally, the results are expressed in common engineering units.

However, it is necessary to select the right dynamic conditions for the application. For example, a V-belt is flexed at constant amplitude, so constant amplitude tests should be used to compare the heat buildup characteristics of compounds. On the other hand, a solid tyre running at low speed is vibrated at constant load and compositions for this application should be studied with constant load tests. The choice of the correct kind of vibration is important because the heat generated per cycle of vibration at constant amplitude is *directly* proportional to the stiffness whilst at constant load it is *inversely* proportional to stiffness.

Dynamic tests can be used to give a fairly accurate prediction of the equilibrium operating temperature of a rubber product in service and the operating temperature will have an important effect on its physical properties. Elevated temperatures generally impair mechanical strength and accelerate the ageing process.

There is a tendency to assume that a compound having high hysteresis will be unsatisfactory but this is not necessarily true. In some applications hysteresis and resilience have little or no bearing upon performance. In certain vibration-damping applications, compounds with relatively low resilience may even be desirable because their damping effect may limit the maximum amplitude developed in service. This may be especially true when the frequency of vibration in service varies over a wide range because destructive amplitudes may develop in a highly resilient compound when the frequency is at or near to resonance. On the other hand, if the frequency and amplitude of vibration produce an excessive heat buildup in a low resilience compound the use of a highly resilient compound is indicated.

The graphs in Figure 11.18 are typical of the results obtained from dynamic testing. Neoprene (CR) and natural rubber (NR) have been compared for their suitability for hydraulic fluid-filled engine mounts by measuring dynamic shear moduli and loss angle at frequencies up to 100 Hz. The graphs show that both materials are equally suitable, the

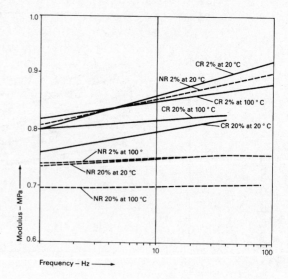

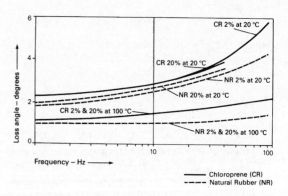

Figure 11.18 *Dynamic modulus in shear and loss angle of comparable Neoprene and natural rubber formulations tested at strains of 2% and 20% and temperatures of 20° and 100°C. The differences in moduli and loss angles are within the generally acceptable limits for control of mass production of automobile mountings and show that properly formulated Neoprene is equivalent to natural rubber for this application*

differences being within the accepted limits of control when mass producing automobile mountings.

There is a regrettable shortage of published quantitative engineering data for elastomers. The most comprehensive is that provided by MRPRA in the form of data sheets covering 46 different natural rubber compounds. In general the test procedures follow the ISO standards but a different approach has been used for the presentation of data on static compression modulus and static and dynamic shear moduli. Disk test pieces are used rather than rectangular and the conditioning deformations are more severe than those required by the standards. For example, when measuring static shear modulus, nine loading and unloading cycles to 100% shear strain, with no delay between cycles, are used and the test piece is taken to failure on the tenth loading. Data from the first and tenth loadings are provided for various strains as chord modulus (see Section 11.1) and tangent modulus (assuming the stress–strain curve is parabolic locally to the chosen strain). The set at the end of the ninth cycle is also provided as a percentage of the thickness. These conditioning strains are roughly double the strains most frequently used in service and therefore provide a safety factor for subsequent calculations.

With the dynamic tests, the rubber is conditioned at 10% strain amplitude for all strain amplitudes up to 10% but with at least several minutes' recovery time after the conditioning.

The approach used by MRPRA is designed primarily for natural rubber compounds, where the crystallization of the polymer can greatly affect the engineering properties, and it will not necessarily apply in all cases to synthetic elastomers. However, there is undoubtedly a need for similar data to cover the whole range of elastomers used in engineering.

12

Transmissibility

Rubber mounts are used to ensure that the dynamic oscillations created in a body are not transmitted to the supporting structure or, conversely, that the vibrations in a support are not transmitted to the body. The mounting for an internal combustion engine is an example of the former and the protective mounting for a delicate instrument an example of the latter. The ability to isolate vibrations is measured by transmissibility (T), which is defined as the ratio of the transmitted force to an externally imposed force for a body supported on a rigid base through the interposition of the mountings. Alternatively, if the base is subjected to periodic oscillatory movements of known amplitude and frequency, the definition becomes the ratio of the transmitted amplitude to the imposed amplitude. Transmissibility can be calculated from the results of dynamic tests and the ability to do so may be built into the software associated with the test equipment. It is common practice amongst vibration and noise-control engineers to express transmissibility in terms of decibels. Strictly a quantity expressed in decibels is a ratio, not an absolute magnitude but by using a reference standard of intensity it can be used to express absolute values.

a) T becomes unity when n/n_o is very small or when it is equal to $\sqrt{2}$;
b) when there is no damping the resonance peak at $n/n_o = 1$ is infinitely high but the height decreases as tan δ increases;
c) T falls quickly as n/n_o is increased, and
d) for any given value of n/n_o greater than $\sqrt{2}$, T decreases with decrease of tan δ,

i.e. the rubber showing the least damping, as conventionally measured, is the most effective in reducing transmissibility in the higher frequency ranges.

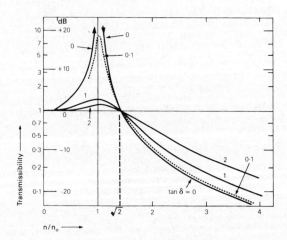

Figure 12.1 *Dependence of transmissibility on the ratio of imposed frequency n to natural frequency n_o for various tan δ. Transmissibility is shown both as a ratio and on the decibel (dB) scale*

12.1 Interpretation

Figure 12.1, taken from a paper presented at the 'Use of Rubber in Engineering' conference held in London in 1966, shows the theoretical relationship between transmissibility and the ratio of imposed frequency, n, to natural frequency, n_o, for various values of tan δ. It can be seen that,

Figure 12.2, taken from the same source, shows the actual curves obtained for a variety of vulcanised rubbers having various degrees of damping. Although natural rubber (NR),

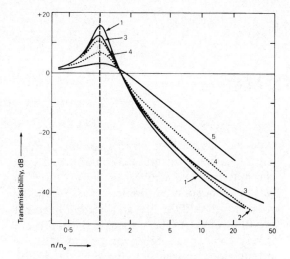

Figure 12.2 *Comparison of transmissibility vs the ratio of n to n_o for elastomers having various degrees of damping: 1. natural rubber (least damping); 2. styrene butadiene rubber (SBR); 3. Neoprene (CR); 4. butyl rubber (IIR); 5. nitrile rubber (NBR) (highest damping)*

styrene butadiene rubber (SBR) and Neoprene (CR) have the least effect on reducing the transmissibility at resonance, they are the most effective at the higher frequencies as the rate of transmissibility change with frequency becomes nearly equal to the theoretical 12 dB per octave. Since an anti-vibration mounting is required to suppress unwanted frequencies well above the resonant frequency these rubbers are evidently the most effective for this purpose. On the other hand, nitrile rubber (NBR), which has the highest damping, inhibits the effect of resonance more successfully whenever there are input vibrations at resonance but is less successful at isolating the system from input vibrations at frequencies significantly above the natural frequency. When selecting a material for a specific design, the optimum will depend on the range of input vibrations of greatest concern.

The addition of carbon black to a rubber compound increases both the dynamic modulus and the phase angle and these properties become amplitude-dependent. At low strain amplitudes, the dynamic stiffness can become unexpectedly high and the phase angle low which result in a much higher natural frequency and higher peak transmissibility than at lower amplitudes. This effect can become very important in the design of anti-vibration mountings that have to deal with a range of input vibrations and emphasizes the need for adequate dynamic testing.

13

Fatigue

Fatigue may be defined as the change in properties that occur in a material after prolonged action of stress or strain. It follows, therefore, that creep, set and stress relaxation in elastomers are in the widest sense examples of fatigue. However, the term is used in this Chapter in the narrower sense to cover the changes that result from crack growth under repeated cyclic deformation (dynamic flexing) or static stress (time-dependent fatigue). Repeated bending or oscillatory movement is applied, for example, in the dynamic tests considered previously in Chapter 11. This type of fatigue causes surface cracking which can eventually propagate. In addition, unless the heat is dissipated, the energy losses due to hysteresis may increase the internal temperature of the elastomer and cause 'heat buildup'. Which effect predominates will depend upon the test piece geometry.

Fatigue cracking is influenced by many factors. First, the type of rubber, the type and degree of cross-linking, additives such as fillers and protective agents (which control the basic crack growth characteristics) and the size of the flaws that are present initially: next, mechanical considerations such as, the shape and size of the article, the nature and magnitude of the deformations and the frequency and form of the cycling: lastly, the environmental factors, temperature, humidity and potentially hostile agents such as atmospheric oxygen and ozone. Ozone, in particular, can cause severe cracking provided the test piece is at least lightly strained. The standard tests for the De Mattia type machines, for example, emphasize that ozone must not be present in a concentration greater than 1 part per 100 million parts of air.

The relative effects of mechanical stresses, oxygen and ozone are illustrated in Figure 13.1, which shows the fatigue life that can be expected for various strains for an unfilled natural rubber vulcanisate containing 1 part per hundred of rubber of antioxidant in various atmospheres. Similar curves would be obtained if fatigue life was plotted against maximum stress. The tests *in vacuo* represent purely mechanical failure and show that the fatigue life approaches infinity when the maximum strain of the fatigue cycle is sufficiently below 100%. A similar fatigue limit is shown by all rubbers but its magnitude will vary.

The results in the laboratory atmosphere, which contained oxygen and a limited amount of ozone, reduced the fatigue limit at 100% strain by about 25% and also reduced the fatigue life at larger strains. When the

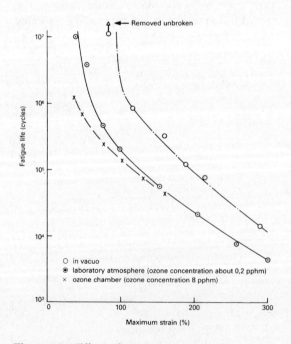

Figure 13.1 *Effects of maximum strain and atmosphere on tension fatigue life for a natural rubber vulcanisate*

ozone concentration was increased the fatigue limit was again reduced at low strains but the ozone had little effect at higher strains where the fatigue life is less. All the results showed that the fatigue life is very dependent upon the maximum strain.

Rubber technologists divide their laboratory tests into two types. The first is known as 'flex cracking', where cracks are induced in fairly thin test pieces without a large increase in temperature. The second type, referred to as 'heat buildup' tests, is made with relatively thick test pieces in which the temperature is allowed to increase until the internal structure of the elastomer breaks down. Neither type provides fundamental information and the results are very dependent upon the test piece geometry and the apparatus used. More precise information about flex cracking is becoming available through the application of the principles of fracture mechanics and these techniques are expected to be used more frequently in the future – see Section 13.3.

13.1 Flex cracking and cut growth

Flex-cracking tests strain the test piece in flexure and were chosen because this is the type of deformation experienced in service by tyres, belting, footwear, etc. Many types of apparatuses have been developed, some of which are only applicable to particular products, but their use is mainly as a form of simulated service test. The De Mattia machine is the most widely used machine in Europe.

The principle of the De Mattia machine is illustrated in Figure 13.2. The upper grip is fixed and holds one end of the test piece stationary. A second grip, which is activated by an eccentric drive, imparts a reciprocating motion to the other end of the test piece. The ISO specifications call for a maximum distance between the grips of 75 mm, a travel of 57 mm and a reciprocating movement of 5 Hz. At least six test pieces and preferably 12, should be accommodated by the machine at one time. The machine may be enclosed in a temperature-controlled chamber for testing at elevated temperatures.

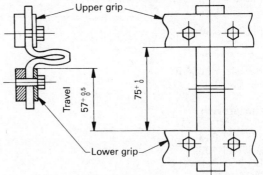

Figure 13.2 *Principle of the De Mattia machine*

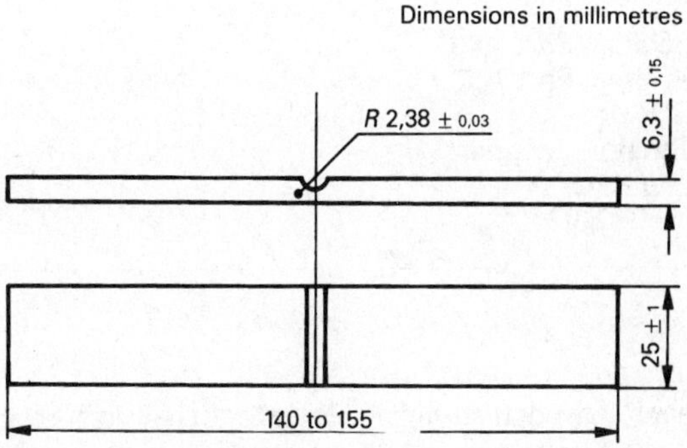

Figure 13.3 *De Mattia test piece*

The test piece is a strip with a moulded groove, as shown in Figure 13.3, which may be moulded individually or cut from a wide slab having a moulded groove. The results of the test are dependent upon the thickness of the test piece and comparisons should only be made with mouldings which are within the tolerances specified.

There are two ISO specifications for flex testing: ISO 132 for the determination of flex cracking and ISO 133 for the determination of cut growth. The test pieces are flexed until the first minute sign of cracking is detected. The number of flexing cycles at this point is recorded. The machine is restarted and stopped for inspection after intervals in which the number of flexing cycles is increased in geometric progression, a suitable ratio being 1.5 on each occasion. The cracking is graded at

Table 13.1 *National and international standards for fatigue testing*

Flex cracking
ISO 132
ASTM D430
BS 903: Part A10
NF T46-015

Cut growth
ISO 133
ASTM D813
ASTM D1052
BS 903: Part A11
NF T46-016

Flexometer testing
ISO 4666/1/2/3
ASTM D623
BS 903: Part A49
BS 903: Part A50

Tension fatigue
ISO 6943
NF T46-021

---

each inspection into six categories, ranging from less than ten pinpricks that can be seen by the naked eye, to a length of the largest crack greater than 3 mm. The weakness of the test is, of course, the subjective nature of the observations.

A variation of the test, known as the 'crack or cut growth' method, is covered by ISO 133. In this case, a 2 mm cut is made through the rubber with a tool of specified geometry and the length of the cut is measured with the aid of a low-powered magnifying glass at frequent intervals. The number of cycles for the cut to extend by 2 mm and then by further extensions of 6 and 10 mm are recorded. Comparison is slightly less subjective than for flex cracking but good reproducibility is difficult.

It is recommended that both flex-cracking and crack-growth tests should be made because certain soft vulcanisates, notably those made from styrene-butadiene rubbers, can show marked resistance to crack initiation yet have a low resistance to crack propagation.

Alternative machines

Although no longer covered by European standards, a number of other machines have been used to test flex fatigue and those which are covered by ASTM standards should be mentioned. ASTM D430 describes the Scott flexer, which is used for testing ply separation in belts, tyres and other articles in which fabric is bonded to rubber and the Du Pont flexer in which test strips are joined together in an endless belt and are driven over and under four pulleys. ASTM D1052 describes the Ross flexer which is used for determining cut growth. This machine controls the maximum and minimum strains rather more precisely than other bending tests and is frequently used for testing shoe soling materials.

13.2 Heat buildup

Heat buildup tests aim to raise the temperature of a relatively bulky test piece by dynamic fatigue and ultimately to cause internal degradation of the polymer, rather than to induce surface cracking. The machines used are known as 'flexometers' but they are quite different from the equipment used for testing flexing. Flexometers can operate in compression, shear or a combination of the two and various designs have been used and standardized over the years. The relevant international standard is ISO 4666 which is issued in three parts: Part 1 Basic Principles, Part 2 The Rotary Flexometer and Part 3 The Compression Flexometer. The rotary unit is derived from an older machine known as the St Joe flexometer and is in some ways similar to the Firestone flexometer described in ASTM D623. The compression unit is similar to the Goodrich flexometer, also covered by ASTM D623. Compression set and creep, when required, are measured at specified times during the tests.

Rotary flexometer

The rotary flexometer test described in ISO 4666/2 is designed to measure the temperature rise and the resistance to fatigue of a rubber compound under rotary shear loading. A cylindrical test piece with its height equal to its diameter, acts as a friction clutch between two chuck plates mounted on parallel shafts which can be rotated. One chuck plate has a fixed bearing and the other has a bearing that can be displaced axially to produce either compression under a constant pre-strain or under a constant pre-stress. The movable bearing can also be displaced transversely with respect to the axis of the test piece. The deflection can be by a predetermined amount (constant cyclic strain) or under the action of a predetermined force (constant cyclic stress)

The chuck plate with the fixed bearing is rotated at a frequency of 14.6 or 25 Hz and the test piece transmits the rotation to the plate with the moveable bearing. The rotation and simultaneous deflection subject the test piece to a sinusoidal form of shear deformation. All movements and forces can be measured and the temperature at the centre of the test piece is taken by means of a needle probe.

Tests are normally undertaken at room temperature and the temperature rise measured after 20 minutes or when equilibrium has been reached. The test conditions for fatigue breakdown measurements should be designed for failure to occur in approximately 30 minutes, the onset of breakdown being indicated by changes in the transverse force or transverse displacement. The test piece is sectioned by cutting in an axial plane after the completion of the test and visually examined for the presence of fine pores, cracks and degradation of the rubber. Tests are repeated for various loading conditions and the resistance to fatigue breakdown is presented graphically as shown in Figure 13.4. The cyclic strain amplitude or the cyclic stress amplitude at which the fatigue life becomes essentially parallel to the log N axis is taken as the limiting fatigue deformability or limiting fatigue stress.

Compression flexometer

The compression flexometer uses cylindrical test pieces 17.8 mm diameter and 25 mm high, which can be moulded or cut from a slab. The test piece is placed between anvils faced with a thermal insulating material. The top anvil is connected to an adjustable eccentric usually driven at an oscillation rate of 30 Hz with a stroke (double amplitude) of 4.45 mm. Strokes of 5.71 or 6.35 mm can also be used when more severe test conditions are required. A compressive load of either 1 or 2 MPa is applied to the test piece through a lever system having high inertia. The temperature is measured with a thermocouple situated in the base of the test piece.

Tests are carried out with a constant applied load and continuous measurement is made of the change in height of the test piece. Permanent set is measured after testing. Tests are made at room temperature, 55° or 100°C. In general, for a medium hardness rubber with ordinary temperature rise characteristics a pre-stress of 1 MPa, a stroke of 5.71 mm and a temperature of 55°C is required for heat buildup tests, the temperature being measured after 25 minutes. More severe conditions are required if breakdown tests are to be completed within a reasonable time.

Fatigue life is expressed as the number of cycles to failure and breakdown is indicated by a sudden rise in temperature or a marked increase in creep and by the onset of internal porosity in the test piece. The use of the

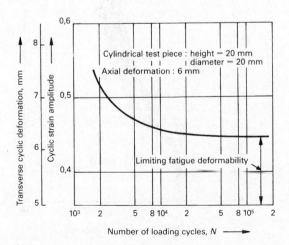

Figure 13.4 *Example of fatigue life curve with predetermined cyclic deformation*

compression flexometer is not recommended for rubbers having a hardness greater than 86 IRHD.

13.3 Fracture mechanics

The mathematical treatment of fracture mechanics began with the classic work of A A Griffith in the 1920s but it was not until R S Rivlin and A G Thomas, working at the Malaysian Rubber Producers' Association (MRPRA) in the 1950s, that the theory was extended to apply to elastomers. The practical application of the theory depends upon determining the energy required to cause unit area of new crack growth, generally termed the 'tearing energy', T. Conversely, T is the internal energy released by the growth of a crack, which under some circumstances can be sufficient to cause rupture of the rubber. The concept of tearing energy is the same as the strain energy release rate or fracture energy as defined for metals and plastics.

Equations for calculating the tearing energy have been developed for several different geometrical shapes and loading directions. For example,

$T = 2 k W c$ (for simple extension)
$T = 0.4 W t$ (for simple shear)
$T = 0.5 W t$ (for uniaxial compression)

where

W is the average stored energy density in the rubber layer,

c is the crack length,

t is the rubber layer thickness, and

k is a constant which varies with strain but lies between 2 and 3.

W is obtained by determining the area under a tensile stress–strain or load deflection curve.

In shear and compression the tearing energy is not a direct function of the crack length. However, in tension, tearing energy depends directly on the crack length, which means that it increases as the crack grows.

This is the reason for avoiding the use of rubber in tension as far as possible and the preference for using rubber in shear or compression.

It has been found that if crack growth rate, expressed as dc/dn (increase in cut length per cycle), is plotted against tearing energy the resulting curves are independent of whether the rubber tested was in tension, shear or compression. There is some scatter with the results but within those limits the relationship between tearing energy and cut growth is a material characteristic and is independent of component geometry. The relationship differs from rubber to rubber but, once derived, the tearing energy can be calculated and the crack growth rate can be estimated from curves, such as those given in Figure 13.5, obtained with strip test pieces in simple extension. At very low tearing energies, i.e. low strains, the failure is by chemical mechanisms and the crack growth rate is very low, 0.1 mm/year or

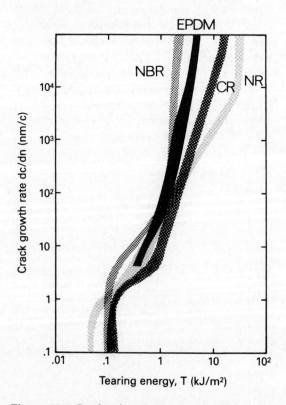

Figure 13.5 *Graphs of tearing energy against crack-growth rate for four engineering elastomers*

less. As the tearing energy increases the crack growth rate increases until finally failure can occur catastrophically from a single cycle. This condition is indicated in Fig.13.5 by dc/dn increasing asymptotically towards the static critical tearing energy. The latter is related to the tear strength as described in Section 10.5.

It can also be seen from Figure 13.5 that no single material will have the best fatigue life in all conditions. For large numbers of very low amplitude cycles, Neoprene will have a better fatigue resistance than natural or nitrile rubber. However, at high tearing energies the situation is reversed. Natural rubber exhibits only moderate crack-growth rates whilst nitrile will fail more quickly and Neoprene and EPDM show intermediate properties. There is also a central region where the fatigue resistance of all the materials is very similar. It should be appreciated that the curves in Figure 13.5 are only examples and different results could be obtained with other formulations.

Tearing energy data enable crack-growth rates to be calculated for rubber components subjected to known loads and/or motions. To apply this information to a component, a load/motion spectrum needs to be determined, either from past experience of similar applications or from engineering analysis. This spectrum can then be converted to a spectrum of tearing energies and the cumulative crack growth calculated. In addition, it is necessary to be able to define the life of a component in terms of crack growth in order to finally complete the analysis. A number of motor manufacturers and specialist consultancy firms are now able to undertake such sophisticated design analyses.

13.4 Fatigue tests in tension

The direct measurement of crack growth in tension provides a useful method of determining tearing energy but there is no international standard covering the technique. A standard servo-hydraulic test machine, such as is used for dynamic testing, can be used and has the advantage that force/deflection data can be collected in situ. This allows tearing energy to be automatically up-dated to take into account set or softening in the test pieces. Great accuracy is possible but only one test piece can be processed at a time.

Multi-station test machines have been developed specifically for these tests and a typical unit is that shown in Figure 13.6, supplied by the Materials Engineering Research Laboratory (MERL) in Hertford, UK. Up to 12 test pieces, each approximately 150 mm long and 25 mm wide, can be mounted in the machine and tested simultaneously. The test pieces are attached to a large plate, six test pieces above and six below, which is driven by an eccentric cam that can produce frequencies between 1 and 50 Hz. The fatigue cycles can be superimposed on a

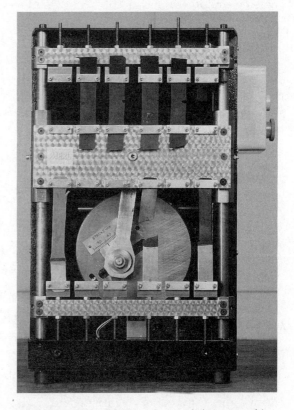

Figure 13.6 *The MERL crack growth fatigue machine provides a means of performing fatigue tests on strips of vulcanised rubber for the purpose of determining the relationship between tearing energy and crack-growth rate. This is an intrinsic property of elastomers and can be used in design calculations*

minimum static strain providing non-relaxing conditions when required. Crack length can be measured to an accuracy of 0.05 mm with a digital-readout travelling microscope or complete monitoring of crack-growth behaviour can be made by video or still camera. Permanent records of any features of interest, at a magnification of up to 30 times, can be recorded by a 35 mm camera. The number of cycles for a specific test may be pre-set on an electronic counter and the machine will automatically stop when this number is reached. Tests may be carried out at any temperature from ambient to 150°C.

The test pieces are normally cut from moulded sheet 1 to 2 mm thick but the results are not in principle dependent upon thickness. Load-deflection or stress–strain characteristics are obtained from standard extensometer equipment and a cut about 0.5 mm long is made in the centre of one edge of the test piece before it is mounted in the machine. The machine stroke is fixed at a chosen value, e.g. 20 mm, to ensure accuracy at higher frequencies. The test piece grip and/or the test piece gauge length can be adjusted to provide any test strain up to about 100%. During the test, the cut length, c, is measured at intervals of cycles, n, corresponding to a 10 to 20% increase in cut length. The rate of growth dc/dn is determined from the difference in cut length divided by the number of cycles between two readings. The tearing energy, T, for this rate is calculated from the average of the two cut lengths and the strain energy density, W, at the maximum strain of the cycle, using the equation $T = 2kWc$. W is obtained from the relevant area under the load-deflection or stress–strain curve. The results are presented graphically in the form shown in Figure 13.5.

The test is stopped when the cut reaches 20% of the test piece width because the theory assumes this ratio to be small. As the cut grows the value of T increases and it is possible to cover about a tenfold change in tearing energy with a single test piece. A different range of T can be covered by cycling to a different maximum strain, thus changing W.

If set occurs during the course of a test the maximum strain will decrease because the maximum extended length remains constant. W must then be obtained by finding the area under the stress–strain curve up to the new but lower strain.

ISO tension tests

Simpler tests, loosely based on the principles of fracture mechanics, have been developed and the method of testing is covered in ISO 6943. Typical machines are the Monsanto fatigue-to-failure and the Wallace-MRPRA fatigue tester. The test pieces can be either dumb-bells similar to those used for tensile testing or rings (Schopper rings) cut from sheet. The dumb-bells are similar to those specified for tensile tests except that they are 1.5 mm thick and generally have a raised bar across each end to aid location and gripping. The rings have a nominal internal diameter of 44.6 mm and an external diameter of 52.6 mm, which gives a nominal radial width of 4 mm. The thickness is again 1.5 mm.

ISO 6943 calls for a fatigue testing machine which can provide a reciprocating motion at a frequency of between 1 and 5 Hz. The Monsanto equipment can test up to 24 dumb-bells and the Wallace–MRPRA 12 rings at one time. The dumb-bells are held in clamps and the rings are slipped over four pulleys as shown in Figure 13.7. The stroke of the machines and the position of the fixed clamps or rollers can be adjusted to provide a range of test strains. In all cases the test piece is relaxed to zero strain for part of each cycle. This is a mandatory requirement for this standard. The machines and testing conditions are designed to avoid heat buildup but the standard does provide for testing at elevated temperatures if necessary. However, excessive set may influence the results at higher temperatures. Even at ambient temperatures, the standard requires set and changes in test piece length to be measured and corrections to be applied as the test proceeds. Tests continue until breakage occurs and the number of cycles is recorded.

Fatigue 63

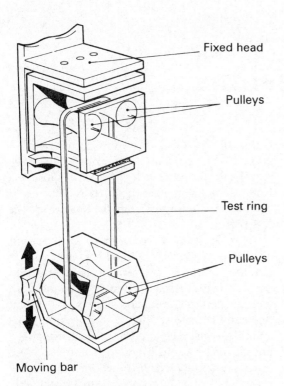

Figure 13.7 *Diagram of the system of pulleys used for ring test pieces on the Wallace-MRPRA fatigue tester. This method of mounting has very low friction and does not depend on the skill of the operator*

It is recommended that a minimum of five test pieces should be used for each measurement and, as it is usual for considerable scatter to occur in the results, that simple statistical analysis methods should be used in their interpretation. For many purposes it is desirable to plot the results in the form of a graph of fatigue life against maximum strain, stress or strain energy density. It is recommended that a logarithmic scale be used for fatigue life and a linear scale for strain. A graph of fatigue life against maximum strain energy density on double logarithmic scales will often give a linear relationship over a considerable range; a similar graph of fatigue life against maximum stress may also give a linear relationship but at a different slope.

13.5 Interpretation

Unfortunately, the test results obtained from the De Mattia tests only have any significance under the exact conditions used and their value is restricted to comparative selection of compounds or production control. They also suffer from the fact that they are partially dependent upon a subjective assessment of the degree of cracking by the operator.

The flexometer tests are also of an arbitrary nature and the exact moment of breakdown is sometimes difficult to determine. Thus the results have little significance apart from under the conditions of the test. They are most often used to measure heat buildup.

The fatigue tests in tension have the advantage that they are not so dependent upon the operator. However, the results can easily be affected by poor preparation of the elastomers and/or test pieces and there is often considerable scatter. The ratio of highest to lowest readings can vary from less than 2 for polyisoprene to greater than 10 for SBR. The ring tests suffer from the problem that abrasion at the contact surface between the ring and the pulley can sometimes cause premature failure.

As tearing energy is a fundamental property of an elastomer the results obtained by the direct measurement of crack growth can be used for design calculations. There have been some successful predictions of the expected life of critical engineering components, not the least of which has been for the structural bearings used in deepwater offshore oil platforms. However, the techniques are relatively new and the interpretation of the practical relevance of the results requires considerable skill.

Temperature, and the presence of oxygen and ozone must always be taken into account when assessing the importance of the results of fatigue testing. In particular, very small quantities of ozone will reduce fatigue life unless the elastomer has been protected by the use of suitable antiozonants.

14

Abrasion

In many applications resistance to wear is one of the most important properties of a rubber compound, yet it is one of the most difficult to analyse and measure. For example, the type of wear experienced by an automobile tyre on a road is quite different from the wear on a hose used for sand-blasting. A tyre is subjected to a grinding action, the severity of which is dependent upon the various types of road surface, whereas the high velocity particles of sharp sand provide a cutting and scouring action on a sand-blasting hose. Although a number of standard tests bear some superficial resemblance to the kind of wear received by a tyre tread they do not generally correlate well with road tests.

In some cases the standard tests can actually be misleading. For example, a rubber compound which under test appears to be unsatisfactory for sand-blasting hose may in practice be excellent. Conversely, it is not safe to assume that a rubber proved to be good for a sand-blasting hose will also be suitable for shoe soling or for use as conveyor belt. Standard methods of abrasion testing should be related to suitable simulated service tests wherever possible.

There are also some purely mechanical testing difficulties that may produce misleading results. For example, an under-cured composition (soft and gummy) or one with a wax or oil content may appear to have excellent wear properties merely because the test machine has been clogged or lubricated by material removed from the surface of the test piece.

Wear is usually considered in terms of abrasion, which is defined as the loss of material that results from mechanical action on a rubber surface. Abrasion resistance is a complicated phenomenon and is dependent on many things, resilience, stiffness, thermal stability, resistance to cutting and tearing, etc. and different applications require these properties in widely varying proportions. Laboratory tests must be completed in a reasonably short time and generally consist of rubbing the test piece against sharp abrasive surfaces of, for example, corundum (aluminium oxide) with heavy loading at high slip velocities and measuring the volume or weight loss. These conditions are difficult to relate to actual service requirements and as a result, many different devices have been developed, some only applicable to one particular application. Many famous names, such as Akron, Dunlop, Du Pont, National Bureau of Standards, Pico and Schiefer and Taber have been associated with abrasion testers. Although the principles, and in some cases the machines, remain the same, international standards organizations now prefer to refer to typical test methods rather than to specific machines by name.

Table 14.1 *National and international standards for abrasion resistance*

ISO 4649 (ISO (DIN) abrader)
ISO 5470 (Taber)
ASTM D394 (Du Pont)
ASTM D1630 (NBS)
ASTM D2228 (Pico)
ASTM D3389 (Taber)
BS 903: Part A9
DIN 53516

14.1 The ISO abrader

The abrader described in ISO 4649 (also known as the DIN abrader) is rapidly becoming the standard machine used in Europe. It is convenient and rapid in use and is well suited

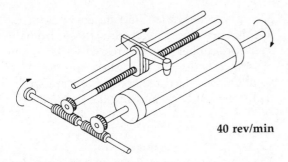

Figure 14.1 *The principle of the ISO (DIN) abrader for measuring abrasion resistance*

to quality control. There is, however, no general close relationship between the results and performance in service.

The principle of the machine is shown in Figure 14.1. A disk test piece in a suitable holder is traversed across a rotating drum covered with a sheet of abrasive cloth, see Figure 14.2. Standard test pieces are cylindrical in shape, 16 mm diameter and not less than 6 mm high. If test pieces of the required thickness are not available a test piece not less than 2 mm thick may be bonded to a base element of hardness not less than 80 IRHD. The holder moves laterally across the drum at a rate of 4.2 mm per revolution of the drum and suitable attachments may be provided to rotate the test piece at the rate of one revolution per 50 revolutions of the drum. The test piece is pressed against the drum with a vertical force of preferably 10, but sometimes 5, newtons by means of weights added to the top of the holder.

The drum has a diameter of 150 mm and a length of about 500 mm and it is rotated at 40 rev/min. A standard abrasive cloth with corundum particles of grain size 60 is specified and it is attached to the drum with double-sided adhesive tape. The test piece is automatically applied to and removed from the drum after an abrasion run of 40 m, which is equivalent to 84 revolutions. In special cases where the mass loss is greater than 600 mg in 40 m, the run may be reduced to 20 m and the results multiplied by two. The test piece is weighed before and after the run. The weight loss is converted to volume loss by dividing it by the density of the material as determined by the method specified in ISO 2781.

Three test runs for each rubber are required but the same test piece may be used if the mass loss is relatively small. The tests are carried out at laboratory temperature but in some cases there may be a considerable increase in temperature at the abrading interface. Such temperature rises are disregarded but the test piece should be allowed to cool to laboratory temperature between test runs.

To overcome the difficulty of maintaining consistent test conditions the results are related to those obtained with a standard rubber. Two standard rubbers are specified, one for use when the results are expressed as relative volume loss and the other when they are expressed as an abrasion resistance index. The two formulae are:

Relative volume loss =

$$200 \times \frac{\text{Volume loss of test piece in mm}^3}{\text{Mass loss of standard rubber in mg}}$$

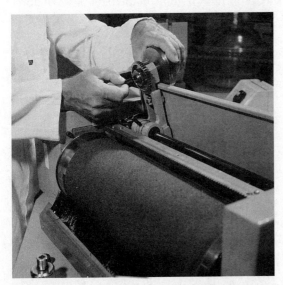

Figure 14.2 *The test piece in the ISO (DIN) abrader is mounted in a holder which can traverse across a drum covered with abrasive cloth which rotates at 40 rev/min. The loss in weight of the test piece is measured after an abrasion run of 40 m, which is equivalent to 84 revolutions. The results are compared with a standard rubber compound and expressed as an abrasion resistance index*

Notes: 1) this formula requires a load of 10 N and a non-rotating test piece.
2) abrasion resistance is the reciprocal of relative volume loss.

Abrasion resistance index =

$$100 \times \frac{\text{Volume loss of standard rubber}}{\text{Volume loss of test rubber}}$$

Note: the rotating test piece method is preferred but both tests must use the same procedure.

The difference between the two methods is not immediately obvious. Relative volume loss is used when a consistent abradant is available that produces an abrasion loss of 210 to 220 mg with the standard rubber. The test method can then be reduced to the measurement of the volume loss of the test rubber. Abrasion index is not dependent upon having such a consistent abradant, although in practice, an abrasive cloth meeting the specification in ISO 4649 is usually used.

14.2 Other abraders

As mentioned previously, many machines have been developed to measure abrasion under specific conditions. Brief descriptions of the most common follow.

Akron

A moulded circular test piece of known weight is mounted on a motor driven spindle in the Akron machine. The periphery of the test piece is brought into contact with the periphery of an abrasive wheel which is also mounted on a spindle. Rotation of the test piece causes the abrasive wheel to rotate and the two are held together under a known force. The axis of the test piece and the axis of the abrasive wheel are not in the same plane which results in a rubbing action. The abrasion resistance is calculated from the loss in weight after a specified number of revolutions of the abrasive wheel. By varying the angle of the test piece relative to the wheel the degree of slip can be varied and its effect studied. Results are compared with those from a standard rubber and expressed as abrasion resistance index as described for the ISO abrader. Standard methods of test are specified in ASTM D394 and BS 903: Part A9.

Dunlop

The Dunlop abrader also uses a specially prepared test piece in the form of a wheel with a bonded steel centre piece which is driven whilst in contact with the side of an abrasive wheel. The degree of slip is varied by means of an eddy current brake acting on the abrasive wheel. The machine and the method of preparation of the test pieces are quite complicated and this abrader is no longer covered by a standard.

Du Pont

The Du Pont machine uses two moulded or built-up test pieces, 20 mm square by 10 mm thick, mounted on a lever arm which has its axis midway between them. A known force holds the test pieces in contact with a flat abrasive disk which is rotated about the axis of the lever arm. The torque in the lever arm is balanced by means of a weight and adjustable spring balance. Tests can be made at constant force or with a force adjusted to give a constant torque on the arm holding the test pieces. The abrasion resistance is calculated from the weight loss after a specified number of revolutions of the abrasive disk and expressed as abrasion resistance index as before. Standards are ASTM D394 and BS 903: Part A9.

National Bureau of Standards

The National Bureau of Standards (NBS) abrader is used for testing soft vulcanised rubber compounds such as those used for shoe soling. Three arms are mounted over a 152 mm diameter metal drum which has abrasive paper attached to it. Each arm is pivoted at one end and each has a weight suspended at the other. The test pieces are mounted on these arms so that they are held against the abrasive surface of the drum by a known force. The drum is rotated at 45 rev/min and the reduction in the thickness of the test pieces is measured by dial gauges. The results are compared with those obtained from standard compounds. The only standard covering this machine is ASTM D1630.

Pico

The Pico abrader uses a pair of blunt tungsten carbide knives which rub the test piece whilst it rotates on a turntable. The force on the test piece and the speed of rotation can be varied. The direction of rotation is reversed at intervals throughout the test and a dusting powder is fed to the test piece surface. The machine is calibrated by five standard rubbers and the results expressed as abrasion resistance index. The test procedure is provided in ASTM D2228.

Taber

The Taber abrader unlike the other machines was not developed by the rubber industry and it is used to test a wide variety of materials, including painted and plastic coated surfaces, textiles, metals and leather as well as rubber. The principle is shown in Figure 14.3. The test piece is a simple flat disk mounted on a rotating platform. The abrasive wheels are attached to the free end of pivoted arms and rotate in peripheral contact with the test piece. The force on the test piece and the nature of the abradant are readily varied and tests can be carried out in the presence of lubricants.

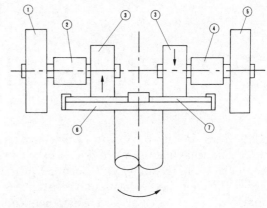

1 & 5 weights
2 & 4 pivoted arms
3 abrasive wheels
6 revolving table
7 test piece

Figure 14.3 *Principle of the Taber abrader*

The Taber machine is particularly suited to testing rubber coated fabrics. Test methods are covered by ISO 5470, ASTM D3389 and BS 903: Part A9.

14.3 Interpretation

The results obtained from abrasion tests are likely to have poor reproducibility, which makes the use of a standard rubber for comparison particularly important. In addition, correlation with service conditions is uncertain and, in general, abrasion tests are not recommended for use in performance specifications.

15
Electrical properties

Elastomers are used extensively in electrical applications because they provide an excellent combination of flexibility and electrical properties. In general, good insulation resistance is required but, if necessary, the elastomers can be made to be antistatic, or even conductive, by incorporating suitable carbon black fillers.

The methods of testing for electrical properties used in Europe are based on the standards developed by the International Electrical Commission (IEC), which are applicable to most materials. National standards organizations have adapted the IEC methods for specific tests but have not always modified them in the same way. The standards listed in Table 15.1 are, therefore, not necessarily identical and individual test conditions need to be checked before comparisons are made.

ISO has published relatively few specifications covering electrical properties of elastomers. However, some national standards exist and BSI, for example, has combined the specifications for elastomers with those for plastics in a five-part standard, BS 903:Part C1 to C5/2782 Part 2. BSI points out that the recommended period of conditioning for vulcanised rubber test pieces is 16 hours (ISO 471) compared with 88 hours for plastics (ISO 291) but they conclude that under the climatic conditions prevailing in the UK, 16 hours is generally adequate for most plastics.

Electrical testing is a complex and specialized subject and will not be covered in detail in this work. However, an understanding of the basic principles is necessary when specifying elastomers for electrical applications. The tests most commonly applied to elastomers are resistivity, insulation resistance, electric strength, tracking resistance, power factor and permittivity. In addition, the elastomer composition must have a proper balance of other properties such as tensile strength, elongation and resistance to heat and water.

15.1 Resistivity

The surface of elastomers may conduct electricity more easily than the bulk of the material and it is usual to distinguish between volume and surface resistivity. Volume resistivity (also known as specific resistance) is the electrical resistance between electrodes, one centimetre square, placed on opposite sides of a hypothetical centimetre cube of the elastomer, assuming negligible surface leakage. The units most commonly used are ohm-centimetres (Ω.cm). Surface resistivity is the electrical resistance between opposite sides of a square on the surface of the elastomer and is expressed in ohms (Ω). The size of the square is immaterial. The following definitions are also relevant. Insulation resistance is the electrical resistance measured between any two particular electrodes on or in an elastomer and hence is a function of both the surface and volume resistivities and of the geometry of the test piece. Conductance and conductivity are respectively, the reciprocals of resistance and resistivity.

Volume resistivity

Volume resistivity tests may be affected by a number of factors including the magnitude and time of application of the applied voltage, the nature and geometry of the electrodes and the temperature and humidity. There are no ISO methods but national standards are based on IEC 93. The tests consist of placing a thin sheet of the elastomer between carefully specified electrodes and applying a direct voltage

Electrical properties

Table 15.1 *National and international standards for electrical properties*

Volume and surface resistivity
(Insulating materials)
IEC 93
ASTM D257
BS 903: Parts C1 & C2
BS 6233
DIN VDE 0303: Part 3
NF C26-215

Volume and surface resistivity
(Conductive & antistatic materials)
ISO 1853
ISO 2878
ISO 2882
ISO 2883
ASTM D991
BS 2044
BS 2050
NF C26-215
NF C26-218

Insulation resistance
IEC 167
ISO 2951
ASTM D257
BS 903: Part C5
DIN VDE 0303: Part 3
NF C26-210

Electric strength
IEC 243
ASTM D149
BS 903: Part C4
DIN VDE 0303: Part 2
NF C26-225

Tracking
IEC 112
BS 5901
DIN VDE 0303: Part 1
NF C26-220
NF C26-221

Permittivity and power factor
IEC 250
ASTM D150
BS 903: Part C3
DIN 53483
NF C26-230

High voltage
IEC 60
BS 923
DIN VDE 0432: Parts 1 to 4
NF C41-101

of either 500 V or 100 V for one minute. Higher voltages and longer times may be used in special cases. Tests using alternating currents are rarely made. The electrodes may be made from graphite, metal foil or film, conducting silver paint or conductive elastomers. The electrodes are backed by metal plates and 'guard' conductors are used to intercept stray currents that might otherwise cause errors. There is always an unknown contact resistance between the electrode and the elastomer interface but this has to be ignored because there is no means of measuring it.

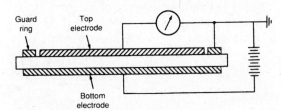

Figure 15.1 *Electrodes, guard ring and circuit for measuring volume resistance of elastomers*

A simplified circuit diagram for measuring volume resistivity is shown in Figure 15.1. In practice, the circuits are more complicated than this and there are a number of practical difficulties in accurately measuring the volume resistance and maintaining a constant applied voltage. The volume resistivity in ohm-centimetres is then given by the formula,

$$\rho = \frac{R_v \cdot A}{h} \text{ ohm.cm}$$

where,

R_v is the volume resistance in Ω,

A is the effective area of the guarded (inner) electrode in cm^2, and

h is the thickness of the test piece in cm.

For circular electrodes, the effective area is taken to be,

$$A = \frac{\pi(d_1 + g)^2}{4} \text{ cm}^2$$

where,

d_1 is the outer diameter of the inner electrode in cm, and

g is the distance between the inner electrode and the inside diameter of the outer (guard) electrode in cm.

Similar formulae exist for square, rectangular, tubular and strip electrodes but circular electrodes are usually used for elastomers.

It is not usually possible to obtain very great precision in measurements of high volume resistivity and results are never quoted beyond two significant figures. Considerable variations between results can often be experienced and two materials would only be considered significantly different if their volume resistivities differed by a factor of 10.

Surface resistivity

The apparatus for measuring surface resistivity is similar to that for volume resistivity but the electrode connections are as shown in Figure 15.2. The outer ring is used as the high potential electrode and the underneath ring as the guard electrode. 500 V or 100 V d.c. voltage is applied for a set time, usually one minute and the surface resistance is measured. Surface resistivity is then given by,

$$\sigma = \frac{R_s \cdot p}{g} \text{ ohms}$$

where,

R_s is the surface resistance in ohms

p is the effective perimeter of the inner electrode in cm, and

g is the distance between the outer diameter of the inner electrode and the inner diameter of the outer electrode in cm.

For circular electrodes,

$$p = \pi(d_1 + g)$$

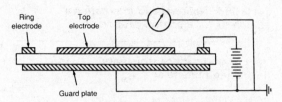

Figure 15.2 *Electrodes, guard ring and circuit for measuring surface resistance of elastomers*

where,

d_1 is the outer diameter of the inner electrode in cm.

As the dimensions of p and g are both length, the surface resistivity is expressed in ohms but it is often referred to as ohms per square. This arises from the fact that the surface resistivity across the two opposite sides of a square of any size is a constant for a given material.

Surface resistance can only be measured approximately because the measured current always includes a component which passes through the bulk of the material. The measured value is also influenced by the contamination on the surface of the test piece at the time of measurement and the humidity and temperature of the ambient atmosphere during conditioning and measurement. Thus surface resistivity is not a fundamental property of the elastomer.

The surface of insulating elastomers is often more conductive than the bulk of the material and there is a measurable difference between volume and surface resistivity. However, with most antistatic and conducting elastomers the surface layer is no more conductive than the bulk and and the current will largely take the 'easy' route through the bulk of the material and surface resistivity has no real meaning. Thus, with lower resistance elastomers it is usual to assume that the resistivity consists solely of the volume constituent.

There is no clear distinction between insulating, antistatic and conductive elastomers. However, in general, elastomers with insulation resistances of up to 10^4 ohms are considered to be conductive; if the resistance is between 10^4 and 10^8 ohms they are said to be anti-static and when the resistance is above 10^8 they are insulating.

15.2 Insulation resistance

ISO 2951 describes empirical methods for determining insulation resistance which do not distinguish between volume and surface resistance. The procedures in the standard can be used for comparing the insulating properties of different elastomers and in suitable circumstances they can be used for testing and comparing finished products.

The specified test pieces can be flat sheets, tubes or rods and the electrodes can be metal bars or made with conductive paint. The resistance is measured after applying a 500 V d.c. supply for one minute in a similar manner as described in Section 15.2 for resistivity. Painted electrodes are spaced 10 mm apart and should either be 100 mm long or the results should be related to 100 mm by multiplying the measured resistance in megohms by the measured length of the test piece in mm divided by 100, the insulation resistance being inversely proportional to the cross-sectional area of the test piece. Similarly, when using bar electrodes, which are specified to be spaced 25 mm apart, the resistance is related to a 25 mm wide test piece by multiplying the measured resistance in megohms by the measured width of the test piece divided by 25. The electrodes may need to be guarded if the resistances are high.

15.3 Conductive and antistatic elastomers

Elastomers are sometimes required to dissipate electrostatic charges and at the same time have sufficient insulation resistance to prevent a person in contact with them receiving a shock. In extreme circumstances, it is also necessary to prevent them igniting in the event of faulty insulation in nearby electrical equipment. Such elastomers are said to be antistatic. Elastomers with insulation resistances which do not meet these safety requirements are termed 'conducting'.

Apart from static electricity, the principal hazard in most buildings and with most electrical equipment is from leakage currents from normal voltage supply mains. To guard against these hazards, it is recommended that the lower limit of resistance for an antistatic elastomer should be 5×10^4 for 250 V mains supplies, that is a maximum current of 5 mA. The limit can be proportionally less for lower voltages.

ISO 1853 specifies a method of measuring the volume resistivity of conducting and antistatic elastomers. The insulation resistance of these materials is very sensitive to strain and temperature. Measurements made on freshly strained material at room temperature and material which has remained unstrained for a short period at 100°C can sometimes vary by a factor of 100 or more. The relationships are complex and arise from the kinetic energy and structural configuration of the carbon particles in the elastomer. In order to make valid comparisons, ISO 1853 specifies that the test pieces shall be conditioned at 70°C for two hours and then at 23°C and 50% relative humidity for not less than 16 hours.

Four electrodes are specified in order to ensure that the contact resistances are kept to a minimum. The principle of the method is shown in Figure 15.3. The strip test piece has metal current electrodes clamped at each end and is connected in series with a voltage source and a means of measuring the current

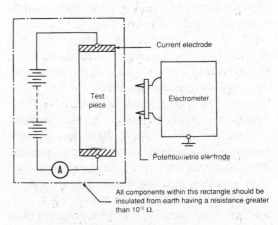

Figure 15.3 *Test circuit for the measurement of the resistivity of conducting and antistatic elastomers*

Table 15.2 Electrical resistance limits for antistatic and conductive elastomeric products based on ISO 2882 and 2883

Product	Electrical resistance, ohms min.	max.
Anaesthetic airways, bellows and face pieces	–	10^6
Anaesthetic tubing	3×10^4 per metre	10^6 per metre
Breathing bags	–	10^6
Flooring material (hospitals)	5×10^4	2×10^6
(explosives handling)	–	5×10^4
(industrial antistatic)	5×10^4	10^8
Footwear (hospitals and antistatic)	5×10^4	5×10^7
(conducting for explosives handling)	–	1.5×10^5
Furniture buffers and feet (hospitals)	–	10^6
Hose	3×10^3 per metre	10^6 per metre
Mattresses and pads (hospitals)	10^4	10^6
Mouldings, small (hospitals)	–	10^6
Sheeting and articles made from sheeting (both surfaces to be tested)	–	10^6
Textile cots and aprons	–	10^8
Tyres, antistatic (solid or pnuematic)	5×10^4	10^7
Tyres for castors and wheels (hospitals)	–	10^4
Tyres for explosives-handling vehicles (Solid or pneumatic)	–	5×10^5

Warning note: products which achieve their antistatic properties by a thin conductive surface coating may lose these properties during use as a result of wear or solvent action.

flowing. The 'potentiometric' electrodes are placed on the test piece between the two current electrodes and the voltage drop between them is measured with a very high impedance electrometer (valve voltmeter). The contact resistance at the current electrodes does not matter and those at the potentiometric electrodes do not affect the measurement if no current is taken by the electrometer. The resistivity is calculated from the measured current and voltage drop and the cross-section and length of the test piece between the potentiometric electrodes. The results are usually expressed as ohm.metres.

There are three ISO standards for measuring the electrical resistance of products made from antistatic and conducting rubbers. ISO 2878 specifies that the electrodes should be formed on the surface by means of a conductive silver lacquer, colloidal graphite or a conductive liquid and defines the positions between which the resistance should be measured. Procedural details are provided for nine products ranging from simple surfaces to footwear. ISO 2882 defines the upper and lower limits of resistance to be applied for 15 products used in hospitals and ISO 2883 does the same for 12 industrial products. A selection of maximum and minimum electrical resistances for a variety of products taken from both standards is given in Table 15.2. These standards are intended for use as production specifications and the methods recommended are suitable for factory inspection and/or service testing.

15.4 Electric strength

The electric strength, sometimes referred to as dielectric strength, is an important property for insulating materials which are to be used in high voltage equipment (e.g. switchgear and transformers), capacitors (also known as condensers) and protective clothing (e.g. electricians' gloves). Electric strength is defined as the nominal voltage gradient (the applied voltage divided by the thickness of the test piece) at which electrical breakdown occurs under specified conditions. The specified test conditions are important because the measured electric strength is not an intrinsic property of the material and varies with the thickness of the test piece, the time for which the voltage is applied, the geometry of the electrodes and the conditioning that has been applied to the material. Electric strength is usually measured using an a.c. voltage at 50 Hz, which is substantially sinusoidal, and is expressed as volts per millimetre or volts per mil (0.001 in), the voltage being measured as a root mean square value (the peak value divided by $\sqrt{2}$).

There is no ISO standard for electric strength and most national standards are based on IEC 243. BS 903: Part C4, for example, specifies two procedures, both using an a.c. voltage at 50 Hz. The first calls for the voltage to be applied at a uniform rate from zero so that breakdown occurs on an average between 10 and 20 seconds. In the second method, known as 'step-by-step', a voltage approximately equal to 40% of the rapidly applied breakdown voltage is first applied. If the test piece withstands this voltage for 20 seconds without failure, the next higher specified voltage and subsequent voltages are applied, each for 20 seconds, until breakdown occurs. Tests are made at 23°C or 90°C with the test pieces in air or immersed in an oil that has adequate electric strength and resistivity (BS 148 and 5730).

The normal test piece for elastomers is a disk of minimum diameter 100 mm or a square of side not less than 100 mm, with a thickness of 1.25 mm. Sheet can be tested at other thicknesses and methods for testing tube, tape, rod and flexible tubing are also specified. Test equipment must be carefully selected and the recommendations given in IEC 60 covering high voltage testing techniques should be followed.

The results obtained by different methods are not directly comparable and the tests are mostly used for comparative selections and for routine, quality control and specification purposes. It should be noted that the electric strength of many materials decreases as the thickness of the test piece between the electrodes increases and with increasing time of voltage application. Also materials with a high electric strength, as measured by the tests, will not necessarily resist long-term degradation processes such as erosion or chemical deterioration in the presence of moisture. However, it is often possible for product specifications, to simply apply a proof voltage for a specified period and the result is either a pass or a fail.

15.5 Tracking

When a discharge over the surface of an insulating material causes local decomposition and produces a charred conducting path, the phenomenon is known as 'tracking'. Non-tracking dielectrics can withstand transient flash-overs without impairment of their insulating properties but where tracking occurs permanent damage results.

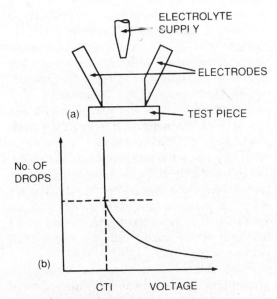

Figure 15.4 *(a) electrode system for tracking test on elastomers; (b) effect of voltage on number of electrolyte drops causing 'failure'. CTI is the comparative tracking index*

Low voltage tests (up to 1000 V) rely on applying a contaminant, such as an electrolyte, between electrodes to induce a flash-over and are generally based on the recommendations given in IEC 112. The number of drops of electrolyte required to cause tracking as indicated by a continuous current passing between the electrodes is noted. The procedure is repeated at different voltages and a graph constructed of drops of electrolyte vs voltage as shown in Figure 15.4. The point at which the voltage starts to increase is taken as the comparative tracking index.

Higher voltage tests, using equipment similar to that for breakdown tests with specified electrode systems, are available but are rarely used for elastomers.

15.6 Permittivity

Relative permittivity, sometimes referred to as dielectric constant or specific inductive capacity, is a measure of an insulation's ability to store electrical energy. It is the ratio of the capacitance of a capacitor having the material as the dielectric, to the capacitance of a similar capacitor having air, or more precisely a vacuum, as the dielectric. Permittivity is therefore a number without units and it is customary to omit the word relative. If measurements are made with an alternating electric field, it is found that at low frequencies, e.g. 50 Hz, the permittivity is similar in value to that obtained under electrostatic conditions but at high frequencies it may differ considerably. It is therefore essential to specify at what frequency the measurements are made, 1 kHz frequently being selected for applications in the audio range. Permittivity changes with temperature and this must also be specified. However, most specifications call for tests to be made only at room temperature.

There is no ISO method for testing the permittivity of elastomers but national standards are based on IEC 250, which is a general standard for insulating materials. Test pieces are either flat sheet or round tube, electrodes for the former being similar to those used for measuring the resistivity of insulating materials (see Section 15.2) but with a smaller guard gap. The apparatus for measuring the capacitance must be sufficiently accurate to measure down to 50 to 100 picofarads, i.e. 10^{-12} farads and sometimes lower. For power and audio frequency tests, transformer bridges are recommended (alternatively, BS 903:Part C3 describes a system known as the Schering bridge with a Wagner earth) but transmission line and resonant cavity techniques must be used at frequencies above 200 MHz. Suitable types of equipment and precautions in their use are described in IEC 250.

15.7 Power factor

In electrical circuits in which current does not flow in phase with an applied alternating voltage, the apparent power absorbed, measured by the product of current and voltage, is greater than the true power (or rate of dissipation of energy) developed and the difference appears as heat. The ratio of true power divided by apparent power is known as the power factor and is expressed either as a decimal fraction or a percentage. The maximum value of the ratio is obtained when the true power is the same as the volt-amperage, a condition represented by current flowing in a pure resistor (power factor = 1). The minimum value relates to the passage of an alternating current through an ideal capacitor (no true power; power factor = 0). For most elastomers the power factor is very small.

The passage of an alternating current through a perfect dielectric would involve no dissipation of energy, since the capacitance current would lead the applied voltage by 90°. In practice all insulating materials exhibit a small in-phase current which makes the phase angle, ϕ, between the total current and the applied voltage less than the ideal value. The cosine of ϕ is equivalent to the power factor (the proof for this will be found in electrical textbooks).

The angle by which the phase difference between the applied voltage and resulting current deviates from 90° is known as the loss angle, δ. The proportion of in-phase current to that 90° out of phase with the voltage is known as the dissipation factor and is represented by the tangent of the loss angle (sometimes referred to as tan δ or loss tangent). However, for most elastomers the component of the current that is 90° out of phase with the applied voltage is so close to the total current that they may be considered to be the same and dissipation factor and power factor can be considered to be identical.

Loss factor is the product of the permittivity and the dissipation factor and, in an a.c. field, the actual power loss in a dielectric is proportional to both the loss factor and to the frequency. Thus, an elastomer to be used at very high frequencies and high voltages should ideally have low permittivity and low power factor combined with high dielectric strength and high resistivity. On the other hand, a dielectric for a capacitor should have a high permittivity in order to keep the physical size of the component small but, as a high loss factor is not normally wanted, a low power factor. A high loss factor may be desirable when it is required to absorb energy, for example in radio frequency or microwave heating, and a high permittivity and power factor combined with low resistivity are then required.

Power factor is usually calculated from the data obtained when measuring permittivity but in some commercial equipment it can be obtained as a direct reading.

15.8 Wires and cables

Thousands of tonnes of elastomers are used every year for primary insulation and for protective sheathing or jackets in wires and cables. For primary insulation the electrical

Table 15.3 *Electrical properties of Du Pont elastomers*

Property	Units	ALCRYN	HYPALON (CSM)	HYTREL (YBPO)	Neoprene (CR)	NORDEL (EPDM)	VAMAC (AEM)	VITON (FKM)
Service temperature (Continuous) (Intermittent)	°C	90 120	90 120	120 150	75 105	90 120	100 130	200 300
Volume resistivity (ASTM D257)	ohm.cm	1.4×10^{10} to 5.0×10^{10}	10^{14} to 10^{10}	1.7×10^{15} to 1.0×10^{17}	10^{12} to 10^{8}	10^{15} to 10^{13}	10^{13} to 10^{10}	5.8×10^{10} to 3.5×10^{15}
Electric strength (ASTM D149)	kV/mm 250 μ 625 μ 3150 μ	– 6 to 18 –	20 to 24 – 8 to 16	60 – 12	5 to 30 – 4 to 6	32 to 40 – 24 to 28	24 to 17 – 8 to 10	– 12 to 20 –
Permittivity at (ASTM D150)	60 Hz 1 kHz 1 MHz	– 8.0 to 11.1 –	– 4.5 to 8.0 –	– 3.7 to 5.4 3.5 to 5.4	– 6 to 9 –	2.3 to 2.8 – –	– 4.5 to 8.0 –	– 8.8 to 10.7 –
Power factor at (ASTM D150)	60 Hz 1 kHz 1 MHz	– 0.17 to 0.4 –	0.03 to 0.07 – –	– 0.008 to 0.02 0.102 to 0.187	0.02 to 0.07 – –	0.0025 to 0.005 – –	0.05 to 0.07 – –	– 0.03 to 0.09 –

Note: Service temperatures for the electrical industry tend to be lower than those for other engineering applications

Table 15.4 *Typical applications for Du Pont elastomers in wires and cables*

	HYPALON (CSM)	HYTREL (YBPO)	NEOPRENE (CR)	NORDEL (EPDM)	VAMAC (AEM)	VITON (FKM)
Aerospace	–	–	–	–	–	□
Automotive	■ □	■ □	□	■	□	□
Communications	–	■ □	–	–	–	–
Computers	–	□	–	–	–	–
Construction	□	–	□	■	□	–
Electromechanical	□	–	□	■	–	□
Electronics	–	□	–	–	–	□
Fibre optics	□	□	□	–	□	–
Mining	□	–	□	■	□	–
Offshore	□	–	□	■	□	□
Power distribution	–	–	□	■	–	–
Power generation	□	–	–	■	–	–
Rail	□	–	□	■	□	–
Ships	□	–	□	■	□	□

■ Insulation
□ Jacket

properties are of greatest importance and these can be modified by suitable compounding. For example, as the carbon black content in a compound is increased, the insulation resistance and electric strength decrease but the permittivity and power factor increase. Most plasticisers have a detrimental effect on electrical properties. In particular, some ester-based plasticisers are poor in d.c. resistivity and significantly lower the insulation resistance of compounds containing them.

When an elastomer is used as a protective sheath or jacket for a cable, its electrical properties are of less importance. This is particularly true if the cable has a metallic shield or if there is a barrier between the cable insulation and the sheath to prevent migration of plasticiser. However, the volume resistivity should be at least 10^{11} ohm.cm in order to prevent current leakage from metal earth screens.

Wires and cables are being increasingly used in hostile environments and there is a continual requirement to save space and weight. Elastomers which retain their mechanical and electrical properties at extremes of temperature and which have excellent resistance to humidity and highly corrosive conditions are therefore in great demand. Fire resistance is also of great importance where the elastomers are to be used in such applications as public transportation, ships and oil rigs. The electrical properties of the high performance elastomers supplied by Du Pont are given in Table 15.3 and some typical applications in Table 15.4.

There are numerous national standards for testing the electrical properties of wires and cables for use in a wide variety of applications. They are mostly based on IEC specifications 227, 245, 502 and 885, the details of which are outside the scope of this work.

16

Thermal properties

Specific heat, heat transfer coefficients and thermal diffusivity are important properties to be considered when processing elastomers because they affect the time required to heat the interior of a rubber product to the vulcanisation temperature. However, they are rarely used for product specifications or in engineering design. On the other hand, thermal conductivity and thermal coefficients of expansion can have significant influences on product design.

Thermal properties and the effects produced by changes in temperature are sometimes confused. The latter have generally been covered in individual sections of *The Language of Rubber* but specialized subjects, such as the effect of low temperatures and heat ageing, will be considered separately.

16.1 Thermal analysis

Elementary thermal analysis (TA) has been used since the early days of the rubber industry. For example, samples of rubber goods were incinerated and the resultant residue weighed in order to check that they contained the requisite amount of rubber polymer. However, the techniques of TA have been considerably expanded since the introduction of thermal analyzers using very powerful micro-processors. These versatile instruments can be used to measure changes in the physical properties of an elastomer as a function of temperature and time. They operate automatically, relatively small samples are required and the data analysis unit assesses the results both accurately and

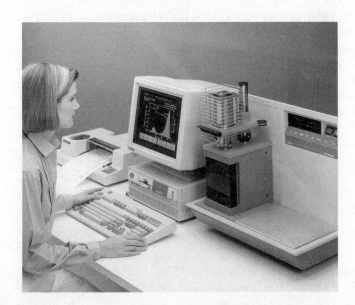

Figure 16.1 *Thermomechanical analysis (TMA) measures linear or volumetric changes in the dimensions of a test piece as a function of time, temperature and force. These data can provide valuable information about the coefficient of thermal expansion, delamination temperature, glass transition temperature, modulus and creep/stress relaxation of elastomers. TMA is widely used for quality control, process optimization and troubleshooting as well as for research and development*

quickly. One example of this type of instrument is the TA Instruments thermal analyzer, shown in Figure 16.1, that was described in Section 11.9. In that case, the specific application was its use for dynamic mechanical analysis (DMA) when investigating the viscoelastic properties of elastomers.

In principle, TA could be used to measure any physical property but in practice it is used most frequently in the rubber industry for:–

1) thermogravimetric analysis (TGA) to measure changes in weight for the provision of data for materials selection and product design,
2) differential scanning calorimetry (DSC) to measure heat flow into or out of a sample, e.g. when measuring specific heat,
3) thermomechanical analysis (TMA) which can be used to measure dimensional changes to provide information about expansion coefficients, glass transition temperatures, creep, etc., and
4) dynamic mechanical analysis (DMA) as mentioned earlier.

16.2 Thermal conductivity

The quantity of heat (Q) is measured in joules, which is the work done when a force of 1 N acts through a distance of 1 m. The joule is equal to 10^7 ergs or 2.778×10^{-7} kW.h. Heat flow rate, or heat flux (ϕ) is the quantity of heat flowing in or transferred to or from a system divided by time and its units are watts (W). The density of heat flow rate (q) or heat flux density, is the heat flow rate divided by area and its units are watts per square metre (W/m^2). Thermal conductivity (λ) is the heat flux density under steady-state conditions divided by the temperature gradient in the direction of heat flow and its units are watts per square metre for one metre thickness and one kelvin difference in temperature,

$$\frac{W}{m.K}$$

where K (kelvin) is the thermodynamic temperature using the Celsius scale. An alternative definition of thermal conductivity is the quantity of heat passing through unit area of a material, when the temperature gradient (measured across unit thickness in the direction of the heat flow) is unity.

The above definitions are universal and the methods of measuring thermal conductivity of elastomers are no different from those used for other insulating materials. Some national standards for measuring thermal conductivity are given in Table 16.1. The thermal conductivities of solid elastomers vary with the amount and the conductivity of each constituent but, in general, they fall between 0.1 and 0.3 W/m.K and the method of measurement selected should be suitable for this range.

Thermal conductivity is important to the designer of rubber products in which heat is generated by vibration, flexing or friction. Rubber mountings, drive belts and sliding seals, for example, are applications where the designer should take thermal conductivity into account. As rubber is a poor conductor, designers must also be especially careful to provide for heat dissipation.

16.3 Thermal coefficients of expansion

The thermal coefficient of expansion is the fractional change in length (area or volume) of

Table 16.1 National and international standards for thermal properties of elastomers

Thermal conductivity
ASTM C177 (guarded hot-plate)
ASTM C518 (heat flow meter)
BS 874
BS 4370: Part 2 (rigid cellular)
DIN 52612 (building insulation)
DIN 52616

Thermal expansion
ASTM D864 (plastics)

Dimensional tolerances
ISO 3302
BS 3734

a unit length (area or volume) of a material per degree of temperature variation. The coefficient of expansion is expressed as per °C, i.e. length (area or volume) increase per unit length (area or volume) per °C. For an isotropic and homogeneous material the superficial coefficient of expansion (area) and the cubical coefficient (volume) are respectively approximately two and three times the linear coefficient. The methods used for measuring the thermal coefficients of expansion of elastomers are the same as those used for other materials but in linear measurements there can be experimental difficulties caused by their low modulus.

The thermal coefficients of elastomers vary with the kind and amount of fillers present; an increase in filler content lowering the coefficient. Elastomers, in general, have linear coefficients approximately ten times that of carbon steel. However, the figures for soft gum stock may be twice those for a hard rubber composition. The variation is so great that moulders of precision products have to take into account the coefficient of the particular compound when designing their moulds. Coefficients also vary with temperature and it is customary to determine average values over a temperature range.

The relatively high coefficient of expansion of elastomers is the principal cause of shrinkage when moulded items cool from the temperature at which they were vulcanised in the mould. This thermal contraction, known as 'mould shrinkage', can vary between 1.5% and 3% (linear) and makes it difficult to mould rubber to close dimensional tolerances. In general, designers should try to ask for tolerances akin to those used when casting metals rather than those used in an engineering machine shop. Recommended tolerances for rubber mouldings are provided in ISO 3302. If the tolerances specified are too tight, the moulders only recourse is to make a lot of mouldings and hope that some of them will be within the required limits. This will obviously increase the cost considerably.

Mould shrinkage is usually measured directly using a standard moulded test bar. The main requirement is that the required accuracy can be obtained. For example, to detect 0.1% shrinkage on a 100 mm bar requires a facility that can measure to 0.1 mm.

17

The Gough-Joule effect

The molecular structure of rubber gives rise to an apparent anomaly which was first observed by Gough in 1805. If a strip of rubber in tension is heated it contracts rather than expands. It follows that if the rubber is held under constant strain it will exert greater stress. This observation was confirmed some 50 years later by Joule and is now known as the Gough-Joule effect. The phenomenon can readily be demonstrated by hanging a pair of scissors by an elastic band from a door knob so that the elongation of the rubber is about 200 to 300%, as shown in Figure 17.1. If the rubber band is gently heated with a match, the scissors rise due to the contraction of the rubber band.

When the conditions for the Gough-Joule effect apply, the modulus of elasticity, or stiffness or ability to carry load of the rubber increases with a rise in temperature. However, the Gough-Joule effect is reversible and it does not apply at temperatures below $-60°C$.

Another related thermoelastic effect is that rubber evolves heat when rapidly stretched and absorbs heat when allowed to contract rapidly. For example, if a rubber band is held to the lips after being stretched rapidly, it feels warm. Conversely the rubber band feels cool when it has been rapidly relaxed but the retraction must be against an applied force. If the rubber is allowed to snap back, kinetic energy is dissipated and no cooling will occur.

There are two apparent inconsistencies associated with the Gough-Joule effect. In Section 16.3 it was stated that rubber expands when heated. In fact, if rubber is under no strain it does expand when heated. The Gough-Joule effect only applies when the rubber is strained first and then heated and it is superimposed on the normal thermal expansion. The two effects balance at about 6% elongation (at normal temperatures) and the force/elongation relationship is then independent of temperature. With a further increase in elongation the Gough-Joule effect predominates. This is illustrated diagrammatically in Figure 17.2 (a) which shows how the relationship between force and length in a

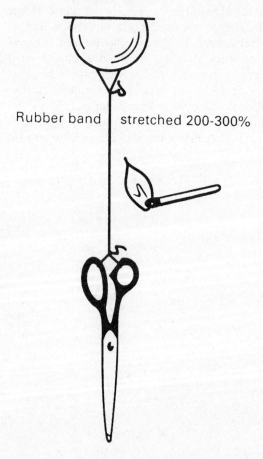

Figure 17.1 *Demonstration of the Gough-Joule effect. The scissors are used to stretch the elastic band by 200% to 300%. On gentle heating the scissors rise due to the contraction of the rubber band*

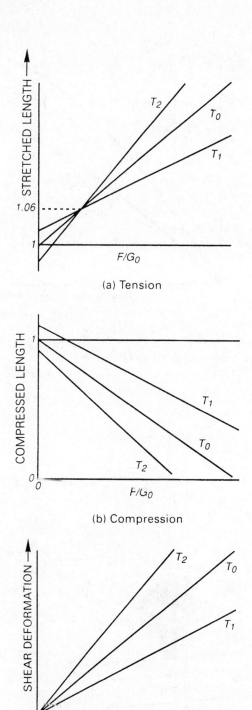

Figure 17.2 *Relationship between force (F) (expressed as F/G_o) and deformation of rubber under strain, at various temperatures, where $T_1 > T_o > T_2$. The slope of the lines is inversely proportional to the absolute temperature*

stretched piece of rubber varies with temperature. For convenience the ratio of force to modulus, G_o, at temperature T_o is used instead of force itself.

Analogous diagrams for compression and shear are shown in Figures 17.2 (b) and (c). In compression the expansion and Gough-Joule effects are seen to be additive and in shear the lines meet at the origin. The slopes of the lines are inversely proportional to the absolute temperature.

The second apparent contradiction arises from the stress–strain curves shown in Figure 17.3(a), which are obtained by running a standard tension test on a rubber compound at room temperature and then at some elevated temperature. These curves indicate that the modulus of elasticity is lower at an elevated temperature, which is contrary to predictions based on the Gough-Joule effect. The explanation is that at higher temperatures plastic flow overrides the Gough-Joule effect. If plastic flow is eliminated by running the test at such a speed that plastic flow does not have time to occur, or by 'conditioning' the rubber by repeated extensions and releases, the positions of the two curves are reversed as shown in Figure 17.3(b). The situation is then as predicted by the Gough-Joule effect. Incidentally, in the standard tension test, the measured stiffness of the rubber may be lower at higher temperatures because the time occupied by the test is insufficient to reach stress/strain equilibrium.

17.1 Practical implications

The Gough-Joule effect has practical implications in the design of O-ring seals for rotating shafts and for torsion springs.

When an O-ring is used to seal a rotating shaft, there is always a temptation to make the ring a little smaller than the shaft, with the idea that it will snap on and make a better seal. This is a fatal mistake. Friction heats the O-ring and since the ring is in tension the Gough-Joule effect applies. The ring tries to contract, creates more friction and more heat

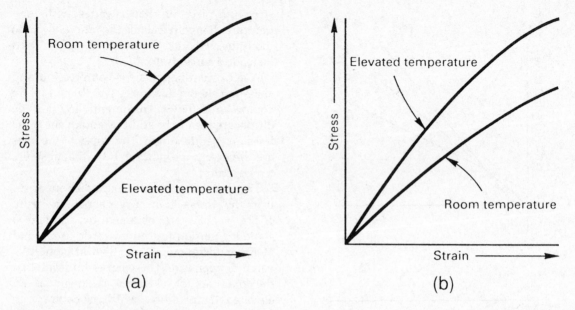

Figure 17.3 *Typical stress vs strain curves at room and elevated temperatures. Graph (a) indicates that Gough-Joule effect does not apply. Graph (b) conforms with Gough-Joule prediction*

develops. The rubber then chars and cracks and finally the seal leaks. On a high speed shaft failure can occur in minutes.

The correct procedure is to make the inside diameter of the ring about 5% greater than the shaft diameter. The seal will then be under peripheral compression in its operating position. The heat curve will level off after a moderate rise in temperature and there will be no further heat buildup. Of course, other factors must also be considered in the design of an O-ring seal but the concept of peripheral compression instead of tension is vital.

Torsional rubber springs, such as that illustrated in Figure 17.4, have a number of advantages. There is no static friction, they are inherently stable and do not require guides and shackles, dust seals are not required and no lubrication is necessary. However, the rubber in a torsional spring is under strain and the Gough-Joule effect causes the rubber to become stiffer as the temperature increases. Thus, if a torsional spring was used in an automobile suspension, a rise in temperature of 40°C could lift the body as much as 3 cm, which would not be acceptable to the

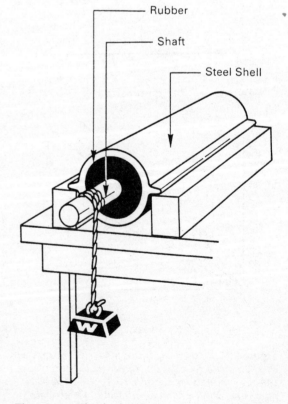

Figure 17.4 *Sketch of rubber torsion spring*

designer. Mechanical means could be added to maintain the car at a constant height but they would add to the cost and complexity of the suspension. The difficulty could also be overcome by reducing the strain in the rubber to below 13% but this would result in a considerable increase in the physical dimensions of the spring, which would also be unacceptable. Torsion springs are therefore not generally used in passenger car suspensions where cost and space limitations would require strains of about 100%.

18

Low temperature properties

All elastomers undergo several kinds of change when they are exposed to low temperatures. Some of the changes occur immediately, others after prolonged exposure. All are reversible; the elastomer regaining its original properties when it is returned to room temperature.

Figure 18.1 shows how the stiffness of a typical Neoprene compound is affected by exposing it to successively low temperatures. As the test piece is cooled through Zone A it becomes progressively more difficult to stretch or bend. This stiffening is gradual until a particular sub-zero temperature is reached. The temperature at which Zone B begins is known as the second order transition temperature, T_g, and depends upon the polymer and on the specific formulation. For example, T_g for Neoprene compounds can vary between $-29°$ and $-40°C$. With a further decrease in temperature into Zone B, stiffness increases sharply. Here a drop of only 11°C may multiply the stiffness a hundredfold. Further cooling into Zone C has little effect.

18.1 Brittle point

At some temperature in Zone B or C the test piece becomes brittle and will shatter on sudden bending or impact. The temperature at which this occurs depends on the rate of application of the load. When determined under certain prescribed testing conditions, this temperature is called the brittle point of the compound. The brittle point will not coincide with the start of Zone C because the stiffness curve is measured at a slow rate of change of load and the brittle point, which is generally measured by an impact test, is at a fast rate. The result of using the higher rate is to effectively move the stiffness curve to the right. Thus, whilst a measurement of stiffness may indicate that the elastomer is still flexible at a given temperature, the impact test could show the compound to be brittle at the same temperature.

18.2 Crystallization

The low temperature changes described above take place quickly – as soon as the test piece has chilled through. Long term exposure to cold produces crystallization which in turn causes stiffening. However, the stiffening only becomes evident after hours, days or even weeks, depending on the temperature and the particular compound involved. While crystallization increases stiffness it does not induce brittleness. Neither does the fact that crystallization has taken place affect the brittle point of the compound. The rate of crystallization is increased by placing the elastomer under strain.

Some elastomers, including natural rubber, butyl rubber and most types of Neoprene are crystallizable. For each elastomer there is a temperature at which crystallization occurs most rapidly. In the case of crystallizable types of Neoprene this temperature is about $-10°C$ and for natural rubber $-25°C$. At higher or lower temperatures crystallization takes place less rapidly. In fact, at extremely low temperatures of about $-46°C$ neither Neoprene nor natural rubber crystallize. The explanation is that the molecules of an elastomer are in constant motion due to their thermal energy and at times may assume a crystal-like relationship to one another. At room temperature, the thermal energy of the molecules is strong enough to overcome the forces that tend to maintain the crystal relationship, so

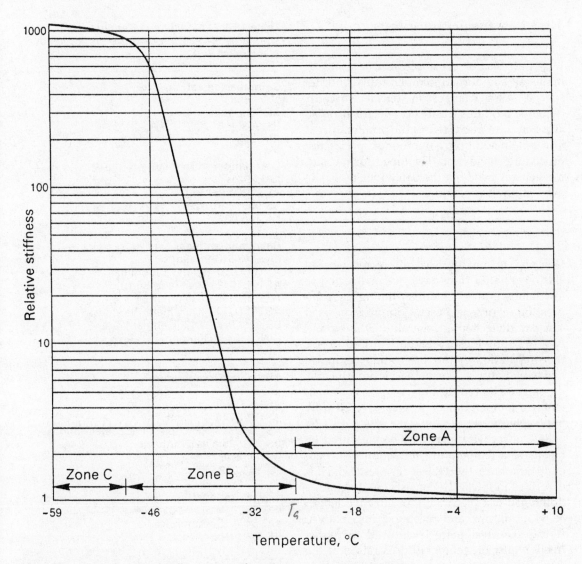

Figure 18.1 *Typical graph of stiffness versus short-time exposure to temperature*

crystallization does not occur. At very low temperatures, the molecular motion is slight and the probability of assuming the proper configuration is nil, so again crystallization does not occur. However, at some intermediate moderately low temperature (0°C for Neoprene) the molecules have sufficient thermal energy to move into a crystal pattern but not enough to break loose again, so the crystal-like structures grow in size and number and the piece of rubber becomes progressively stiffer.

Certain ester-based plasticizers are sometimes used in Neoprene compounds to depress the brittleness point and to reduce stiffening at low temperatures. These in effect displace the short-term stiffness curve in the direction of the lower temperature because they 'lubricate' the vulcanisate internally. However, they also accelerate the crystallization rate. This lowers the temperature below which crystallization would not otherwise occur and these plasticizers should be used sparingly.

18.3 Low temperature tests

It is possible for most physical tests to be made at sub-normal temperatures and for special cases changes in tensile strength, dynamic modulus, resilience electrical resistance, etc., are measured as the temperature is lowered. In addition, a number of specific test procedures have been developed for measuring general trends in the behaviour of elastomers at low temperatures.

Rate of recovery

One way to measure the rate of recovery of elastomers after they have been subjected to low temperatures is by determining the compression set or tension set. Compression set at constant deflection is generally favoured and ISO 1653 provides a standard procedure for the preparation of test pieces and testing at low temperatures. The method is, in principle, the same as that for testing compression set at ambient temperatures and was covered fully in Chapter 6.

Rate of recovery may also be measured by means of a temperature retraction test, generally known as the TR test. The test, which is described in ISO 2921, consists of stretching a dumb-bell test piece with an effective length of 50 or 100 mm and placing it in a bath at −70°C. The test piece is allowed to retract freely whilst the temperature is raised at the rate of 1°C per minute. A suitable apparatus is shown in Figure 18.2. The upper test piece holder is fitted with a locking device which enables the test piece to be stretched to a maximum of 350% and it is counterbalanced to maintain a slight tension of between 10 and 20 kPa on the test piece. It is essential that the cord and pulley system is virtually friction free. An elongation of 50% should be used if the effect of crystallization is to be minimized. If the effect of both crystallization and low temperature is to be studied an elongation of 250% (or half the ultimate elongation if 250% is unobtainable) is recommended. If the ultimate elongation is greater than 600% an elongation of 350% should be used. The stretched test piece is immersed in the bath at −70°C for 10 minutes and the locking device is then released to allow the test piece to retract freely. At the same time the temperature of the bath is raised at the rate of 1°C per minute. Should the elongated test piece retract to the original length at −70°C another cooling medium and a lower temperature should be used. Readings of the retracted length are taken every two minutes until the retraction reaches 75%.

Table 18.1 *National and international standards for low temperature properties of elastomers*

Compression set
ISO 1653
ASTM D1229
BS 903: Part A39
DIN 53517

Low temperature testing (TR test)
ISO 2921
ASTM D1329
BS 903: Part A29
NF T46-032

Change in stiffness
ISO 1432 (Gehman)
ASTM D797 (Young's modulus)
ASTM 1043 (Clash and Berg)
ASTM D1053 (Gehman)
ASTM D2240 (Durometer hardness)
BS 903: Part A 13 (Gehman)
DIN 53548 (Gehman)
NF T46-025 (Gehman)

Brittleness temperature
ISO R 812
ASTM C509
ASTM D746 (Impact)
ASTM D2137 (Flexibles and coated fabrics)
BS 903: Part A25 (Impact)
DIN 53546
NF T46-018

Crystallization
ISO 3387 (Hardness)
ISO 6471 (Compression)
BS 5294 (Hardness)
DIN 53541
NF T46-027

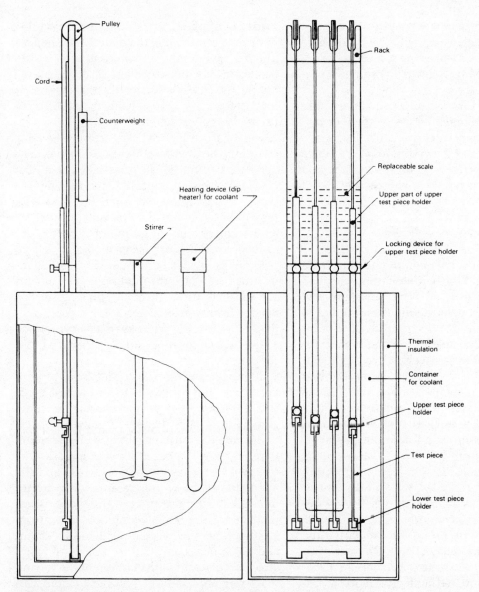

Figure 18.2 *Diagram of the apparatus recommended for the temperature retraction test (TR test). Dumb-bell test pieces, with an effective length of 50 or 100 mm, are stretched to up to 350% in a bath held at −70°C. The test pieces are allowed to retract freely whilst the temperature is raised at the rate of 1°C per minute*

The retraction is calculated from the formula,

$$\text{Retraction \%} = \frac{l_s - l_r}{l_s - l_o}$$

where,

l_s is the length of the test piece in its stretched condition

l_r is the length of the test piece at the observed temperature, and

l_o is the length of the test piece in its unstretched condition.

The percentage retraction is plotted against temperature and the temperatures which show retractions of 10%, 30%, 50% and 70% are determined (TR10, TR30, TR50 and TR70).

The standard points out that tests made at different elongations do not necessarily give the same results. Also, as the test procedure is arbitrary it must be followed closely if reproducible results are to be obtained.

The TR10 and TR70 figures are of particular interest. The difference betweeen them increases as the tendency to crystallize increases. Further, TR70 has been found to correlate with low temperature compression set and TR10 with brittle points in vulcanisates based on polymers of similar type.

Change in stiffness

The hardness of rubber compounds increases with stiffness and the hardness tests described in Chapter 4 can be adapted for use at low temperatures. In determining changes due to simple temperature effects or second order transitions, a test piece is conditioned for short times at successively lower temperatures and the durometer hardness is recorded as a function of temperature. To determine changes due to plasticiser-time effects or to crystallization, the test piece is conditioned at constant temperature and hardness is plotted as a function of time.

Although the test is simple, it suffers from the limitations associated with all durometer measurements. Nevertheless, when they are made by experienced rubber technologists, durometer tests are useful for comparing the low temperature behaviour of different compositions. Dead-load measurements of hardness are standardized and give more reproducible results. They are particularly useful for investigations into crystallization and are described in greater detail in Section 18.5. One difficulty which applies to both durometer and dead-load tests is that test pieces at low temperatures frequently display 'drift' so that the instantaneous reading is followed by a drop in hardness. The rate and magnitude of the decrease vary widely with different compositions but are often substantial.

A stiffness test using a simple bar-shaped beam test piece is described in ASTM D797. The test piece is supported by anvils at each end and the deflection is measured when loads are applied centrally. Young's modulus in flexure can then be calculated and changes due to low temperatures compared. The beam test has the advantage of being simple but the modulus data can only be measured at low deflections whilst rubbers are usually used at large deflections.

Torsional tests are very convenient for studying the change in stiffness as the temperature is reduced. In principle, a strip test piece is twisted through 180° and the angle to which it has recovered after 10 seconds is measured. The test is repeated at various temperatures and a graph of temperature against torsion (i.e. shear) modulus is prepared. The results are quoted as the temperatures at which the modulus is 2, 5, 10 and 100 times its value at room temperature. The results from torsional tests are more useful than those from a simple beam test in as much as the deflections more closely resemble those of service.

Originally, in the Clash and Berg type of apparatus, the torque to twist the test piece was provided by a system of weights, cords and pulleys. However, the Gehman apparatus, where the torque is provided by a torsion wire, is now favoured internationally. The Gehman test is covered by ISO 1432 and the apparatus is shown diagrammatically in Figure 18.3. A strip test piece, 40 mm long, 3 mm wide and 2 mm thick is moulded or cut from sheet and held between a fixed bottom clamp and an upper clamp that is capable of being attached to a stud at the bottom of the torsion wire. A pointer attached to the torsion wire moves over a moveable protractor to indicate the degree of twist. A number of test pieces can be accommodated in the rack and each can be brought in turn to be attached to the torsion wire. The rack and the test pieces can be lowered into an insulated container which can be cooled by a liquid or gas heat transfer medium. Care must be taken to ensure that the medium has no effect on the rubber being tested.

The first test is made at room temperature by moving the torsion head quickly through 180° and noting the pointer reading after 10 seconds. If the reading does not fall within the

Low temperature properties 89

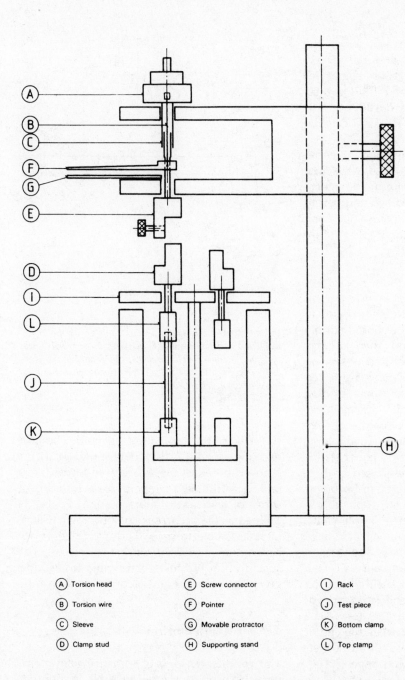

- (A) Torsion head
- (B) Torsion wire
- (C) Sleeve
- (D) Clamp stud
- (E) Screw connector
- (F) Pointer
- (G) Movable protractor
- (H) Supporting stand
- (I) Rack
- (J) Test piece
- (K) Bottom clamp
- (L) Top clamp

Figure 18.3 *Diagram of apparatus recommended for the determination of the stiffness characteristics of elastomers at low temperatures (Gehman test). A strip test piece is held between a fixed bottom clamp and an upper clamp attached to the bottom of a torsion wire. A pointer attached to the torsion wire moves over a protractor to indicate the degree of twist*

range 120° to 170° the torsion wire must be changed to another of different torsional characteristics. The standard specifies three torsional wires, each 65 mm long, having torsional constants of 0.7, 2.81 and 11.24 mN.m. The temperature is then adjusted to the lowest level desired and measurements made either at 5°C intervals with increasing temperature or in step changes at five-minute intervals with a continuous temperature increase of 1°C per minute, from which a graph of angle of twist against temperature can be drawn. The relative modulus at any temperature is the ratio of the torsional modulus at that temperature to the torsional modulus at room temperature. The results are expressed as the temperatures at which the relative modulii are 2, 5, 10 and 100 respectively. A table showing the twist angles for relative moduli of 2, 5, 10 and 100 for twist angles at room temperature from 120° to 170° is provided in the standard to avoid calculation.

Although in theory this method gives an absolute measure of shear modulus the actual result is dependent on details of the procedure and should be regarded as an apparent modulus. Also, the results from a Gehman apparatus are not always comparable to those obtained with a Clash and Berg.

Brittleness temperature

The temperature at which brittleness occurs in a rubber compound can be measured simply by cooling a test strip and then quickly bending it around a mandrel and observing whether it cracks. New specimens are tested at successively lower temperatures until a temperature is reached at which the test piece breaks into two or more separate pieces. However, simple bending tests do not give reliably reproducible results and better standard procedures have been developed.

The method generally adopted in Europe is that described in ISO R812. A strip test piece is clamped at one end to form a simple cantilever which, after cooling, is struck at a controlled speed by a striker head as shown in Figure 18.4. The test piece can be a strip or a 50 mm dumb-bell test piece, such as is used for tensile testing, with one end tab removed. The critical dimensions are the test piece thickness and the distance between the end of the grip and the point of impact of the striker, specified as 2 mm and 6.4 mm respectively. The striker radius is specified as 1.6 mm. With these dimensions the maximum surface strain in the test piece is held to about ±10% and with the velocity of the striker controlled to between 1.83 and 2.12 metres per second, the rate of straining is constant at about ±17% and adequate reproduciblity is obtained for most purposes.

The standard illustrates two suitable apparatuses. One uses an electric motor to drive

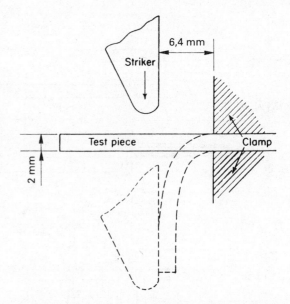

Figure 18.4 *A strip test piece is clamped at one end to form a cantilever for the low temperature brittleness test for elastomers. The broken lines show the position of the striker and test piece after impact when the test piece has not been broken*

the striker and the other a solenoid. The heat exchange medium may be a liquid or a gas. Four or more test pieces are used for each test and the brittleness temperature is found by a series of tests until failure occurs. Failure is defined as the occurrence of any crack, fissure or hole visible to the naked eye or complete separation into two or more pieces. For specification purposes, it is usual to test at a given temperature and record a pass or fail.

18.4 Interpretation

The various methods of measuring low temperature properties are applicable to different situations and do not have equal relevance to any particular application. Some workers have found good correlation between brittleness point and TR10 and brittleness and T10 (Gehman). Reasonable correlation has also been found between modulus (hardness) and modulus (Gehman) and T10 and TR10. However, the relationship between test results and actual performance in service is less secure.

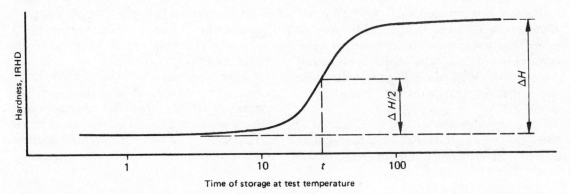

Figure 18.5 *Recommended method of reporting crystallization effects measured by change in hardness. The time for half the hardness increase to occur between initial and final hardness is determined from the smooth curve obtained by plotting hardness against time of storage at the test temperature*

18.5 Crystallization tests

In principle, any of the low temperature tests can be used to study crystallization effects by conditioning the test pieces at the low temperature for much longer times. However, a hardness test has been standardized in ISO 3387 for studying materials which have an initial hardness, at the test temperature, within the range 10 to 80 IRHD. The test is applicable to raw, unvulcanised but compounded, and vulcanised rubbers and determines the progressive stiffening occasioned by storage at a specified temperature.

The test pieces should be 8 to 10 mm thick with a side length or diameter of at least 42 mm and should be de-crystallized immediately before testing by heating them in an oven at 70°C for 45 minutes. They are then conditioned at room temperature for at least 30 minutes but no more than 60 minutes before testing. Tests should be made at one of the specified nine temperatures between 23° and −70°C.

Hardness measurements are carried out in accordance with ISO 48 (30 to 85 IRHD), ISO 1400 (85 to 100 IRHD) or ISO 1818 (10 to 35 IRHD), as described in Chapter 4. The same hardness gauge should be used throughout any one test, even if the ISO specification would require a different range instrument to be used because the rubber had become harder. Hardness measurements are taken at sufficient intervals to provide a graph of hardness against time. If possible, the test should continue until the final maximum hardness has been reached, as shown in Figure 18.5, but if this time is inconveniently long, the test may be terminated after 166 to 168 hours. The test results should preferably be reported as the time required for half of the total hardness increase between initial and final hardness to occur. Alternatively, the increase in hardness from the initial reading after a defined storage time may be quoted.

The above test is made with the rubber in an unstrained condition. An alternative way of studying crystallization is described in ISO 6471. The method is to measure the recovery from a low compression strain (with no crystallization) and from a high compression strain (after crystallization) and calculating the degree of crystallization from the difference between the two. The test is similar in principle to that for compression set as described in Section 6 but the test pieces for the high strain are compressed by as much as 60%. The apparatus, which must be capable of being kept in a low temperature bath or enclosure, consists of two parallel, flat, highly polished stainless steel plates, between the faces of which the test pieces are compressed. The compression device must be able to apply the force within 30 seconds and to release it whilst the apparatus remains at the low temperature.

If the time dependence of crystallization is to be studied, the tests may be carried out at any convenient temperature, degree of compresssion and holding time. The half-time to crystallization is then determined from a graph of recovery against the logarithm of the holding time.

For the rapid determination of the relative tendency of rubbers to crystallize and for comparing rubbers of similar composition, a table of recommended temperatures, nominal compressions and holding times, for various types of rubbers is provided in the standard.

19

Heat ageing

The properties of an elastomer will generally change after prolonged exposure to high temperatures. Natural rubber, for example, will become soft and gummy whilst Neoprene will harden slowly under the same conditions. The extent to which either softening or hardening is undesirable will depend upon the particular service required. The rate at which the properties of an elastomer change increases logarithmically with the temperature. Relatively small changes in temperature may, therefore, cause large differences in the degree of deterioration.

Tests for heat ageing are carried out for two purposes. Firstly, there are tests to establish the changes in physical properties at elevated service temperatures. Secondly, there are accelerated ageing tests at high temperatures which attempt to predict the long-term life at lower temperatures. The international standard for both is ISO 188 which specifies methods using an air oven or an oxygen pressure chamber. The tests consist of ageing test pieces for a given period at a given temperature and then measuring the physical properties that are considered important. In the absence of any specific requirements, the standard suggests measuring tensile strength, stress at an intermediate elongation, elongation at break and hardness. Comparison is then made with similar test pieces that have not been aged.

In the oven method, the test pieces are exposed to air at atmospheric pressure in either the usual single chamber oven or a multi-cell oven. The latter has the advantage that dissimilar materials can be tested without the danger of exchange contamination by plasticisers and other chemicals. The total volume of the test pieces in a single chamber oven should not exceed 10% of the free air space of the oven. The air flow in both types of oven must be steady and at a rate that provides between 3 and 10 complete changes of air per hour. It is also important that no copper or copper alloys are used in the construction of the oven because they could accelerate the ageing process. The temperature of the test can be as required but 70° or 100°C are the most commonly used for general purposes. However, many special-purpose elastomers, such as those supplied by Du Pont, operate and are tested at higher temperatures. The length of test is recommended as 1, 3, 7, 10 or a multiple of 7 days.

In the oxygen pressure chamber method it is recommended that the test pieces are exposed to oxygen at 70°C and at a presssure of 2.1 MPa. Otherwise the procedure is similar to that for the oven method.

Although not specified in the standard, it can often be advantageous to age for a series of times and plot a graph of the property against time. In some cases, it may also be useful to draw a graph of the logarithm of the rate of change of the chosen property (taken to be equivalent to a reaction rate) against the inverse of the absolute temperature – known as an Arrhenius plot. If a straight line is obtained, it is assumed that the rate of change at a lower temperature may be estimated by extrapolation.

A variation of these procedures for a typical heat-resistant Neoprene is illustrated in Figure 19.1, where the temperature at which the elongation of a test piece falls to 100% is plotted against the logarithm of the exposure time in an air oven. As the result is a straight line, it is assumed that it provides a means of estimating the service life of this compound at any intermediate temperature. The relevance to industrial practice is justified because, for most applications, a resilient material must

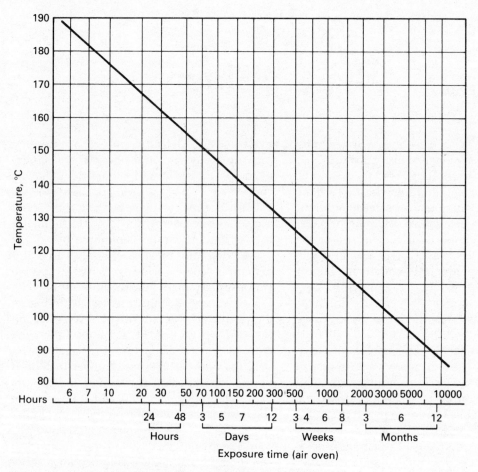

Figure 19.1 *Heat ageing characteristics of Neoprene*

have an elongation of at least 100% to remain serviceable. It is interesting to note that in this case an increase in temperature of 10°C, causes a decrease in service life of 50%, which agrees with the polymer engineer's rule of thumb that a difference of 10°C halves or doubles the life of a component.

19.1 Interpretation

The main use of ISO 188 is for quality control but the detailed requirements of the procedures must be followed closely to obtain good reproducibility. No universal correlation has been found between accelerated tests and natural ageing and the greater the disparity between ageing and service conditions, the less reliable any correlation becomes. Experimental difficulties, of course, arise because of the time required to obtain natural ageing data and the extent to which extrapolation may be required. Rubber products can and do last 50 years or more!

Ageing is also affected by the thickness of the rubber. As the deterioration is mainly due to oxidation, the rate of oxygen diffusion into the bulk of the rubber can have a serious effect on the results. If the rate is too low, during a test or in use, the bulk of the material may experience little or no ageing, even though the surface of the same rubber is very badly aged. The rate of oxygen diffusion will become even more significant when the temperature is increased. Comparisons should, therefore, only be made with results obtained from test pieces of similar shape and size.

Table 19.1 National and international standards for heat ageing properties of elastomers

Heat ageing
ISO 188
ASTM D454
ASTM D572
ASTM D573
ASTM D865
BS 903: Part A19
DIN 53508
NF T46-004
NF T46-005
NF T46-006

Stress relaxation
ISO 6914

It should also be noted that heat ageing tests on thin test pieces in the presence of air can give gross underestimates of the useful life of a product. For example, bridge bearings, which have a small area exposed to the air compared with their total volume, will last many times longer than would be indicated by short-term tests on thin sections.

19.2 Stress relaxation

The collection of multi-point ageing data is very time consuming and the use of separate test pieces for each point in the time and temperature sequence can introduce errors. Age testing by means of stress relaxation measurements can reduce some of the difficulties. The stress in a test piece is measured whilst it is subjected to an ageing procedure. The stress may be measured under conditions of continuous or intermittent strain. In the former, the test piece is held stretched throughout the test and with the latter the test piece is aged in the unstressed state and then stretched to a fixed length for a short time at periodic intervals and the stress measured. The second method is, of course, a measure of the change in modulus as a function of time. Both methods are specified in ISO 6914.

The continuous strain method requires a special apparatus, such as the Wallace-Shawbury self-recording age tester, which is a miniature stress relaxometer. The test piece is held between two grips in such a manner that it can be totally immersed in a cell of a multi-cell oven. The upper grip is supported by a horizontal beam which is mounted on, and balanced about, high grade miniature ball-bearings. The downward force exerted by the strained test piece is balanced by an upward force applied by a vertical spring. As the test piece relaxes the spring lengthens and the beam tilts. This movement energizes a small electric motor which adjusts the position of the spring anchorage to reduce the force applied by the spring until a condition of balance is restored to the beam. Attached to the spring anchorage is a pencil which records on a chart the reduction in the force applied to the spring. The unit is calibrated to allow a reading of 100 at the commencement of the test with full force and a reading of zero at zero force. The graph drawn by the pencil then indicates the percentage of force initially applied plotted against time.

The above apparatus may also be used for the intermittent strain test, a small air cylinder being used to release the tension. However, one of the advantages of the intermittent test is that normal testing equipment can be used.

19.3 Interpretation

Stress relaxation measurements can be interpretated in terms of changes in network structure by means of the kinetic theory of rubber elasticity. This states that at constant elongation and temperature the force in a rubber test piece is proportional to the number of active chains in the network. As a first approximation, any part of a polymer molecule joining two crosslinks can be considered as an active chain.

During the ageing of rubber vulcanisates crosslinking reactions occur as well as the breakdown of the network. Thus both continuous and intermittent measurements are of interest and give different results. During continuous measurement the test piece is always strained and crosslinks form in the stretched test piece. These can be expected to form between pairs of units in the polymer chains so that the newly formed crosslinks are not under strain. Consequently, the tension in the test piece will not be changed by their formation. It follows that continuous relaxation measurements do not show an effect from crosslinking and only measure the degradation reactions.

In the intermittent test, crosslinks are formed in the unstretched test piece and show their full effect only when the test piece is strained and the load measured. This type of measurement gives the combined effect of both crosslinking and degradation. If the crosslinking outweighs the force an intermittent test result will show an increase in force whereas the continuous test result must always show a decrease. Strictly, the intermittent test does not measure stress relaxation and ISO 6914 does not use the expression 'stress relaxation'. However, the term is in common use in the literature and in industry.

If continuous and intermittent tests are made simultaneously it is possible to obtain separate estimates of the two different reactions.

The stress relaxation method gives better information on network ageing than tests based on tensile strength; it is quicker to perform and therefore costs less. The amount of material required is small and the test piece can be sufficiently thin to avoid difficulties that might arise from surface reactions. Finally, the accuracy is higher than for measurements of ultimate properties, so more subtle effects can be investigated.

20

Adhesion

The term 'adhesion' is used in the rubber industry to cover a number of different aspects of the subject. The adhesion between unvulcanised elastomers is important to the rubber manufacturer when fabricating composite structures, such as tyres and conveyor belts. This form of adhesion is known as 'green tack' or just 'tack'. It is of no interest to the engineer when preparing product specifications and will not be considered in this work.

The adhesion between elastomers and other substrates, such as metals and fabrics, is very important to engineers. Rubbers are bonded to both metals and fabrics in the construction of tyres; rubber-to-metal bonding forms the basis of most rubber suspension and support systems and rubber-to-fabric bonding is all important in the manufacture of belting, hose, inflatable boats and similar fabricated items. The actual adhesion may be made by direct vulcanisation of the rubber to the substrate, sometimes with the addition of primers and/or bonding agents to the substrate. Alternatively the vulcanised rubber may be bonded to the substrate, using primers and/or pre-treatments and suitable adhesives.

The term adhesion is also used in connection with rubber-based adhesives themselves. These can be solutions of the rubber in solvents or water-based emulsions of the rubber as, for example, in the adhesives based on natural or synthetic rubber latices. This is a separate industry which is beyond the scope of the present work.

Finally, there is the unwanted adhesion between vulcanised rubbers and metals that mainly arises during long-term storage. This adhesion is associated with the corrosion of the metal caused by the proximity of the rubber compound and a number of simple contact tests have been developed to assess the extent of the problem. These are of particular importance for military applications but will not be given further consideration in this work.

The testing of composite materials and of individual products is also outside the scope of this section of *The Language of Rubber* but there are a number of tests that are used for production quality control and for evaluating bonding techniques that may be considered as rubber tests. Some of these tests, of course, may also be used for assessing the effectiveness of rubber-based adhesives.

20.1 Adhesion to metals

Very strong bonds to metals are necessary for applications such as rubber mountings and in most cases the bond needs to be, and is, stronger than the rubber itself. It is desirable to measure the bond strength by testing the actual product but this is not always possible. However, the test selected should emulate the working conditions as far as possible. There are three basic conditions: peeling, direct tension and shear.

Peel tests

The peeling test is specified in ISO 813, where it is referred to as the 'one-plate' method. The test fixture shown in Figure 20.1 is mounted in a tensile testing machine, which preferably should be of the low-inertia type, capable of operating at a constant rate of extension. The test piece consists of a strip of rubber 6 mm thick, 25 mm wide and 125 mm long bonded to to a metal strip to provide a bonded area 25 mm square as shown in Figure 20.2. The rubber is vulcanised in contact with the metal strip. If primers and/or bonding agents are

Table 20.1 *National and international standards for adhesion*

Adhesion to metals

Peel
ISO 813 (one-plate)
ASTM D429 (method B)
BS 903: Part A21 (method B)
DIN 53531: Part 1
NF ISO 813

Tension
ISO 814 (two-plate)
ISO 5600 (conical ends)
ASTM D429 (methods A & C)
BS 903: Part A21 (method A)
BS 903: Part A40 (conical ends)
DIN 53531: Part 2 (conical ends)
NF ISO 814 (two-plate)

Shear
ISO 1747
DIN ISO 1827
NF T46-020

Adhesion to fabrics

Peel
ISO 36
ASTM D413
ASTM D2630 (strap)
BS 903: Part A12
DIN 53530
NF T46-008

Tension
ISO 4637
BS 903: Part A27

Adhesion to cord

Textile
ISO 4647 (H-test)
ASTM D2138
BS 903: Part A48
NF ISO 4647

Metal
ISO 5603
ASTM D2229 (steel)
ASTM D1871 (single strand)

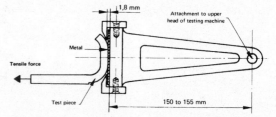

Figure 20.1 *Test fixture for adhesion-to-metal peel test*

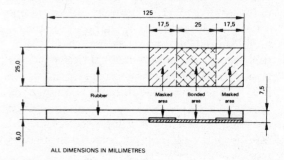

Figure 20.2 *Test piece for adhesion-to-metal peel test*

used they must be specified in the report of the test.

The test piece is placed symmetrically in the fixture with the separating edge towards the operator. Before the load is applied the rubber is stripped from the metal plate for a distance of approximately 1.5 mm using a sharp knife. The test piece is then placed in the grip and the heads are separated at 50 mm/min until separation is complete. The peeling angle is 90°. The maximum force required to cause separation over the distance of 25 mm is recorded, preferably by means of an automatic graphical recorder. If the rubber stock starts to tear the operator must cut the rubber back to the metal before continuing the test. The adhesion is calculated by dividing the maximum force recorded by the width of the test piece and is reported as newtons per millimetre of width.

The type of failure is reported in this and other similar tests as:–

R   for failure in the rubber
RC for failure at the interface between the rubber and the cover cement
CP for failure in the interface between the cover cement and the prime cement, and
M for failure at the interface between the metal and the prime cement.

The percentage failure of each type is included in the report. In practice it is not always possible to distinguish between RC and CP and in some cases only a single coat bonding system may be used.

Adhesion 99

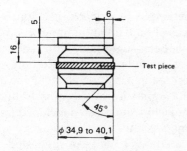

ALL DIMENSIONS IN MILLIMETRES

Figure 20.3 *Test piece for adhesion-to-metal tension test*

Most of the ISO rubber testing standards call for conditioning of the test pieces before the test is made but this is particularly important when testing bond strengths. The time between vulcanisation and testing should not exceed six days and comparable tests should be made with equal time intervals after vulcanisation. Test pieces or components should be held at the test temperature for at least 16 hours before testing, unless otherwise specified. It is also important that the exposed surfaces of the rubber and the metal are kept free from dust, moisture and foreign matter and are not touched by hand during assembly.

Tension tests

The method of testing the adhesion of rubber to metal by direct tension, also known as the 'two-plate' method, is specified in ISO 814. The test piece consists of a disc between 35 and 40 mm in diameter and 3 mm thick, bonded to two parallel metal plates not less than 9 mm thick, as shown in Figure 20.3. The plates are approximately 0.08 mm less in diameter than the rubber in order to prevent the rubber tearing from the edges of the metal during the test. The test piece is prepared by vulcanising the rubber in a suitable mould.

The test piece is mounted in a fixture in a tensile testing machine, care being taken to ensure that it is accurately aligned in order to ensure that the tension is uniformly distributed over its cross-section. Any misalignment will tend to introduce a peeling action. In practice, the stress at the rubber/metal interface does not remain even because shear forces are introduced as the rubber deforms under tension. The bond strength therefore depends on the shape factor of the rubber disk (see Section 10.3), the strength increasing with decreasing thickness. The jaws of the testing machine are separated at 25 mm/min and the result is calculated by dividing the maximum force applied by the cross-sectional area of the test piece and expressing it as meganewtons per square metre. The type of failure is expressed as described above for the peel test.

The bond strength is often stronger than the rubber and a tension test has been developed with cone-shaped metal end pieces in order to encourage failure at the interface between the rubber and the metal. The method is covered by ISO 5600. The geometry of the test piece is shown in Figure 20.4. The unvulcanised

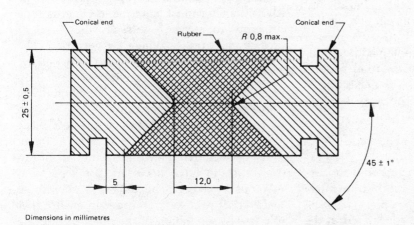

Figure 20.4 *Test piece for cone-shaped metal-to-rubber tension test*

rubber is bonded to the metal in a suitable mould, using primers and cover cements if required. The test piece is mounted in a tensile testing machine and the jaws separated at 50 mm/min. The result is simply expressed as the maximum force recorded in newtons. If the failure is in the rubber, it is recorded that the bond strength is greater than that of the rubber. The percentages of the types of failure are expressed as described previously. The stress distribution is concentrated at the tips of the cones and involves peel and shear forces rather than pure tension. The results are generally lower than those obtained with a plain disk test piece of similar diameter and about the same as the results of peel tests.

Shear tests

Quadruple element test pieces are recommended in ISO 1747 for testing bond strengths in shear. The test piece is similar to that used in ISO 1827 to measure shear modulus as shown in Figure 20.5 and described in Section 10.4. There is a proposal to combine the two

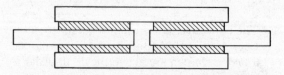

Figure 20.5 *Test piece for metal-to-rubber shear test*

standards by allowing the test for shear modulus to be taken to the point of failure but at present ISO 1747 is still applicable. In ISO 1747 the test piece is strained at a rate of 50 mm/min and the result obtained by dividing the maximum force by the total bonded area of one of the double sandwiches and is expressed in megapascals. The type of failure is reported as described previously.

Dynamic tests

In service, bonds between metals and rubbers may be subjected to impacts (high strain rates) or to repeated dynamic cycling (fatigue). Neither of these factors are considered in the standard methods described previously. Impact tests based on Izod and falling weight apparatus have been suggested, with the test piece receiving a single blow. In addition, a high speed test using a double element shear test piece and a sophisticated servo-hydraulic universal test machine has been developed for testing automobile bumpers. However, no standard test methods have been suggested.

Tests based on flexometers and many other specialized pieces of equipment have also been developed for dynamic testing but, again, there are no standards.

20.2 Adhesion to fabrics

Tests for adhesion between elastomers and fabrics are usually carried out in peel or direct tension. Peel tests are the most common but tension tests are useful when testing thin coatings where the elastomer is too thin or too weak to withstand a peel test.

Peel tests

The recommended peel test is that given in ISO 36 which can be used for two plies of fabric bonded with rubber or a rubber layer and a fabric bonded together. A 25mm wide strip test piece long enough to permit separation over at least 100 mm is used. The testing machine must be power driven and equipped with a suitable dynamometer capable of maintaining a substantially constant rate of traverse of the moving head during the test and fitted with an autographic recorder. The dynamometer must be of the low inertia type in order to measure correctly the series of peaks and troughs which are recorded. The peaks are caused by the stripping force building up to the point where a separation occurs and the troughs by the falling off of the force immediately afterwards.

The minimum thickness of the test piece constituent components, or any one of them, must be such that the weakest component can transmit the force necessary for separation without breaking. The thickness should be reduced, if necessary, in order to ensure that the line of separation of the plies, during the test, lies as close as possible to the plane of the axis of the strips of the test piece held in the grips. The fabric and rubber are separated by hand over a length of about 50 mm and the body of the test piece placed in the non-driven grip and the ply to be separated in the power-driven grip. The angle of separation should be approximately 180°. It is important to ensure that the axes of the strips of the test piece are in the same plane. The rate of travel of the power-driven grip should be 50 or 100 mm/min so as to give a rate of ply separation of 25 or 50 mm/min.

The median peak force during the test is calculated from the continuous trace of force against time using one of the three methods described in ISO 6133. The method used depends upon whether there are less than 5, 5 to 20 or more than 20 peaks. In essence, the peak forces are arranged in ascending numerical order and the middle value, if there is an odd number of peaks, or the arithmetic mean of the two middle values if there is an even number of peaks, is the median force. When there are 5 to 20 peaks the outer 10 percentages of the traces are ignored. When there are more than 20 peaks, the outer 10 percentages are again ignored and 10 equally spaced vertical lines are drawn on the remainder of the trace. Only the peak values situated nearest to each of the vertical lines are considered. In each case, the range is taken as the difference between the values of the lowest and the highest peaks. The adhesion strength is the median peak force divided by the width of the test piece, expressed in newtons per millimetre.

The type of failure is reported as:

R   for failure in the rubber layer
RA for separation between the rubber layer and the adhesive
AT for separation between the adhesive and the fabric
RB for failure in the rubber bond between two fabric plies
T   for failure in the fabric, and
RT for separation between the rubber and the textile when no adhesive is present.

Direct tension tests

A direct tension test is described in ISO 4637 which more nearly measures the true adhesion between fabric and rubber. The test piece consists of two metal cylinders, 25 mm diameter, between which a test piece, 32 mm square, is cemented with a cyanoacrylate adhesive as shown in Figure 20.6. The cylinders are gripped in a tensile test machine and separated at 50 mm/min. The maximum force recorded at failure, in kilonewtons, is the adhesion strength. The mode of failure is expressed as an approximate percentage using the following designations:

C   for cohesive failure within the rubber, and
RF for adhesive failure between the rubber and fabric.

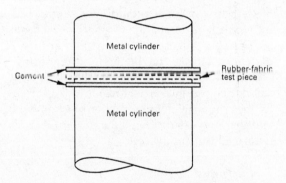

Figure 20.6 *Test piece for rubber-to-fabric direct tension test. The test piece is square and extends beyond the edges of the metal cylinders*

The test is simple but the preparation of the test piece requires a great deal of care. The metal surfaces must be accurately lapped and degreased and the assembly has to be cemented together using a special jig specified in the standard. However, comparative results can be less misleading than those obtained from peel tests.

Dynamic tests

The measured bond strength of a rubber/textile system may be different if the test is carried out under static or dynamic conditions. It is surprising, therefore, that there is no European test for measuring dynamic ply separation. In principle, any dynamic fatigue flexing test, such as described in Chapter 13, could be used but most of the existing standards are intended to induce cracking rather than to test ply separation.

In the USA, a strap peel test is described in ASTM D2630 in which the test piece is optionally flexed using the Scott flexer detailed in ASTM D430 (see Section 13.1). The latter standard also suggests that the Du Pont flexer can be used for testing composites.

20.3 Adhesion to cord

The adhesion of textile and metal cords to rubber is an important factor in the manufacture of tyres, belting and other products. Static tests consist of moulding a single cord into a block of rubber and measuring the force required to pull it out. The results are very dependent upon the method of moulding, which is specified in great detail in the standards. The measured force is also affected by the way in which the rubber block is held and by the extent to which the rubber is deformed during the test.

Textile cord

The 'H-pull' or 'H-block' test is described in ISO 4547. The test piece consists of two pieces of rubber, each 6.4 mm wide and 3.2 mm thick, interconnected by a single cord moulded into both end pieces. Cotton fabric is used to reinforce the outer sides of the rubber end pieces. The assembly resembles the letter H from which the test derives its name. The pieces of rubber are held in special clamps, which are separated at a constant rate of 100 mm/min in a low-inertia testing machine. The maximum force to separate the cord from the rubber is recorded to the nearest newton. At least eight test pieces should be used and the arithmetic mean of the results is taken to be the adhesion value. The appearance of the cord indicating whether the rubber has remained adhered to it or not is also reported.

Wire cord

Wire cord and single wires are tested by similar methods which are described in ISO 5603. Again, the method of moulding the test pieces is critical and is described in detail. In this case a series of wire cords are moulded into a single rubber block, the sides of which are reinforced by a strip of sheet metal or steel cord fabric. The rubber block is held in the upper clamp of the testing machine and the free end of the wire cord is held directly in the lower clamp. Each wire cord is tested separately, the rubber block being passed successively through the upper clamp. The jaws are separated at between 50 and 150 mm/min and the force at which the cord parts from the rubber is recorded. the adhesion force is calculated by dividing the maximum force at separation by the embedded length of the cord into the test piece and is expressed as newtons per millimetre or kilonewtons per metre to the nearest integer. At least 10 test cords must be tested and the mean value and standard deviation determined for each test condition. Each test piece should be examined and the type of adhesion failure expressed as:

- R   to indicate failure in the rubber
- M   to indicate that the failure is in the interface between the cord and the rubber and that the bare cord surface is visible.

The results should be reported in percentages, to express the 'percentage coverage' in steps of 25%. For example, 25R/75M means that 75% of the cord surface is visible.

Dynamic tests

Tyres are very definitely fatigued during use and dynamic tests should also be used to assess the bond efficiency between cords and rubber. There are no standards but a wide variety of procedures have been reported in the literature. Some workers have used the same or a similar test piece to that used in static tests and applied a cyclic tensile stress or strain. Others have used some form of fatigue tester operating in compression/shear to repeatedly stress the cord/rubber composite or even to flex samples in the form of a belt.

20.4 Interpretation

Unlike most other properties, adhesion is not strictly a function of the composition of the rubber part. The method of fabrication, the kind of metal or fabric and the cleanliness of the bond surfaces can all have their effects. In addition, variations in testing technique can frequently cause identical materials to give different results on different occasions. It is, therefore, necessary for the conditions of the test methods to be agreed in detail between the supplier and the user if standards are to be used in product specifications. However, the standard tests can be used successfully for the comparative testing of bonding systems, providing experienced operators are used and close control of the test conditions is maintained.

21

Permeability

Permeability is a measure of the ease with which a liquid, vapour or gas (i.e. fluids) can pass through an elastomeric film or laminate. The process is one of absorption and diffusion. The fluid dissolves into the elastomer on one side of the film and then the dissolved material diffuses through the film to the opposite side, where evaporation takes place.

The definition of permeability is the rate of flow of the fluid, under steady state conditions, between opposite faces of a unit cube of the elastomer, when subjected to unit pressure difference and controlled temperature. Permeability is a fundamental property of the elastomer but for vapours and liquids it is more usual to measure the rate of transmission under defined conditions of temperature and humidity, expressed as grams of fluid per square metre per 24 hours ($g/m^2 d$). Rate of transmission measurements are often wrongly called permeability.

Permeability is an important factor in many applications for elastomers. Linings for reservoirs, flexible fuel tanks, gaskets and seals, diaphragms for pumps and valves, hose, inner tubes, balloons, inflatable boats and airships are some of the applications in which permeability must be kept within reasonable limits. Permeability measurements are also of interest in the study of diffusion and gas solubility characteristics in relation to polymer structure.

21.1 Permeability to gases

Gas permeability is usually measured by setting up a pressure differential across the test piece and measuring, by change of pressure or volume, the amount of gas passing to the low pressure side of the system.

Constant volume method

The constant volume method for the measurement of permeability is covered by ISO 1399. The apparatus consists of a metal cell having two cavities separated by the test piece. The high pressure cavity is filled with the test gas at the required pressure, which must be measured to an accuracy of 1%. The low pressure side is connected to a pressure measuring device, usually a capillary U-tube manometer with an adjustable height reservoir. The test cell must be maintained to within ± 0.5°C of the required temperature because the permeability of gases is extremely sensitive to temperature.

The test piece is a disk between 50 mm and 65 mm diameter and 0.25 mm to 3 mm thick, with a free testing surface of 8 to 16 cm^2. After the cell and test piece have been assembled the high pressure side is filled with gas at the test pressure. The increase in pressure on the low pressure side is then measured as a function of time, the manometer being adjusted to ensure that the measurements are taken at constant volume. Steady state conditions are indicated by a linear relationship between pressure change and time and may take at least an hour to be established.

The permeability is calculated from the formula given in the standard and is expressed as metres squared per second per pascal ($m^2/s.Pa$) and typical figures for a natural rubber gum stock are of the order of 9×10^{-17}. The apparatus has a useful working range of 0.1×10^{-17} to 15×10^{-17} $m^2/s.Pa$.

Constant pressure method

The test cell for measuring permeability at constant pressure is described in ISO 2782 and is similar to that described above. However,

Table 21.1 *National and international standards for permeability*

Permeability to gases

Constant volume
ISO 1399
BS 903: Part A17
DIN 53536
NF T46-037

Constant pressure
ISO 2782
BS 903: Part A30
DIN 53536
NF T46-034

Permeability to vapours

Water vapour
ISO 2528
ASTM E96
DIN 53122: Part 2
NF ISO 2528

Volatile liquids
ISO 6179
BS 903: Part A46
DIN 53532

the low pressure side is connected to a device to measure the volume increase as the gas diffuses to the low pressure side whilst maintaining constant pressure. In ISO 2782 a graduated capillary tube, of accurately known cross-section, is used to measure the volume change. The capillary tube can be arranged as a vertical U-tube with a reservoir, which is effectively the same as the apparatus used in ISO 1399, or as a horizontal capillary with a single drop of liquid which is pushed along as the volume increases as shown in Figure 21.1. The operation of the apparatus is similar to that for the constant volume method, the readings of the meniscus in the capillary tube being plotted against time. The permeability is then calculated from the formula supplied in the standard.

Carrier gas methods

When greater sensitivity is required carrier gas methods may be used but they are not covered by international standards. The test gas flows on one side of the test piece and a second gas, the carrier gas, flows on the other

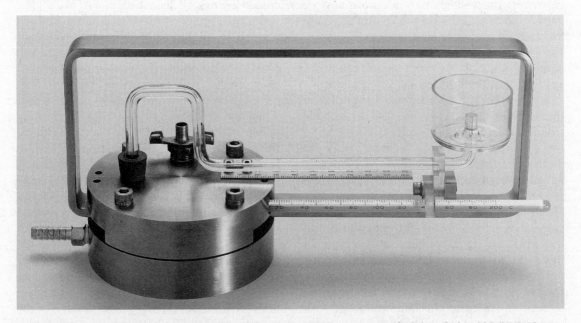

Figure 21.1 *Gas permeability tester for rubber using the constant pressure method described in ISO 2782. Gas permeability is of importance in the manufacture of inner tubes and liners for tubeless tyres, hose, seals, balloons, etc.*

side and is quantitatively analysed to determine the quantity of test gas that has passed through the test piece. As there need be no pressure differential across the test piece there is less likelihood of leaks and it is easier to support the test piece. The greater sensitivity also enables less permeable materials to be tested and the different transmission rates of the components of gas or vapour mixtures can be measured. Thermal conductivity, gas chromatography and similar methods of analysing the carrier gas have all been reported. Carrier gas methods are of particular value for packaging applications because the test method can relate closely to actual working conditions.

21.2 Vapour permeability

The most usual method of measuring the rate of vapour transmission, especially for water vapour, is the dish method described in ISO 2528. This standard can be used for many sheet materials including, paper, board, plastics films and laminates as well as for elastomers and fabrics coated with elastomers. However, the test should not normally be used for materials thicker than 3 mm or where the transmission rate is less than 1 g/m^2d. Nor is it applicable to materials that are damaged by hot wax or that shrink to an appreciable extent under the test conditions.

The method consists of using a thin disk of the material under test to seal a suitable circular dish containing a dehydrating agent – usually anydrous calcium chloride. The edges of the dish are sealed with a wax mixture which is specified in the standard. A template is used to define accurately the effective area of the test piece. The dish assembly is placed in a cabinet at a controlled temperature and humidity and it is weighed at intervals to measure the amount of water vapour transmitted through the test piece and absorbed by the dessicant. The test is continued until the increase in weight is substantially linear with time and the system is therefore in equilibrium. The result is expressed as a transmission rate with the units of g/m^2d. The internal depth of the dish below the plane of the test piece should not be less than 15 mm for transmission rates above 100 g/m^2d or 8 mm for lower rates. The effective area of the test piece should be about 50 cm^2.

In general, the transmission rate is not a linear function of temperature or relative humidity and the test conditions should therefore be chosen to be as close as possible to those required in service.

An alternative procedure is to contain the water within the dish and determining the quantity of water vapour transmitted out into a dry atmosphere by measuring the weight loss of the dish.

If the dish assembly is inverted, it is possible to measure the transmission rate when the water is in contact with the test piece but a more general procedure applicable to all volatile liquids is described below.

21.3 Volatile liquids

A method for determining the transmission rate of volatile liquids through elastomers and coated fabrics is described in ISO 6179. The apparatus, which is illustrated in Figure 21.2, consists of a lightweight aluminium container with a screw-on collar which retains the test piece. The rotating part of the collar applies pressure to the clamp ring through ball bearings so that the test piece is not distorted when the collar is tightened. The two filling valves allow the liquid to be changed during the test without disturbing the test piece. This is recommended when a mixture of liquids is used which are not transmitted at the same rate, thus changing the properties of the liquid left in the container. When testing materials without a reinforcing fabric, a circular piece of stainless steel wire mesh is mounted with the test piece in order to support it on its outer surface. The exposed area of the test piece should be about 10 cm^2 and its thickness between 0.2 mm and 3 mm. The volume of the container should be between 60 and 100 cm^3.

After assembly, approximately 50 cm^3 of the liquid is introduced into the container and the

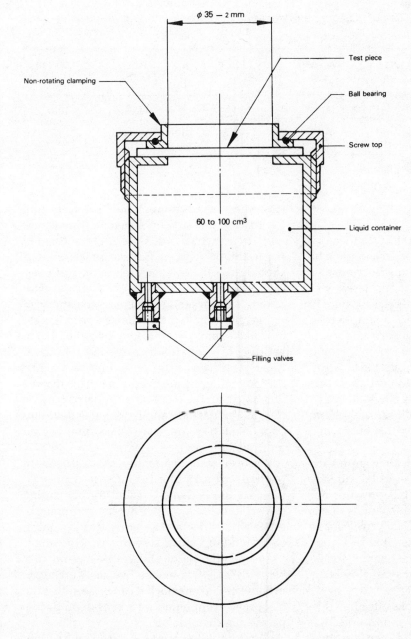

Figure 21.2 *Cell for measuring the transmission rate of volatile liquids through elastomers. The cell is sealed with the elastomer under test and the volatile liquid is introduced into the cavity. The cell is then inverted and the loss in weight under controlled conditions is measured. The rate of transmission is expressed as grams per square metre per 24 hours*

container is weighed. The assembly is then inverted so that the liquid is in contact with the test piece and maintained at the test temperature for 24 hours. The assembly is cooled to room temperature and weighed again. The test is repeated with the liquid in contact with the test piece for a further 72 hours. For Method A in the standard, the container is then emptied and re-filled at 24 hour intervals until the weight loss per 24 hours is effectively constant. In Method B, which is only used for single component liquids, the weight loss is simply determined after the second exposure of 72 hours but in this case it should be noted that equilibrium conditions may not have been reached.

Table 21.2 *Relative permeabilities of elastomers to gases at 25°C. (Natural rubber arbitrarily set to 100)*

	Hydrogen	Oxygen	Nitrogen	Carbon Dioxide	Methane
Natural Rubber	100	100	100	100	100
Neoprene	26	17	13	19	11
SBR	78	72	73	93	73

21.4 Interpretation

Many factors affect permeability but one of the most important is that the system must be in equilibrium when the measurement is made. For some vapours, particularly with thick films, it can take several days or more to reach equilibrium. Even then, a simple relationship that assumes that permeability is directly proportional to the solubility and diffusion constants is only applicable to 'ideal' gases. For other gases and for vapours, permeability is not necessarily a constant. Measurements made before equilibrium has been reached will give a smaller apparent permeability.

Also, for some gases and all vapours the rate of transmission may not be directly proportional to the differential pressure nor inversely proportional to the thickness of the test piece. This can be especially true if the elastomer is not homogeneous. It is, therefore, very important to select the test parameters to match the service conditions as closely as possible.

Another important variable affecting permeability is temperature. For example, the permeability at 50°C may be four or five times that at 15°C. Also, different compounds of the same type of elastomer will have different permeabilities. It follows that the permeability of a product cannot be predicted merely from a knowledge of the type of elastomer it contains.

Further complications can occur if the elastomer is distorted during the test. If the elastomer is restrained and it swells, an hydrostatic pressure or change of dimensions will occur which will affect the measured permeability. Also, if the elastomer is stretched during the test, the area of the test piece will increase and its thickness decrease. Both of these changes tend to increase the apparent permeability unless the dimensional changes are taken into account. On the other hand, because of molecular orientation, stretched films may have a lower permeability than corresponding unstretched films. This may be illustrated by the fact that an inflated balloon does not deflate as rapidly as published rates of transmission for unstretched films would indicate.

With so many variables it is difficult to relate test results to actual service. However, the tests are extremely useful for comparing the relative merits of different compounds, especially if they are tested against a control compound of known service characteristics.

The relative permeabilities of natural rubber, SBR and Neoprene to five gases are given in Table 21.2 from which it can be seen that the Neoprene compounds are considerably less permeable than the others.

Some liquids are more soluble in certain elastomers than others. Therefore, in considering permeation by volatile liquids, the ease with which the liquid dissolves in the elastomer becomes of prime importance. Hydrocarbons, for example, are far more soluble in natural rubber than in Neoprene and, as a result, Neoprene is much less permeable to these liquids.

22

Resistance to weathering

Deterioration in physical properties can occur when elastomers are exposed to the weather. This deterioration can be observed as cracking, peeling, chalking, colour changes and other surface defects and ultimately by failure of the product in service. Most of these defects can be prevented by suitable compounding and many of the special synthetic elastomers are inherently more resistant to deterioration than natural rubber. The inclusion of carbon black, in particular, will increase the resistance of most elastomeric compounds to attack by ultraviolet (UV) radiation.

By far the most important cause of deterioration by weathering is the presence of ozone. Less than one part per hundred million (pphm) of ozone in the atmosphere can severely attack non-resistant rubbers if they are in a slightly strained condition. However, the concentration of ozone in the atmosphere is such a variable parameter that it is very difficult to reproduce actual working conditions for test purposes. Laboratory tests are therefore normally used and are discussed in detail in Chapter 23.

Sunlight, oxygen, moisture and temperature also affect elastomers. Short-wave UV light is the most damaging of the sun's radiations and can cause surface hardening which can lead to crazing, chalking and gradual erosion of the surface of the elastomer. Oxygen can cause oxidation of the polymer and leads to loss of mechanical properties and elasticity and this effect is accelerated at high temperatures. Moisture can be present more often than is generally supposed. A study in the USA showed that materials at six locations in North America were wet for about 30% of the time and that the predominant source of the moisture was dew rather than rain. The water itself may not have been particularly destructive but the study concluded that the dew contained a high percentage of oxygen which penetrated the material and caused internal oxidation. The continual presence of moisture can also allow time for soluble additives to be leached from an elastomer.

In addition to the above, deterioration of elastomers can be caused by the presence of gases and chemicals peculiar to a particular locality. Salt spray in marine locations is but one example.

22.1 Assessment of deterioration

The assessment of the deterioration after weathering can be a difficult task. Measurements of physical properties on test pieces that have been subjected to weathering is fairly straightforward but it is necessary to ensure that the results are not affected by cracks or similar damage. More difficult is the fact that it is frequently necessary to make a visual assessment of the changes in physical appearance. This is necessarily a subjective process, the results from which can vary between operators.

ISO 4665/1 provides a basis for assessing the changes in properties after test pieces have been subjected to natural weathering or artificial light. Physical tests are made with the standard test pieces recommended for the particular property. Control test pieces are kept in the dark for the same period as the exposed test pieces and tested at the same time as the exposed test pieces. The deterioration is calculated as a simple percentage change in the property compared with the initial value. Visual assessments of the

Table 22.1 *National and international standards for weather resistance*

ISO 4665 Parts 1, 2 and 3
ASTM D518
ASTM D750 (Weather-Ometer™)
ASTM D1171 (Ozone)
DIN 53386
DIN 53387
DIN 53388
NF T46-040
NF T46-041
NF T46-042

changes in appearance and surface properties are made on the following scale:

0 none
1 barely perceptible
2 moderate
3 substantial

which is necessarily arbitrary and can be difficult to interpret.

Changes in colour are assessed on the 'grey scale' following the recommendations in ISO 105-A01 and -A02. Grey scales vary from 1 to 5; grade 1 corresponding to the strongest contrast and grade 5 to zero contrast, i.e. two samples with identical colour. The exposed and the reference test pieces are compared and the rating of colour change is the grade on the grey scale which shows an equivalent contrast. If the contrast lies between two grades, both are reported, e.g. a 3–4 rating means that the contrast is greater than that of rating 4 but less than rating 3. The type of colour change may also be recorded. Thus any changes in hue, brilliance, lightness or any combination of these changes is described by adding terms listed in the standard. Grey scales can be purchased from a number of international standards organizations.

ISO 4551/1 also mentions that colour and changes in colour may be measured by instrumental means and promises that suitable means will be published in a future international standard.

22.2 Natural weathering

Natural weathering tests are generally long-term tests in areas which relate to the climatic conditions under which the products will be used. In some cases, a form of accelerated test is obtained by exposing the test pieces in areas, such as Florida in the USA, where the percentage of UV light is greater than normal. Deterioration may also be encouraged by placing the test pieces under strain, either by extending them in special racks or by bending them in a loop or around mandrels. Mandrels are frequently used when testing electric cables and extruded sections.

Recommendations for the type of exposure equipment, the preparation of test pieces, the test conditions and the procedures to be used are provided in ISO 4665/2. The methods and test pieces for determining ozone resistance are the same as for laboratory tests and are considered in Chapter 23. Exposure racks for most purposes are simple flat frames but it is recommended that brass, steel and copper should be avoided in their construction. The exposed surfaces should preferably be at 45° to the horizontal facing the equator but the angle may be varied to suit specific requirements. For example, vertical exposure may be required to reproduce the conditions on the outside of buildings. If maximum annual irradiation of the test piece is required the angle to the horizontal should be the exact latitude angle of the site. It can be seen, therefore, that the test conditions can be very varied.

Solar radiation is measured by the use of blue dyed wool standards as specified in ISO 105-B01. These are pieces of woollen fabric that have been dyed to a range of seven standard blue colours. On exposure to light the colours fade and are then compared with unexposed samples using the grey scale described above. As the blue dyed wool standards can be affected by rain, ozone, etc., in a different manner from the elastomers, they are exposed under suitable transparent protective covers.

The blue dyed wool standards are usually exposed until the colour comparison agrees with grey scale contrast number 4. Thus, exposure stage 1/1 is reached when standard 1 gives a contrast equal to 4 on the grey scale; 2/1 when standard 2 shows similar contrast and so on in the same manner to stage 7/1. The duration of stage 7/1 is about one year in natural daylight in temperate climates. If longer test periods are required it is recommended that when the exposure stage has reached 7/1 the exposed standard be replaced with a new blue standard 7. When this new standard reaches grey scale 4 the exposure stage is designated 7/2. This process is repeated as many times as is necessary. The standard points out that this procedure is far from satisfactory and should only be used when no better alternative, such as the use of suitable instruments, is available. However, blue dyed wool standards are readily available and the fund of data based on their use is so large that they continue to be used.

22.3 Artificial weathering

Natural weathering is necessarily a long process and many units for performing accelerated tests have been developed. The light source must have a spectrum similar to that of sunlight but with an increased intensity in the UV wavelengths. Early units used carbon or mercury arcs, later units xenon arcs and now fluorescent tubes are becoming popular. ISO 4665/3, whilst specifing methods of exposing elastomers to artificial light, only considers xenon arcs for the light source. However, there is an appendix containing information about fluorescent tubes which points out amongst their advantages that they are less costly to use and that, as they produce little infra-red radiation, there is generally no problem with overheating.

This part of the standard specifies test conditions, the preparation of test pieces and the method of assessing the deterioration using blue dyed wool standards, etc., with reference to the recommendations in Parts 1 and 2 where necessary. The main consideration is the effect of light but the use of water sprays to simulate rain is included. Generally the test pieces are not strained during the test and ozone is not present. There are also useful appendices to ISO 4665/3 which include the correlation between artificial and natural

Table 22.2 *Weathering of ALCRYN window gaskets*

Colour	White	Red	Green	Maroon
Process	moulded	moulded	molded	moulded
Initial properties				
Tensile strength (MPa)	9.3	9.6	8.9	8.8
Elongation (%)	395	360	390	370
Modulus at 100% (MPa)	3.3	4	3.4	3.6
After 1000 hrs in Q.U.V				
Tensile strength (MPa)	10.2 (110%)	9.1 (95%)	8.5 (96%)	8.1 (92%)
Elongation (%)	380 (96%)	411 (114%)	440 (113%)	405 (109%)
Modulus at 100% (MPa)	4.0 (120%)	3.5 (87%)	3.1 (92%)	3.1 (86%)
Surface degradation	none	none	none	none
Colour change	none	none	none	none
After 2000 hrs in Q.U.V				
Tensile strength (MPa)	7.3 (80%)	9.5 (99%)	9.0 (101%)	9.4 (107%)
Elongation (%)	370 (94%)	366 (102%)	405 (104%)	388 (105%)
Modulus at 100% (MPa)	3.1 (94%)	4.2 (105%)	3.6 (106%)	3.9 (108%)
Surface degradation	none	none	none	none
Colour change	none	none	none	none

The percentage retention figures are shown in brackets.

weathering and the spectral distribution of simulated solar radiation.

A typical example of the use of an accelerated weathering test to assess a new product was that carried out by Du Pont at their Customer and Technical Service Centre at Hemel Hempstead, UK. Four coloured compounds of ALCRYN, melt-processable halogenated polyolefin rubber, were tested on a QUV accelerated weathering unit for their suitability for use as window gaskets. Test pieces were subjected to a 12 hours weathering cycle; eight hours exposure to UV light having an average wave length of 340 ångström units at a temperature of 70°C, followed by four hours exposure to condensing water at a temperature of 40°C. It can be seen from Table 22.2 that there were only minor changes in properties after 1,000 and 2,000 hours. In addition, there was virtually no discoloration after 2,000 hours. Samples of ethylene-propylene terpolymer (EPDM) elastomer were tested in the same manner and there were signs of chalking after only 500 hours.

22.4 Interpretation

The number of parameters, the considerable variation in conditions from site to site and seasonal differences make it very difficult to measure natural weathering in absolute terms. It is, therefore, desirable to include test pieces of known weathering characteristics in any series of tests and to consider only comparative performances. Thus, the best correlation that can be expected is that a series of tests exposed in, say, Florida have the same relative ranking as those exposed in any other place. This in itself can provide valuable information and can often be used to eliminate compounds that fail quickly. It is often recommended that exposure trials of any new product should begin as early as possible. The experience and data will then always be ahead in time of actual use and may be used to give advance notice of any possible trouble.

The results of accelerated weathering tests are equally difficult to correlate with natural weathering and actual performance. Accelerated tests tend to exaggerate the UV region of the sun's spectrum and the temperature during the test is often higher than that experienced in practice. Part of the degradation is then due to the high temperature rather than to the action of the UV radiation. Also the absence of ozone during the tests eliminates the most serious factor that affects weather resistance. Again, the best that can be expected is that the relative ranking of test pieces correlates with that obtained by natural weathering.

Weathering tests with thin test pieces can be misleading. When the elastomer is relatively thick, for example in bridge bearings, the degradation may only affect the surface and the mechanical properties of the bulk of the elastomer will remain satisfactory. In such cases, practical experience must take precedence over laboratory or site tests.

23

Resistance to gases

Three gases to which elastomers are regularly exposed are air, oxygen and ozone and they are the only ones for which internationally recognized standard tests exist. The exposure of elastomers to air and oxygen is considered in detail in Chapter 19 and the effect of air and oxygen is part of the weathering tests discussed in Chapter 22. The permeability of elastomers to vapours and gases is a separate subject and is covered in Chapter 21.

If data for other vapours or gases are required the test pieces are usually exposed in chambers through which the gas can be circulated. If the gas can be readily maintained in its liquid state, tests are sometimes made with the liquid rather than the gas.

23.1 Resistance to ozone

Ozone is generated from the oxygen in the air by the action of the UV light from the sun or by high voltage discharge (corona) and is universally present in the earth's atmosphere.

Table 23.1 National and international standards for resistance to ozone

Ozone
ISO 1431 Parts 1 and 2
ASTM D470 (cables)
ASTM D1149
ASTM D1171
ASTM D3395 (dynamic)
BS 903: Parts A43 and A44
DIN 53509
DIN 53509 Part 2
EDIN 53509 Part 1
NF T46-019
NF T46-038
NF T46-039

The concentration can vary from less than 1 to 5 parts per hundred million (pphm) in rural areas to as much as 50 pphm in heavily polluted districts. Although ozone has no effect on unstrained rubbers, almost all elastomeric products are under some strain and can be severely attacked by ozone at concentrations as low as 1 pphm.

The effect of ozone is to produce clearly visible and mechanically very damaging cracking of the rubber surface and testing involves assessing the extent of the cracking and/or its effect on mechanical properties. The resistance to ozone can be improved by the use of anti-ozonants and/or waxes which form a protective surface bloom. Laboratory tests have great importance for the evaluation of such additives.

The international standard for testing ozone resistance is ISO 1431. Part 1 covers static and Part 2 dynamic testing. A third part to cover the methods of measuring ozone concentration is in preparation but is held up because of the difficulty in obtaining correlation between the various available procedures.

Equipment

The test equipment, shown diagrammatically in Figure 23.1, consists of a closed chamber held at a constant temperature, through which is passed ozonized air at a known concentration. All parts in contact with the ozone must be constructed of materials, such as aluminium or glass, which do not decompose ozone. The tests must be made in the dark but an inspection window and a light source for inspecting the test pieces during the test may be incorporated. The dimensions of the chamber are not critical but the flow rate and velocity of the ozonized air do affect the severity of attack and must be controlled. The

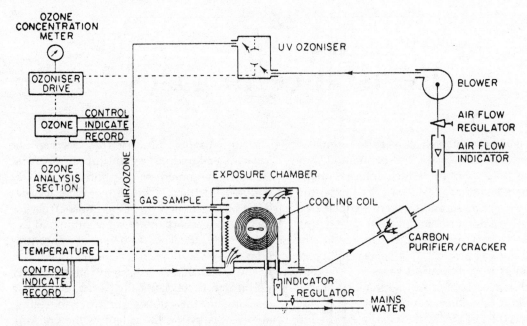

Figure 23.1 *Circuit diagram for an automatic ozone exposure apparatus*

ozonized air must be evenly distributed through the chamber, either by a fan or a diffuser and a cooling coil is usually supplied to help maintain a constant temperature.

The standard also recommends that the test pieces should be mounted on a mechanically rotated carrier that moves the test pieces, with their length parallel to the gas flow, in a path that covers at least 40% of the available cross-sectional area of the chamber. Each test piece should visit the same position within the chamber every eight to 12 minutes.

The source of the ozone may be either an ultraviolet lamp or a silent discharge tube. The latter is usually necessary if very high concentrations of ozone are required but oxygen must be used instead of air in order to avoid the formation of nitrogen oxides which can affect the cracking. The ozonized oxygen or air is then diluted with air to attain the required ozone concentration.

As most ozone tests are carried out in the range 25 to 200 pphm a very sensitive method of measuring the concentration is necessary. The traditional chemical methods use the reaction of ozone with potassium iodide to produce iodine but these are not suitable for continuous monitoring or automatic control. In practice, electrochemical or UV absorption methods are used but they are difficult to correlate with each other and with the chemical methods. The test results therefore tend to differ from machine to machine and between laboratories. This is not a problem when comparative tests are used to distinguish between different compounds but it means that inter-laboratory comparisons are difficult.

The test pieces can be strips of not less than 10 mm width, 2 mm thick and not less than 40 mm between the grips of the holders before stretching. Alternatively, they may be dumb-bell shaped, 5 mm wide and 50 mm long with the tab ends 12 mm square. Further variations allowed by the standard are T50 dumb-bells as used for tensile testing, annulus-shaped pieces that produce a continuous range of extensions when mounted and stretched and rectangular shaped pieces bent in a loop which also provide a gradation of extensions within one test piece. Extrusions may be wrapped around mandrels but the resulting strain may be less well-defined than it is with strips.

The test pieces are always mounted in a stretched condition between suitable clamps. The areas to be clamped are preferential sites for cracking and are usually covered with an ozone resistant paint, such as one based on Du Pont HYPALON. The clamps themselves should be 'soaked' in ozone prior to use.

The surface of the test piece should be as smooth as possible and it is recommended that the sheet should be moulded between polished aluminium foil which is left on the rubber until the test piece is required. This provides protection against handling and ensures a fresh test surface at the time of testing.

Test procedures

The test pieces should be stored in the dark in an essentially ozone-free atmosphere and should be conditioned before use in a strained state for between 48 and 96 hours. This enables the anti-ozonants and/or waxes to form a protective surface bloom. If these additives are not used, the conditioning period is unnecessary.

ISO 1431/1 provides for a series of possible ozone concentrations and test temperatures but unless otherwise specified it recommends a concentration of 50 pphm and 40°C. High humidities can affect the rate of ozone attack and the standard recommends that the relative humidity should not exceed 65% at the test temperature, unless the test is designed to evaluate the effect of higher humidities.

Ideally a series of strains should be used but if for specification or quality control purposes a single strain is used it is usually 20%. The test pieces are examined at regular intervals for cracking by using a lens of × 7 magnification. For most purposes a simple 'cracked' or 'not cracked' is all that is required but if a 'degree of cracking' is wanted an arbitrary scale has to be used which is necessarily very subjective. The most widely used scale is 0 to 3, where 0 is for no cracking, 1 is for cracks only seen under magnification, 2 is for very small cracks and 3 is for anything worse. Even this simple rating system falls down when there are only one or two large cracks.

The standard provides for three procedures. Procedure A requires the test pieces to be strained at 20% elongation and for them to be examined after 72 hours. The results are usually reported simply as 'no cracking' or 'cracking' but estimates of the degree of cracking may be made if they are required.

With Procedure B the test pieces are strained at one or more of the elongations specified and are examined after exposures of 2, 4, 8, 16, 24, 48, 72 and 96 hours and, if necessary, at suitable intervals thereafter. The time for the first appearance of cracks is noted for each elongation and this is taken as a measure of the ozone resistance at the specified strain.

Procedure C uses four or more of the standard elongations and the test pieces are examined at the same time intervals as for Procedure B. The time of the first appearance of cracks is recorded at each elongation. The 'threshold strain', which is defined as the highest tensile strain at which a rubber can be exposed at a given temperaure to air containing a given concentration of ozone without ozone cracks developing on it after a given exposure period, can then be estimated. For some elastomers, a graph of strain against time to first apperance of cracks will produce a curve which becomes asymptotic to the time axis. This represents the 'limiting threshold strain' which is defined as the tensile strain below which the time required for the development of ozone cracks increases very markedly and can become virtually infinite. In other words, it is the strain below which the elastomer is unaffected by ozone.

Another method of presenting the results is to plot the logarithm of strain against the logarithm of the time to first cracking, recording both the longest time at which no cracks are seen and the earliest time when cracks are observed as shown in Figure 23.2. For some rubbers the curve may approximate to a straight line but this should not be assumed since it can lead to large errors in estimating the threshold strain. Unless otherwise specified, the threshold strain at the longest test period should be reported.

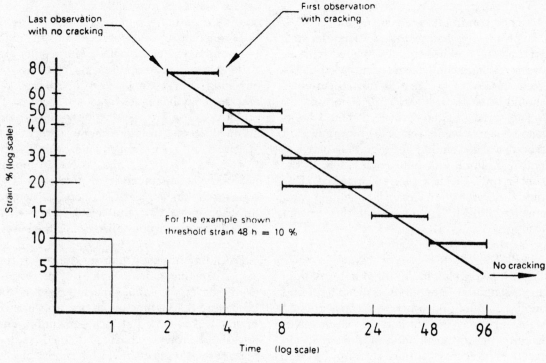

Figure 23.2 *Method of presenting results in a graphical form as recommended in ISO 1431/1*

Dynamic testing

Many products are subjected to cyclic strain in service and the mechanical movement can remove the surface bloom provided by protective wax additives. It is, therefore, advisable in some instances to use a dynamic test. A standard method is given in ISO 1431/2. The basic construction of the equipment is similar to that described in ISO 1431/1 but the test pieces are supported in a dynamic unit instead of a rotating carrier. One end of the test piece is held in a fixed position and the other is reciprocated by an eccentric driven by a constant speed motor. The travel is set so that the initial minimum distance between the grips gives zero strain and the maximum distance gives the specified maximum strain. The motion must be in a straight line and in the direction of the common centreline of each opposing pair of grips. Both sides of the test piece must be in contact with the ozonized air and its length must be in the direction of the air flow. The recommended frequency is only 0.5 Hz in order to reduce any effect due to fatigue.

The exposure and expression of results is generally similar to that for the static standard but either continuous cycling or a sequence of dynamic cycle interspersed with periods of static strain is specified. The results are expressed either as the presence or absence of cracks after a given time, or the time (or number of cycles) to the first appearance of cracks. T50 test pieces have also been used, with complete fracture as the criterion of failure. The tests take more time but the assessment is much easier and not subjective.

It has also been suggested that stress relaxation tests, similar to those discussed in Chapter 9, could be used to provide a quantitative assessment of ozone cracking but, although commercial units are available, they are not in general use.

23.2 Interpretation

Cracking on the surface of an elastomer may make it difficult to measure some changes in physical properties. This is particularly true for ozone resistance where cracks are the first manifestation of deterioration. However, if tension strength, elongation and hardness remain unaffected by ozone it is probable that the other mechanical properties will also be satisfactory.

The pattern and severity of ozone cracking varies according to the magnitude and nature of the applied strain. In service the strain may vary from a minimum at one point, which need not necessarily be zero, to a maximum at some other point. It is therefore, necessary to consider the pattern of cracks at all extensions in the range. However, the relationship between strain and severity of cracking is not simple. A few cracks, often large, are found at strains slightly above the threshold and the cracks will become more numerous and smaller at progressively higher strains. It is quite possible for the cracks at very high strains to be so small as to be invisible to the naked eye. As the exposure time increases, numerous very small cracks may coalesce to form larger but relatively shallow cracks. Hence it is possible that a non-resistant rubber at high strains could be more suitable in some applications than a more resistant rubber just above its threshold strain. It is therefore dangerous to have a specification that just requires a rubber to pass a standard single strain and period test.

It should also be remembered that the standard tests are made with thin test pieces deformed in tension. In practice, the finished products may be quite thick and subject to minimum strains and minor cracks may not have any significance on the overall performance. A further consideration is that the deterioration due to surface cracking will not proceed at a constant rate – or even at an easily predictable rate.

Under dynamic strain conditions a distinction needs to be made between ozone cracking and the cracking resulting from fatigue failure. Ozone attack is the sole cause of crack initiation at cyclic strains below a characteristic strain, known as the mechanical fatigue limit. Once this limit is exceeded, the rate of crack growth increases rapidly and is mainly the result of mechanical fatigue, assisted in many elastomers by the presence of atmospheric oxygen. In this region the effect of ozone is small and becomes increasingly negligible at higher strains. Mechanical fatigue can also occur at low strains once ozone cracks reach a certain size. For these reasons, the ranking order of different elastomers can vary according to the magnitiude of the strain, so that the test conditions used should, as far as possible, match those anticipated in service.

Finally, there is very little correlation between the results from laboratory tests and service performance. The tests are, therefore, usually only used for comparing and rating different elastomers and compounds, where they are extremely useful to the rubber technologist. Nevertheless, ozone tests are sometimes included in product specifications in order to eliminate compounds that might fail rapidly from ozone attack. In these cases, the simple 'cracking'/'no cracking' procedure may be all that is required.

24

Resistance to liquids

The action of liquids on elastomers may result in the absorption of liquid by the elastomer, extraction of soluble constituents from the elastomer or chemical reaction with the elastomer. In some cases two, or all three, of the reactions can take place. Absorption is usually greater than extraction and there is a net increase in volume (generally known as 'swelling') but this is not always true. For some products a decrease in volume or dimensions could be more serious than swelling and if there is a significant chemical reaction a low swelling may hide a large deterioration in physical properties. Consequently, although degree of swelling provides a good general indication of resistance it is also important to measure the change in other properties.

The international standard for resistance to liquids is ISO 1817 which specifies how to determine the change in volume, extracted soluble matter and tensile stress–strain and hardness properties of the elastomer after immersion and after drying out the test liquid. Choice of liquid, conditioning, temperatures and duration of test are specified. Although the tests basically consist of simply measuring properties before and after immersion, the recommendations need to be followed closely if reproducibility is to be achieved.

Table 24.1 *National and international standards for resistance to liquids*

ISO 1817
ASTM D471
ASTM D1460
ASTM D3137
BS 903: Parts A16 and A18
DIN 53538 Parts 1 to 5
DIN 53521
NF T46-013

Attack by liquids usually proceeds quickly at first and then reaches either a maximum or a slow rate of further change. The time to reach equilibrium or maximum swelling will be approximately proportional to the square of the thickness of the test piece and directly proportional to the viscosity of the liquid. Care should, therefore, be taken to ensure that the test results have reached equilibrium and it is sometimes advisable to take several readings and plot a graph of swelling or absorption against time.

Change in volume and dimensions

The size of the test piece when measuring change in volume should be 1 to 3 cm^3, have a uniform thickness of 2 mm and should not be longer or broader than 50 mm. The test piece for measuring change in dimensions should be rectangular, 50 mm long and 25 mm wide and also 2 mm thick. The test consists of completely immersing the test piece in the liquid in a stoppered glass bottle or tube at the required temperature. The test piece must be freely exposed at all surfaces without restraint. This can be achieved by suspending the test piece with wires or by resting it on glass marbles. If the test temperature is near the boiling point of the liquid it may also be necessary to fit a reflux condenser to miminize evaporation. The volume of the liquid must be at least 15 times the volume of the test piece.

The change in volume may be measured gravimetrically or by measuring the change in dimensions. The former is preferred. The test piece is weighed in air and in a liquid (usually water) before and after immersion and the volume change calculated on the basis that the volume is proportional to the weight in air minus the weight in water. Care must be taken to exclude air bubbles when weighing in water and a trace of detergent can be helpful. If the elastomer is less dense than water a sinker must be used, just as it is when measuring density. The standard stipulates how the excess liquid is to be removed and the test piece dried.

If the change in volume is to be calculated from the change in dimensions a dial gauge is recommended for measuring thickness and an optical system for length and width.

The results are expressed as percentage change in volume or percentage change in length, width or thickness. Sometimes, especially for purposes of quality control, the change of mass obtained by weighing in air before and after immersion is used.

Change in physical properties

The procedure for measuring the change in tensile stress–strain and hardness follows the recommendations in the relevant standards (see Chapter 4 and 5). Two procedures are specified: one immediately after immersion and the other after immersion and subsequent drying. The exposure procedure is similar to that described above, dumb-bells or rings being used for tensile measurements and a piece of sheet for micro hardness tests. The normal hardness test piece is too thick for equilibrium to be reached in a reasonable time. The cross-section of the tensile test piece is measured before immersion but the gauge length for elongation measurement is marked after immersion. This is convenient but means that the stress at a given elongation becomes an arbitrary measurement suitable for comparison purposes only.

Extraction

ISO 1817 specifies two methods for measuring the soluble matter extracted by a liquid that is 'readily volatile', by which is meant liquids that boil below 100°C. The first method involves drying the test piece after immersion and comparing its mass before and after immersion. Errors can be introduced if the elastomer is oxidized by air during immersion, especially if the temperature is high. The second involves evaporating to dryness the liquid used in the test and weighing the non-volatile matter. This can be unsatisfactory if the elastomer contains extractable volatile additives, for example plasticisers, that will be evaporated with the test liquid. Neither method is very accurate but extraction is rarely included in product specifications.

One surface tests

In many applications, such as hoses and flexible containers, only one side of the elastomer is exposed to the liquid. ISO 1817 provides for a relatively simple test using the apparatus shown in Figure 24.1. The test piece is a disk of about 60 mm diameter which can

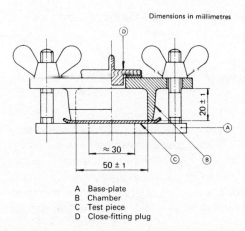

Figure 24.1 *Apparatus for testing the effect of liquids on one surface of an elastomer*

be as thick as the material to be used in service. The test piece is weighed in air and placed in the apparatus. The chamber is filled to a depth of approximately 15 mm and the closing plug inserted. The apparatus is then maintained at the required test temperature for the test period. After cooling to ambient temperature the test piece is removed, dried and re-weighed. The result is expressed as change in mass per unit area.

24.1 Oil and chemical resistance

A large proportion of elastomeric products are used in the automobile industry and many others are frequently in contact with oils of various types. Ideally, an elastomer should be tested with the oil with which it will be in contact during its life. However, it may be in contact with more than one oil and with oils that vary in composition. ISO 1817, therefore, specifies a range of four simulated fuels, three mineral oils and two fluids, that can be used in standard tests. These oils and fluids are available from various authorative organizations and can be expected to be reproducible.

Chemicals can only be tested individually and over the years a great deal of information has been accumulated. Du Pont, for example, can provide data about chemical resistance, for all their elastomers, which cover well over 2,000 chemicals and temperatures, much of which can be substantiated by service experience.

24.2 Water resistance

All elastomers will absorb water but the penetration is very slow compared with most organic liquids and a long time is required for the standard test pieces to reach equilibrium. Also, the deterioration in properties due to the absorbed water is not great. The most important consideration is the effect on the surface and water absorption is often expressed as change of mass per unit area of surface, rather than by percentage change in weight or volume.

The compounding ingredients used can be more important in judging water resistance than the elastomer itself. Hydrophilic compounding materials should be avoided if the amount of water absorbed is an important consideration. The use of water-soluble compounding ingredients may also result in degradation of physical properties as well as contamination of the water or water solution. In certain cases, undesirable tastes or odours may be imparted by water-extractable materials.

24.3 Interpretation

Although, in some respects, the accelerated test methods may closely simulate service conditions, there may not be any direct correlation with actual performance. The thickness of the elastomer must always be taken into account. The rate of penetration is time-dependent and the bulk of a thick product may remain unaffected for the whole of the projected service life. Also, the action of a liquid on an elastomer may be markedly affected by the presence of atmospheric oxygen. However, the accelerated tests are used extensively for control purposes and for comparative tests when developing elastomers for resistance to particular liquids.

Storage of rubber products

Rubber products may be stored for long periods, especially if they are for military use, and care must be taken to ensure that they do not deteriorate. Accelerated tests may be of some use in estimating life in storage but it is also necesary to follow good warehouse practice. Rubber products should be stored in a cool and preferably dark indoor warehouse. The relative humidity should be kept below 80% and the temperature between 10° and 30°C. If these conditions are not possible, precautions should be taken to avoid at least direct sunlight. Different rubbers should be stored separately to avoid migration of their constituents.

Rubber products should not be stored under strain. For example, extrusions should not be hung over a single projection, conveyor belting should not be stored in close folds and hose should not be rewound on to a tighter coil to save space. In addition, rubber products should not be stored close to electric motors, switchgear or any other equipment capable of generating ozone.

If unsuitable storage conditions are unavoidable, it may be possible to improve the situation by compounding. Some proprietary surface coatings may also help but, ultimately, use should be made of elastomers, such as Neoprene, ALCRYN, HYPALON and other speciality polymers, which are inherently stable. The first cost may be greater but the final cost will be less. Failure in a military application could be catastrophic and no saving in initial cost could be justifiable.

26

Which elastomer?

The number of possible rubber compounds is very large indeed. The type of polymer, the fillers and extenders used in compounding, the vulcanization system, etc., can all be varied to provide small, and often desirable, differences in the properties of the finished article.

One important variable that has an effect on the properties is the crosslink density of the base polymer. This is a measure of the chemical bonds within the polymer and, in general, if the crosslink density is increased there will be an increase in hardness, modulus, extrusion resistance and resistance to blistering under rapid decompression. However, at the same time there will also be a decrease in elongation. Properties such as tear strength, fatigue life and toughness, which pass through optimums with change in crosslink density, may be improved or worsened. A pictorial illustration of the changes which can take place and which must be taken into account is provided in Figure 26.1. Generally, it is preferable for the supplier of the rubber part to make this decision.

26.1 ASTM D2000/SAE J200

The American Society for Testing and Materials (ASTM) and the Society of Automotive Engineers (SAE) formed a joint committee in the early 1940s to produce a classification system based on the physical properties of vulcanised rubbers rather than their compositions. Their aims were to avoid unnecessary duplication of grades of elastomers and to eliminate inconsistent requirements in specifications. The system is now published as ASTM D2000 *Standard classification system for rubber products in automotive applications* and as SAE J200. Although the standard was created for the very large automotive market, it is equally applicable to other industries.

ASTM D2000 is used extensively in Europe but is being replaced by ISO 4632 which, although it differs from the American standard in detail, is similar in principle. Yet another slightly different version is published as BS 5176. Only the ISO standard will be considered here but it must be emphasized that the three standards are not interchangeable – each must be used on its own.

The classification system is based on the premise that the properties of all vulcanized rubbers can be arranged into characteristic material designations. These designations are determined by Type, based on resistance to heat ageing; by Class, based on resistance to swelling in oil and by Group, based on low temperature resistance. The combined use of Type, Class and Group and the values for basic and additional properties are designed

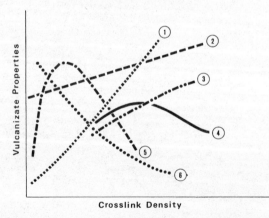

Figure 26.1 *Vulcanisate properties vs crosslink density. 1. Static modulus, extrusion resistance and blister resistance; 2. High speed dynamic modulus; 3. Hardness; 4. Tensile strength; 5. Tear strength, fatigue life and toughness; 6. Hysteresis, permanent set, coefficient of friction, elongation and compression set.*

Table 26.1 *National and international standards for classification systems for vulcanised rubbers*

ISO 4632/1
ASTM D2000/SAE J200
BS 5176

---

to permit the complete description of the quality of any vulcanised rubber. The American and British standards do not include the Group classification.

The use of the system leads to what is known as a 'line call-out' and tables are provided in the standard to identify a wide range of properties. A typical example of a line call-out taken from ISO 4632 is:

BCD 6461-3 A14 B14 EO14 EO34 K21

The first three letters indicate the Type, Class and Group. In this case B means that the test temperature must be 100°C; C that the percentage volume swelling in oil No. 3 (as defined in ISO 1817) must not exceed 120% and D that the maximum brittleness temperature (as defined in ISO 471) must be −40°C.

The next figures define the four basic physical properties. The 6 requires the hardness to be between 56 and 66 IRHD; 4 that the minimum tensile strength will be 10 MPa; 6 that the minimum elongation at break will be 300% and 1 that the maximum compression set after 22 hours at the Type temperature of 100°C should not exceed 80%. The 3 preceded by a hyphen indicates the grade number, and shows that additional properties are required. These follow as a letter to

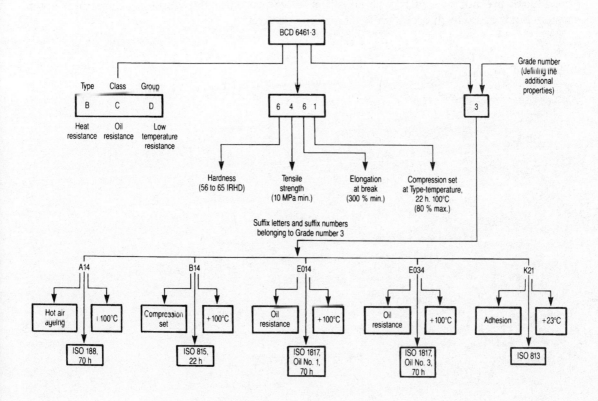

Figure 26.2 *Detailed example of a 'line call-out' taken from ISO 4632/1*

indicate the property and a two figure number to indicate the method and temperature of test (as defined by the original Type letter). A detailed explanation of this example is provided in Figure 26.2.

It can be seen that the 'line call-out' can define the properties required in as much detail as is required. However, no more detail than necessary should be included. In some cases it is sufficient to simply define the Type and Class. The line call-out is not an alternative to simulated service tests. It simply lists the properties that the rubber manufacturer must use to maintain production quality and reproducibility. Neither should these properties be used as design data, especially with respect to heat resistance.

The American and British specifications have appendices that list some of the elastomers that satisfy the conditions of a number of combinations of Types and Classes. These appendices are not part of the classification system but they do form useful guides for the selection of suitable elastomers. ISO plan to publish a similar list as a Technical Report, to be designated ISO 4632/2.

The classification system can be used to produce a graphical comparison of elastomers, as was shown in Figure 3.1. The various elastomers are positioned in accordance with their Type and Class as specified in ASTM D2000. At one extreme are KALREZ perfluoroelastomer and VITON fluoroelastomer, both of which have excellent heat and oil resistance. In the other corners are the polysulphide rubbers, having minimum swell and heat resistance and natural rubber and SBR with relatively poor resistance to both swelling and heat. Although ISO 4632/1 does not cover thermoplastic elastomers, ALCRYN, HYTREL and Olefinic TPE have been included in the graph for comparison. Note 2 with the graph is also important. The fact that the temperature of testing in ASTM No. 3 oil varies with the Type is sometimes overlooked.

Part 2

Simulated service tests

Introduction

It has been shown in Part 1 that although the laboratory tests used by the rubber technologist may not provide a reliable indication of performance, they are essential to the manufacturing process if consistent product quality is to be maintained. Also, it is frequently true that two elastomers with different physical properties (as measured by the rubber technologist) will give finished products with equal performance – perhaps with one of them at a lower price. It is, therefore, desirable that a purchase specification should be based on the ability to perform rather than on a number of arbitrary physical properties.

Ideally, every product should undergo a life test under the actual service conditions but this would slow up, if not prevent, the introduction of new ideas. An alternative is to develop simulated service tests which duplicate the effects of the actual service and produce a recognizable and, if possible, a measurable failure in a relatively short time. Care must be exercised to ensure that the means of accelerating the test do not materially alter the performance of the elastomer. For example, an increase in temperature may decrease the time to failure but it might also cause a degradation of the polymer which would not be encountered in practice. A change in concentration of an aggressive oil or fluid might have a similar effect.

It is not always necessary to have a simulated service test which leads to a total failure of the product. A significant loss of usefulness may be just as useful, especially if it can be measured. What is important is that simulated service tests should be carefully designed and constantly reviewed to provide results that consistently accept proven products and reject those found to be unsuccessful in actual service. In practice, the development of an adequate simulated service test is not as difficult as it might at first appear. Engineers working in the field are thoroughly familiar with the service conditions that their products encounter and they can usually develop a test that provides the information required.

27.1 Purchase specifications

Once a satisfactory simulated service test has been developed it can be used to identify products which meet the performance requirements and to establish a purchase specification. A provisional specification should then be given to the rubber supplier and he should be allowed to submit formulations and/or designs that he considers are suitable. If one of these is shown to be satisfactory by the simulated service test, the purchaser and supplier should jointly decide which standard laboratory tests are required to ensure a consistent product. These tests are then added to the purchase specification.

If the above recommendations are followed, the rubber manufacturers are free to use their expertise to offer an economical product and the purchaser is assured that the part will meet his performance requirements. As mentioned previously, it is frequently found that two suppliers can offer products which, although equally suitable for the application, require a different set of parameters for production control.

Failure to follow the above procedure may lead to unrealistic specifications. If the specification is too tight it may narrow the field of competitive bidding and encourage suppliers to incorporate safety factors in calculating their costs of manufacture. Of course, it is possible to refine purchase specifications based on laboratory tests to the point where they establish legitimate price competition and also provide an adequate product. The joint efforts of the rubber and automotive industries which led to the publication of ASTM D2000 (see Chapter 26) is the classical example of an attempt to do his. However, the refining process is costly in time and effort and the amount of material involved must be large enough to make it worthwhile.

28

Automobile coolant hose

The Volvo Group manufactures approximately 400,000 passenger cars and 50,000 commercial vehicles and buses each year and it has an enviable reputation for quality and reliability. The Volvo Car Corporation, which is responsible for the Group's car operations, has a large research and development laboratory at Gothenburg, Sweden, devoted entirely to polymers and its method of specifying rubber components closely follows the recommendations given in Chapter 27. The specifications are based on extensive field research and simulated service tests yet leave the suppliers free to adapt their formulations to suit their own technical knowledge.

The field research consists of identifying the worst possible conditions under which a vehicle is likely to be driven. For example, in Europe a passenger car might be driven at high speed on motorways for well over half of its life, city driving may account for a moderate mileage and holiday driving towing a caravan in mountainous areas for a short time. The engine temperatures for each condition are then measured and account is taken of the 'heat sink' effect which causes a rise in temperature immediately after an engine is switched off. All this information is statistically analysed to provide test temperatures and times which accurately reflect the working life for the component being studied.

28.1 Simulated service test conditions

The hose is mounted with standard metal and plastic couplings so that the ends are fastened in the identical positions that will be used on the vehicle. The hose is filled with a coolant consisting of equal parts of glycol and water and connected with an expansion tank fitted with a pressure relief cap. The whole assembly is placed in a heating chamber equipped with a hot air circulation system.

In the case of passenger car cooling systems the hoses are first pre-aged. The hose is kept under pressure for a number of weeks with the coolant slightly above the normal working temperature. The hose is next subjected to a dynamic pressure pulsation test with the coolant fluid temperature at approximately its normal working value. A fatigue test follows in which the conditions are similar but at the same time one of the ends of the hose is vibrated vertically at 11 Hz with an amplitude of ±20 mm about the neutral point, see Figures 28.1 and 28.2. Changes of frequency are allowed if the test frequency happens to coincide with the natural resonance of the assembly and the fatigue test may be omitted if the assembly is not subjected to vibrations in practice. Finally, the equipment is allowed to cool and the test is repeated at room temperature for five minutes.

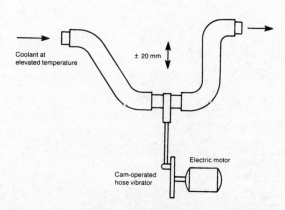

Figure 28.1 *The hoses used in the vibration test are usually mounted in pairs. One end connector in each case will be made of plastic if plastic headers are to be used in the car*

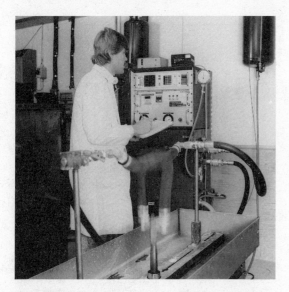

Figure 28.2 *Test conditions are closely controlled during simulated service tests. The hoses are mounted in pairs with standard metal and/or plastic couplings so that the ends are fastened in the identical positions that are used on the vehicle*

28.2 Overall specifications

No leakage, cracks or other defects must occur in the hose or the couplings at any time during the period of the simulated service test. The working life of the vehicle is thus simulated in a test lasting under 10 weeks. In practice, it has been found that correctly manufactured rubber hoses usually pass this test. If failure does occur, it is usually at the couplings. It is of interest to note that KEVLAR para-aramid fibre is the favoured reinforcement for the coolant hoses in current use.

In addition to the simulated service tests, the hose has to meet a number of Volvo quality assurance tests. Some of these are based on ISO specifications and others on tests devised by Volvo itself. They include, hardness, permanent set (slightly modified from ISO), low temperature set, chemical and fluid resistance, adhesion of fabric reinforcement, bursting pressure, resistance to vacuum and ageing.

Elastomeric seals

Elastomeric seals are an essential part of modern machinery and they are made in a wide variety of shapes and sizes. They can be divided broadly into two categories: **Static seals**, where sealing takes place between surfaces which do not move relative to one another and **Dynamic seals**, where sealing takes place between surfaces which have relative movement. O-rings and flat gaskets are examples of the former and seals where there is a rotary movement of a shaft relative to a housing or a reciprocating movement of a rod or piston in a cylinder of the latter. In addition there are many sub-divisions within the two general categories. For example, flexible couplings for pipes may be described as **semi-static** or **flexible-static** because they can accommodate swivelling motion. Also, bellows and gaiters, where the main function is to prevent the ingress of dirt, dust or other harmful contaminants, are known as **exclusion seals**.

Published specifications for seals cover the dimensions of the housing rather than the details of the elastomeric components but they often include recommendations for quality, storage, handling, fitting and some include methods for performance rating. O-ring seals are the exception and ISO 3601 covers dimensions, tolerances, etc., for the elastomeric rings themselves.

Industry and the seal manufacturers have developed a number of simulated service tests but they are mainly used for establishing geometrical configurations and for the selection of elastomers. In the end, there is no substitute for testing the seals in prototype units working under the actual operating conditions.

One of the largest manufacturers of seals, with subsidiaries all over the world, is Carl Freudenberg in Germany. It has a large test facility at its headquarters in Weinheim (near Frankfurt) with over 200 test rigs for testing dynamic seals. The test methods used are generally those given in DIN standards, principally to DIN 3761 (ISO 6194 is equivalent but not identical). A typical installation is that

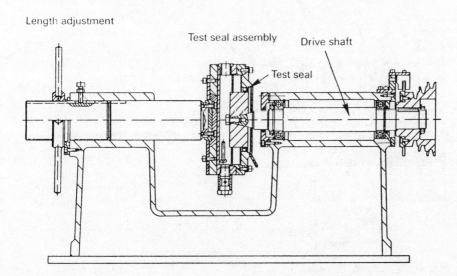

Figure 29.1 *Typical test rig for dynamic testing of elastomeric shaft seals*

shown in Figure 29.1. Shaft speeds can be up to 8,000 rev/min and tests last for periods selected to meet the customers' requirements as specified in DIN 3761.

Freudenberg uses a number of simulated service tests to determine the effect of seals operating in dusty and dirty water conditions. For the dust test a standard Arizona dust with specified particle size is added to a test rig containing hypoid gear oil as shown in Figure 29.2. The rig arrangement for testing with dirty water, as shown in Figure 29.3, simulates

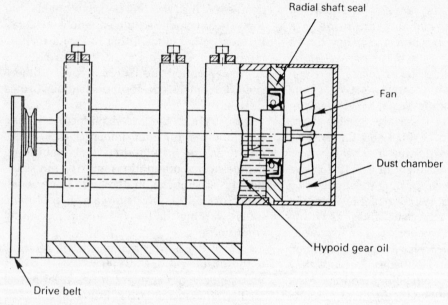

Figure 29.2 *Rig for testing shaft seals in dusty conditions*

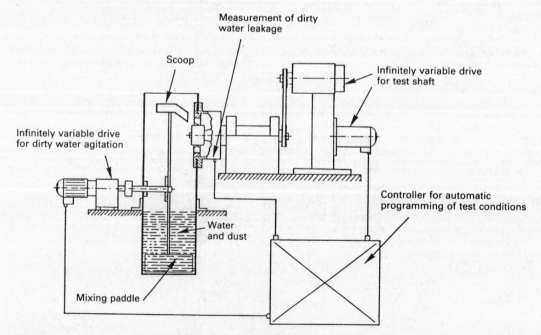

Figure 29.3 *Rig for testing shaft seals with dirty water*

the conditions under which a wheel bearing is expected to operate. The test conditions vary widely and are designed to meet the individual requirements of the end-user so there are no absolute test specifications.

The company also devises simulated service tests using its customers' own equipment. For example, complete rear axle assemblies have been tested.

29.1 Valve stem seals

The company has developed a number of simulated service tests for valve stem seals. In most cases the customer's own cylinder head assemblies are adapted for the test as illustrated in Figure 29.4. The camshaft is turned by a variable speed electric motor and oil is circulated at the required temperature through the assembly. A vacuum of about 0.6 bar is induced inside the inlet and exhaust ports. The oil lubricates the camshaft, bucket tappets and the valve stems as it would in an engine but it is not possible to simulate the presence of the hot combustion gases, petrol-air mixture and the varying conditions of pressure, ageing of the oil, etc., that would be present in practice. The oil that leaks past the valve guides is collected in measuring cylinders situated under each set of valves.

A typical test will be made for 500 hours with a cylinder head temperature of 120°C, which is approximately equivalent to 50,000 km on the road. Some specifications call for the test to be intermittent in order to simulate more closely the operating condition of a car. The oil leakage is recorded at regular intervals and the results plotted graphically. A leakage of 20 ml per valve is considered to be acceptable but some customers will accept and may require leakages of up to 100 ml. The wear that occurs between the valve stem and its guide is also measured after each test.

Light metal cylinder heads and smaller and more powerful engines are increasing the operating temperatures required for both petrol and diesel engines. Polyacrylate elastomers can be used at temperatures up to 130°C but some types of fluoroelastomers, such as VITON which can be used at temperatures up to 200°C (if there are no problems with charcoal deposits in the area of the sealing lip) are increasingly being specified. VITON is also better able to withstand the attack of the more aggressive additives that are being used in modern lubricating oils.

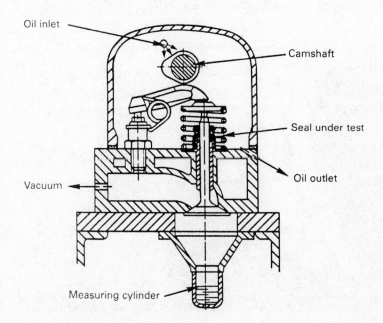

Figure 29.4 *Rig for testing valve stem seals*

29.2 Bellows

Elastomeric bellows or boots are exclusion seals and when they are used for sealing drive shafts in automobiles they have to withstand extremely stringent conditions. They have to operate over a temperature range from −40°C to 100°C and in the future may have to operate at up to 150°C; they must not be affected by the lubricating greases used in the bearings they protect; they must have good mechanical properties to stand up to impact from stones and also have high elongation at break to take up the relative movement between the shaft and the differential gear; they require rotational stiffness and they must retain their impermeability and tightness under all these conditions. Some of the properties required can be met by the physical design of the convolutions and others can be established by the laboratory tests described in Part 1 of this work. However, simulated service tests are required to establish the dynamic response under extremes of temperature.

Freudenberg has developed extensive facilities for cyclic dynamic testing of bellows. The bellows are first immersed in a standard drive shaft grease for 72 hours at 100°C to establish that the change in volume of the elastomer is within acceptable limits – generally less than +10%. The boots are then fitted onto shafts which can be rotated and which can also be set at various angles, as shown in Figures 29.5 and 29.6. The specifications for the tests depend upon the customers but a typical cold test cycle would be for the assembly to be held at a temperature of −40°C and a bend angle of 18° for two hours followed by cycles of 15 minutes at 1,100 rev/min (reached after a maximum of five seconds) then a 60 minute stationary period. The number of cycles before failure is recorded and the arithmetic average of at least three separate tests is taken to be the final life cycle.

Figure 29.5 *Elastomeric test boot in free state*

Figure 29.6 *Elastomeric test boot in fitted position*

A typical high temperature test would be to raise the temperature of the assembly to between 100° and 150°C and to rotate the shafts at 140 rev/min for 240 hours. Acceptable bellows would show no signs of cracking after this test.

Neoprene has for many years been the standard elastomer for the manufacture of automobile bellows and it remains the favoured material when good flexibility at temperatures up to 120°C is of paramount importance. If operation at temperatures up to 150°C is required ethylene oxide-epichlorohydrin (ECO), hydrogenated nitrile or polymers of ethylene-methyl acrylate with carboxyl groups (VAMAC) are used. The ease with which HYTREL engineering thermoplastic elastomer can be fabricated into bellows is encouraging its use in applications where low temperature flexibility is of importance.

Flue duct expansion joints

The sulphur that is always present in the coal used in power stations is converted into sulphur dioxide (SO_2) during combustion and some is oxidized to sulphur trioxide (SO_3), which dissolves in water vapour to form sulphuric acid (H_2SO_4).

When untreated flue gases are discharged into the atmosphere, the sulphur oxides react with water vapour in clouds or directly with rain to generate weak sulphurous and sulphuric acids. This is the notorious 'acid rain' which can acidify rivers and lakes and progressively attack trees and buildings.

The amount of sulphurous oxides in the gases can be substantially reduced by the installation of wet scrubbers before the exhaust stack. These scrubbers, however, not only clean the flue gases to a great extent, but also cool the gases to between 50° and 70°C – well below the acid condensation dew-point. Thus, highly aggressive condensates are formed inside the flue ducts, the reheater and the chimneys, potentially leading to serious corrosion problems.

A power station will have expansion joints throughout the entire duct system from the boiler to the chimney to accommodate movement and relieve the stresses caused by thermal expansion and contraction. These expansion joints were formerly complex metal bellows that were prone to corrosion by acid and erosion by fly ash. Soft multilayer expansion joints also suffered from corrosion and abrasive fly ash. In addition, they did not seal; the leaking condensates then led to corrosion on the outside of the site ductwork.

Suitably designed and constructed elastomeric expansion joints, such as is shown in Figure 30.1, are inherently more resistant to corrosion and erosion and give much greater tolerance to vibration, lateral displacement and misalignment. They are also lighter and easier to install. An additional very valuable benefit is that elastomeric expansion joints provide an excellent seal against wet corrosive condensates.

VITON fluoroelastomer is one of the best materials for this application and two types of elastomeric flue duct expansion joints are in use. The first is the moulded reinforced fluoroelastomer belt joint which has a total joint thickness between 3 and 6 mm. Typical reinforcement materials are acid-resistant glass fibre fabrics and stainless steel wire mesh. Asbestos reinforcement has fallen from favour because of the dangers now associated with its use. The second type, is the composite

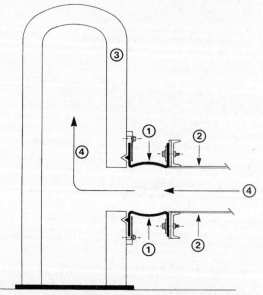

Figure 30.1 *The excellent resistance to attack by corrosive chemicals at high temperatures enables flue duct expansion joints made from VITON reinforced with wire mesh (1) to be clamped directly on to the chimney. The hot flue gases (4) pass through the horizontal metal ductwork (2) into the vertical concrete chimney (3). This construction is leak-proof, easy to install and requires less space than previous designs*

or insulated joint with a sandwich construction which, in general, will tolerate higher temperatures for longer periods. The inner ply is usually made of coated glass fibres surrounded by one or more layers of insulation. An outer layer of reinforced fluoroelastomer provides the gas seal. Internal metal baffles are often used to protect the glass fibre from abrasion by hot fly ash.

30.1 Simulated service test rig

The long time required to test expansion joints and the difficulty and cost of 'on site' testing led Du Pont to set up a simulated service test rig in the USA. The rig, shown in Figure 30.2, consists of a closed loop recirculating ducting system with a sample chamber capable of holding 48 flat test pieces. The test pieces are

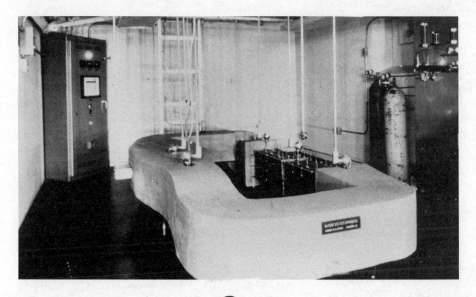

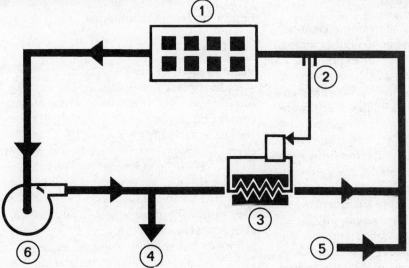

1) Test chamber
2) Temperature control
3) Heater
4) Sample point
5) Gas and acid make-up
6) Blower

Figure 30.2 *A view of the Du Pont simulated flue gas test unit with schematic diagram of the installation*

exposed by clamping them to openings along the two sides and the top of the sample chamber. There are no test positions in the bottom of the chamber. One side of each test piece is exposed to a synthesized flue gas which is circulated through the apparatus by a blower. Fresh gases and acids are added and the stream is purged continuously during a test. The temperature of the gas can be regulated between ambient and 315°C by a programmable heater/controller which can be set to provide cycling between two control temperatures on a timed schedule.

The composition of the synthetic flue gas is chosen by Du Pont to represent that of a power station burning a coal containing 5.5% by weight of sulphur, with a 20% excess of air. The target specific composition of the gas is:

Constituent	Volume %	Weight%
moisture	12.60	7.80
carbon dioxide	11.50	17.40
oxides of nitrogen	0.50	0.80
nitrogen	71.60	69.00
oxygen	3.20	3.50
sulphur dioxide	0.60	1.40
sulphur trioxide	0.03	0.08
hydrogen chloride	0.05	0.06

The gas velocity within the sample chamber is adjusted to an average of 165 linear metres per minute. The pressure at all 48 positions is normally slightly positive and the exposure of each position is reasonably equivalent. Temperatures and gas and acid flows are monitored daily and analyses of oxygen and carbon dioxide are made with an Orsat apparatus at regular intervals. In addition to these routine measurements, complete analysis of the flue gas (including SO_2 and SO_3) is carried out periodically by an independent analytical laboratory. The results obtained have been proved to be consistent and the synthesized gas is considered to provide a close analogy to the flue gases in a power station.

Test pieces of VITON are made by press curing but, in accordance with the practice adopted by the manufacturers of flue duct expansion joints, they are not post cured. The physical properties measured are hardness, 100% modulus, elongation and tensile strength using the methods described in Part 1. The thicknesses of the test pieces have been measured before, during and after the tests and it has been shown that changes in thickness due to erosion are not significant. After exposure, the physical properties are measured again and compared with those of the original.

30.2 Test conditions

The exact conditions for a test depend upon the application but Du Pont mainly use four standard specifications. All tests are carried out on a continuous 24 hours per day basis.

Test A

The theoretical acid dew-point for the synthetic gas composition given above is about 154°C and this test consists of exposing the joint material to a gas temperature alternating between 143° and 166°C. Each cycle consists of raising the gas temperature from 143° to 166°C and returning it to 143°C once an hour. The gas thus passes through the acid dew-point temperature 48 times a day. The duration of the test is six months.

Test B

The objective of this test is to expose the samples to test conditions severe enough to differentiate the effect of polymer fluorine content on performance. The temperature of the gas is varied from 121° to 260°C and back every 12 hours and the test is run for eight weeks. These conditions are very severe because the acid condensate, which is formed when the temperature drops below the acid dew point, is subsequently heated to 260°C. This is not typical of the service conditions for the joints but the upper temperature limit of 260°C is an 'excursion' temperature sometimes specified by customers.

Test C

The objective of this test is to simulate a peak load situation where extra boilers are brought on stream for a short term peak demand and then shut down. The gas is heated from 71° to 166°C over six hours, held at 166°C for six hours, cycled back to 71°C over six hours and held at 71°C for six hours. One complete cycle takes a day and the test is run for one year. The gas temperature is not lowered to ambient because such low temperatures do not contribute appreciably to the degradation of VITON but the temperature of the gas drops below its acid dew-point of 154°C twice a day.

Test D

Fluoroelastomer flue duct expansion joints are usually rated for continuous service at 200°C with temporary short periods at higher temperatures, such as might occur if a forced draught fan should fail. This test simulates that condition by exposing the joint material to a gas temperature of 315°C for 48 hours. This is not a difficult test for VITON because the thermal gradient in the test piece results in a lower effective temperature at its air side. The gas side of the test piece stands this short term excursion without significant damage to its physical properties. Also, there is no liquid acid concentrate present at this high test temperature.

30.3 Results

Extensive tests over many years encourage Du Pont to believe that its simulated test rig satisfactorily reproduces the chemical conditions existing in practical flue duct systems and correlation with field performance is good.

The test rig has two main purposes. Firstly, it aims at defining the best possible compound formulation regarding the type of VITON crosslinking system and the metal oxide that is used to accelerate crosslinking. In many tests over some seven years VITON B, crosslinked with a bisphenol C cure, proved to be the best formulation. Secondly, the test facility has also enabled all the major manufacturers of expansion joints to compare the quality of their commercial products with laboratory samples made under ideal conditions.

The test does, however, have certain limitations. The test pieces are not flexed or strained in any way and they are not subjected to the erosion caused by fly ash or the gypsum which is formed during the scrubbing process. Nevertheless, the results show VITON to be a preferred material for this application and this conclusion has subsequently been confirmed by many successful installations throughout the world.

31

V-belts

Endless V-belt drives are a widely used method of transmitting power. Their construction is complex and usually consists of a top filler layer of elastomer – often Neoprene chloroprene elastomer – textile cords situated at a neutral axis to take the tension, sometimes with cross-cord layers above and below and a base of cushion elastomer. The whole assembly is surrounded by a fabric jacket and is then moulded to form the finished belt. Alternatively, the belts are built by coiling the different layers on to a drum to form a slab. The slab is cured and individual (bandless) belts are then cut out of it.

The belts are trapezoidal in cross-section and one or more belts are run in grooved pulleys to transmit the power. The grooves are designed so that only the sides of the belts are in contact with the pulley and the depth of the trapezium can vary between about 45 to 95% of its width, depending on the construction. The methods of construction, the materials used and the technical specifications vary widely.

There are also many other types of elastomeric drive belts including, toothed belts that provide synchronous or positive non-slip engagement; belts used in pulleys with movable flanges to provide variable speed drives and V-ribbed belts in which the under-cord part has a number of small V-sections moulded into one piece as shown in Figure 31.1.

The Gates Corporation of Denver, USA, is one of the world's major manufacturers of V-belts and its subsidiary, Gates Europe, manufactures no less than 12 million V- and Micro-V belts every year in its plants at Erembodegem in Belgium and at Balsareay in Spain. Production volumes of this magnitude require rigorous quality assurance systems and test procedures and the Gates Corporation has extensive test facilities in all of its plants, the results from which are centrally coordinated and analyzed in Denver.

Stress–strain analysis and testing of V-belts to destruction have led to mathematically-based design data which enable V-belts to be accurately designed in the first place. The belts then have to be accepted by the customer for the particular application and the production quality is maintained by the use of international standard quality systems and test procedures. However, in addition, there is an increasing demand, particularly from the automotive industry, for simulated service tests which reflect more closely the actual operating conditions. These require test equipment that can use actual engines, the running of which can be computer-controlled to follow the operating conditions specified

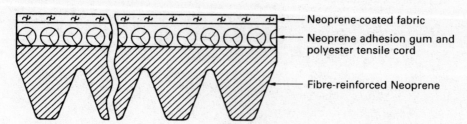

Figure 31.1 *Cross-section of a Micro-V ribbed belt. The extremely low profile is very efficient and enables one belt to drive a large number of accessories*

by the customer. Experiments with this type of equipment are running at the Gates plant at Dumfries, Scotland, for testing synchronous and Micro-V ribbed belts and a fully-equipped laboratory for simulated service tests will subsequently be installed in its plant at Aachen, Germany.

31.1 Industrial V-belts

Industrial V-belts are generally designed to last at least three or four years and the powers transmitted by such drives are frequently very large – up to 1,500 kW. Accelerated test methods are therefore essential and it is desirable to recover as much of the energy that is put into the test rig as possible.

Belt fatigue can be accelerated by decreasing the pulley diameters, increasing the speed of rotation, the belt tension and the power transmitted and by raising the operating temperature. All these methods are used but it is essential to establish that the method of acceleration does not also affect the pattern of the failure.

Belts for use at large power ratings are tested on motor-generator dynamometers. The belt transmits power from a d.c. electric motor to a d.c. generator, which in turn supplies most of the electricity required to drive the motor. The net power consumed is, therefore, only that lost by the friction in the system. In this way, many kW can be used for testing at a net cost of only a few kW.

The generator is supported on a pivoted cradle, as shown in Figure 31.2, so that belt tension and torque can be accurately controlled. The torque is derived from the reading on a scale which measures the reaction of the generator directly in kilograms. The forces registered are multiplied by the lever arm (distance between the axis of the generator and point of scale attachment) to give the torque value in newton metres.

The V-belt is mounted on a 1:1 pulley ratio to simplify calculations and it is run at a fixed torque, speed and tension until it eventually fails. If possible, conditions are selected so that failure occurs in 150 to 200 hours. Necessary adjustments to motor speed and generator torque are made by electrical controls on a panel where belt speed and elapsed testing time are also registered.

The dimensions and certain properties of V-belts and of some of the test equipment are

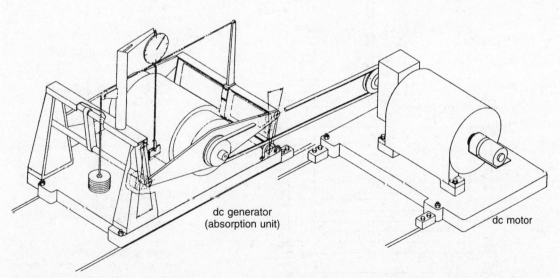

Figure 31.2 *Electric dynamometer V-belt tester. The generator is mounted in low friction bearings which allow limited rotation about its shaft. Thus, reactive torque (test torque) can be measured with a calibrated scale. A balanced cradle system maintains constant belt tension over several centimetres of centre distance movement. The motor and generator are track-mounted to accommodate various belt lengths*

specified by international standards but the method of testing the belts is left to be arranged between the manufacturers and the customers. Belts for use in the automotive industry have special requirements and are considered in Section 31.2.

Tests on V-belts for industrial applications which only require low power transmission are undertaken with a wide variety of test equipment. The main requirements are to provide a suitable method of applying the necessary torque and either to recover or remove the energy consumed. One typical test rig uses a water brake to apply the load. A series of metal disks or a rotor is turned in a water bath and the power absorbed at a given speed can be varied by altering the amount of water in the casing. The resistance to rotation of the shaft can be measured by allowing the casing to swing about the shaft axis and the balancing torque by weights on a lever arm. The heat generated can be removed by allowing the water to flow through the case. Another system uses a magnetic clutch as an electric brake.

Dead-weight testers, as shown in Figure 31.3, are used to obtain additional data at low cost. The equipment consists simply of two sheaves of small diameter, the lower of which has attached a known weight to provide a constant tension. No power load is applied to the belt but the small diameter pulleys induce flex fatigue.

31.2 Automotive V-belts

One of the largest markets for V-belts is in the automotive industry and comprehensive standards exist which cover belt and pulley

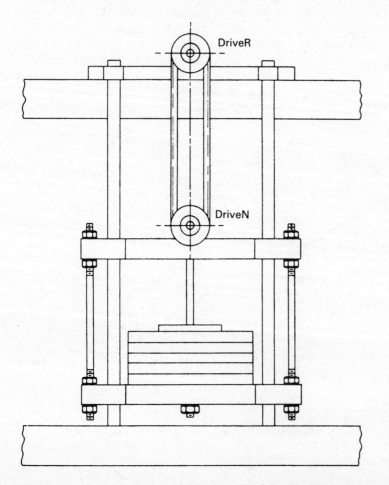

Figure 31.3 *The dead-weight tester is a simple and inexpensive piece of equipment that applies a constant tension to the belt. No power is transmitted to the belt but the small diameter pulleys induce flex fatigue*

dimensions. ISO 2790, SAE J636 and DIN 7753/3 are the standards in most general use in Europe. In addition, standards for testing V-belts (ISO 5287, SAE J637 and DIN 7753/4) have been developed which specify the testing equipment and the methods of testing but leave the tester free to specify the power to be transmitted. Thus, although the tests are not true simulated service tests, they are widely used for product development, source approval and quality verification.

Test equipment

ISO 5287 specifies both two- and three-pulley test machines but the three-pulley is used most frequently. The equipment is shown diagrammatically in Figure 31.4 and consists

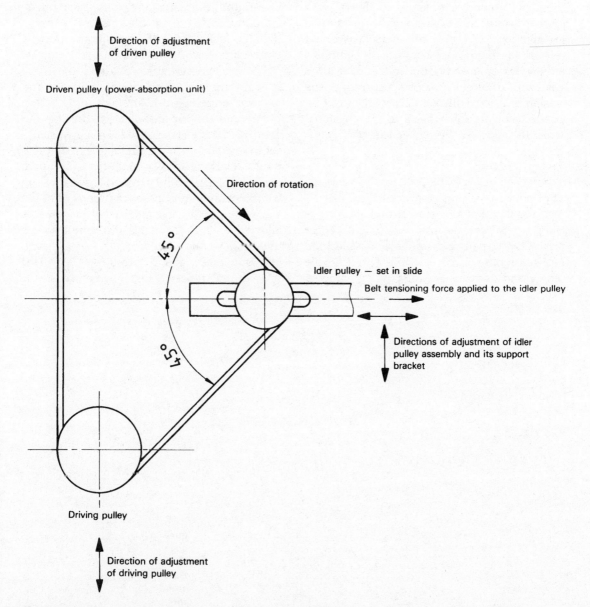

Figure 31.4 *Diagrammatic representation of the equipment for three-pulley testing of automotive V-belts*

of a driving pulley, a driven pulley with a suitable power-absorption unit, a method of applying a known tension to the belt and a method of determining belt slip. The power-absorption unit is usually an electric or water brake as described previously. The slip may be measured in a number of ways but Gates uses proximity switches that count the number of gear teeth passing by each second. The gear wheels have 60 teeth so that a speed of 1 rev/min gives a count of 60 teeth per minute or 1 tooth per second. The speed is shown on a digital display. Slip is calculated by a formula given in ISO 5287 but is automatically worked out by a microprocessor which computes either slip or speed ratio and allows for a variety of different gear or speed ratios. The system is fitted with an alarm which can be pre-set with thumb wheels to give a signal when the allowed slip is reached.

Test procedure

The outside circumference of the belt is measured without tension before the belt is installed in the testing equipment. After mounting the belt on the pulleys, the specified belt tensioning force is applied to the idle pulley and (leaving the idler pulley support bracket free to move in its slide) the drive is brought up to specified rotational frequency. The relevant load is then applied to the drive pulley as quickly as possible. The drive is run for five minutes, not including the starting and stopping time. The machine is stopped and allowed to stand for at least 10 minutes. The drive is then turned manually for several revolutions of the belt and a dial indicator mounted in contact with the idler pulley support bracket, is used to determine the maximum limits of travel of the idler pulley support bracket. The idler pulley is immediately locked in the position midway between the two limits of travel.

The machine is re-started and the drive brought up to the specified rotational frequency, the test load is applied to the driven pulley and the slip between the driving and the driven pulleys measured. The slip alarm is set to operate when the slip is 4% greater (8% total if working to SAE J637) than the initial measurement and the drive is allowed to run continuously under these conditions until the alarm stops the machine. The machine is allowed to cool for at least 20 minutes. The idler pulley support bracket is re-set as described previously and the test repeated. This procedure is repeated until the belt fails.

Part 3

Vocabulary of the rubber industry

ability (softening ability)

A

Nr	ENGLISH	FRANÇAIS	DEUTSCH	ITALIANO
1	ability (softening ability)	capacité f de ramollissement	Weichmachungsfähigkeit f	capacità f di rammollimento
2	abrader	abrasimètre m	Schleifapparat m, Abriebsgerät n, Abriebsprüfgerät n	abrasimetro (utensile m)
3	abrader (Du Pont)	abrasimètre, m Du Pont	Abriebprüfer (Du Pont-) m	macchina f Du Pont per prove d'abrasione (abrasimetro Du Pont)
4	abrasion	abrasion f	Abrieb m	abrasione f
5	abrasion index	abrasion (indice d')	Abriebindex m	abrasione (indice d')
6	abrasion resistance	résistance f à l'abrasion	Abriebbeständigkeit f	resistenza f all'abrasione
7	abrasion service test	essai m pratique pour la mesure de l'abrasion	Abriebprüfung f in der Praxis	prova f pratica di abrasione
8	abrasion (Taber abrasion)	abrasion f Taber	Abrieb m nach Taber	abrasione f Taber
9	abrasion test (general term)	essai m d'abrasion	Abriebprüfung f	abrasione (prova f di)
10	abrasion test (on the road)	usure (essais m d') sur route	Abnutzungsprüfung f	prova abrasione su strada
11	abrasion tester	abrasimètre m, machine f pour essais d'abrasion	Abriebprüfgerät n	macchina f per prove di abrasione
12	abrasive	abrasif m	Schleifmittel n	abrasivo m
13	absorption	absorption f	Absorption f	assorbimento m
14	absorption capacity	absorption (capacité f d')	Absorptionsvermögen n	assorbimento (capacità f di)
15	absorption (moisture absorption)	absorption f d'humidité	Feuchtigkeitsaufnahme f	assorbimento d'umidità
16	absorption tester	absorption (machine f d'essai d')	Absorptionsprüfer m	assorbimento (apparecchio m per prove di)
17	absorption (water absorption)	absorption f d'eau	Wasseraufnahme f	assorbimento m d'acqua
18	acceleration	accélération f	Beschleunigung f	accelerazione f
19	accelerator	accélérateur m	Beschleuniger m	accelerante m
20	acceptor	accepteur m	Akzeptor m	accettore m
21	acceptor (acid)	accepteur m d'acide	Säureakzeptor m	accettore m di acidi
22	accumulator (hydraulic accumulator)	accumulateur m hydraulique	Akkumulator m (hydraulischer)	accumulatore m idraulico
23	accumulator (storage battery accumulator)	accumulateur m	Akkumulator m	accumulatore m, batteria
24	acetyl value	indice m d'acétyle	Acetylzahl f	valore m di acetile
25	acetylene hose	tuyau m à acétylène	Acetylenschlauch m	tubo m per acetilene
26	acid corrosion	corrosion f par les acides	Säurekorrosion f	corrosione f da acidi
27	acid hose	tuyau m pour acides	Säureschlauch m	tubo m per acidi
28	acid (inorganic)	acide m inorganique	Säure f (anorganische)	acido m inorganico
29	acid (organic acid)	acide m organique	organische Säure f	acido m organico
30	acid proof	acides (résistant aux)	säurefest	acidi (resistente agli)
31	acid resistance	résistance f aux acides	Säurebeständigkeit f	resistenza f agli acidi
32	acid resistant, acid resisting	acides (résistant aux)	säurebeständig	resistente agli acidi
33	acid (stearic acid)	acide m stéarique	Stearinsäure f	acido m stearico
34	across the grain	perpendiculaire au granulé	quer zur Faser	perpendicolarmente alla venatura
35	activation	activation f	Aktivierung f	attivazione f
36	activator	activateur m	Beschleuniger m, Aktivator m	attivatore m
37	active ingredients	ingrédients actifs	aktive Bestandteile	ingredienti m attivi
38	additive	additif m	Zusatz m, Zusatzmittel n	additivo m
39	adhere (to)	adhérer à	haften, kleben, verkleben	aderire
40	adherent (adj)	adhérent	haftend, klebend	aderente
41	adhering (adj)	adhérent	haftend, klebend	aderente
42	adhesion (attraction between molecules)	adhésion f, collage m	Adhäsion f, Bindungskraft f	adesione f (attrazione molecolare)
43	adhesion (bonding)	collage m	Haftung f, Bindung f, Adhäsion f	adesione f (attacco m)
44	adhesion force	force f de collage	Adhäsionskraft f, Klebkraft f	forza f di adesione
45	adhesion (peel adhesion)	adhésion f, résistance f au pelage	Schälfestigkeit f	adesivo superficiale (potere m)

A

Nr	ESPAÑOL	PORTUGUÊS	SVENSKA	NEDERLANDS
1	capacidad f de ablandamiento	capacidade f de amolecimento	mjukgöringsförmåga	vermogen om week te maken
2	abrasímetro m muela	abrasímetro m	slipapparat	slijtageapparaat
3	abrasímetro Du Pont	máquina f de abrasão Du Pont	Nötningsprovmaskin (Du Pont)	slijtageapparaat (Du Pont)
4	abrasión f	abrasão f	avnötning	slijtage
5	abrasión (índice de)	abrasão (índice de)	avnötningsindex	slijtage index
6	resistencia f a la abrasión	resistência f à abrasão	nötningsbeständighet	slijtageweerstand
7	ensayo m de abrasión en servicio	ensaio m prático de resistência à abrasão	praktiskt avnötningsprov	praktijkproef voor slijtagebepaling
8	abrasión f de Taber	abrasão f Taber	avnötning enligt Taber	slijttest volgens Taber
9	abrasión (prueba f de)	abrasão (ensaio m de)	avnötningsprovning	slijtageproef
10	desgaste por fricción (ensayo m de)	abrasão (ensaio m de) na estrada	avnötningsprov	slijtageproef (op de weg)
11	abrasión (máquina f para ensayos de) , abrasimetro m	abrasímetro m, equipamento m para ensaio de abrasão	avnötningsprovmaskin	slijtageapparaat
12	abrasivo m	abrasivo m	slipmedel	schuurmiddel
13	absorción f	absorção f	absorption	absorptie
14	absorción (capacidad f de)	absorção (capacidade f de)	absorptionskapacitet	absorberend vermogen
15	absorción de humedad	absorção de humidade	fuktupptagning	vochtopname
16	absorción (ensayador m de la)	absorção (equipamento m para ensaio de)	absorptionsmätare	absorptietester
17	absorción de agua	absorção de água	vatten absorption	wateropname
18	aceleración f	aceleração f	acceleration	versnelling
19	acelerante m	acelerador m	accelerator	versneller
20	aceptor m	receptor m	acceptor	acceptor
21	aceptor m (de ácido)	ácido receptor	syraacceptor	zuuracceptor
22	acumulador m hidráulico	acumulador m hidráulico	ackumulator (hydraulisk)	accumulator (hydraulische)
23	batería de acumuladores	acumulador m	ackumulator	accu
24	índice de acetilo	índice m de acetilo	acetylvärde (-tal)	acetylgetal
25	tubo m para acetileno	tubo m para acetileno	acetylenslang	acetyleenslang
26	corrosión f ácida	corrosão f pelos ácidos	syrakorrosion	aantasting door zuren
27	tubo m para ácidos	mangueira f para ácidos	syraslang	zuurbestendige slang
28	ácido m inorgánico	ácido m inorgânico	oorganisk syra	zuur (anorganisch)
29	ácido m orgánico	ácido m orgânico	organisk syra	zuur (organisch)
30	prueba acida	ácidos (à prova de)	syrafast	zuurbestendig
31	resistencia f a los ácidos	resistência f aos ácidos	syrabeständighet	zuurbestendigheid
32	resistente a los ácidos	resistente aos ácidos	syrabeständig	zuurvast
33	acido esteárico	ácido esteárico	stearin syra	stearine zuur
34	transversalmente al grano; a través del grano	perpendicular ao grão m	korn	dwars op de draad
35	activación f	activação	aktivering	activering
36	activador m	activador	aktivator	activator
37	ingredientes activos	ingredientes activos	aktiva beståndsdelar	actieve bestanddelen
38	aditivo m	aditivo m	tillsats	toevoeging
39	adherir	aderir	vidhäfta	kleven aan
40	adherente	aderente	vidhäftande	klevend
41	adherente	aderente	vidhäftande	klevend
42	adhesión f	adesão f	vidhäftning (bindning)	aantrekkingskracht
43	adherencia f, pegado	aderência f	vidhäftning	hechting
44	fuerza f de adhesión, fuerza f de adherencia	fôrca f de adesão	vidhäftningskraft	kleefkracht
45	pelado (resistencia al)	delaminação (força de)	avskalningsvidhäftning	striphechting

A

adhesion power, adhesion strength

Nr	ENGLISH	FRANÇAIS	DEUTSCH	ITALIANO
46	adhesion power, adhesion strength	pouvoir m d'adhésion, force f de collage	Haftvermögen n	potere m adesivo, forza f di adesione
47	adhesion tester	adhésion (appareil m pour essais d')	Adhäsionsprüfer m	adesione (apparecchio m per prove di)
48	adhesive	adhésif m	Klebstoff m, Kleber m	adesivo m
49	adhesive (adj)	adhésif	klebend	adesivo
50	adhesive fabric	tissu m adhésif	Gewebe n (selbstklebend)	tessuto m adesivo
51	adhesive (pressure-sensitive adhesive)	adhésif m sensible à la pression	Haftkleber m	adesivo (autoadesivo)
52	adjustment	ajustage m	Einstellung f, Justierung f	regolazione f
53	adsorption	adsorption f	Adsorption f	assorbimento m
54	ageing (ageing apparatus)	appareil m de vieillissement	Alterungsgerät n	apparecchio m per prove d'invecchiamento
55	ageing (dynamic ageing)	vieillissement m dynamique	Alterung f (dynamische)	invecchiamento m dinamico
56	ageing of rubber	vieillissement m du caoutchouc	Gummialterung f	invecchiamento m della gomma
57	ageing (oil ageing)	vieillissement m dans l'huile	Ölalterung f	invecchiamento m in olio
58	ageing oven	étuve f de vieillissement	Alterungsschrank m	stufa f per prove d'invecchiamento
59	ageing (oven ageing)	vieillissement m en étuve	Ofenalterung f	invecchiamento m in stufa
60	ageing resistance	résistance f au vieillissement	Alterungsbeständigkeit f	resistenza f all'invecchiamento
61	ageing test	essai m de vieillissement	Alterungsprüfung f	prova f d'invecchiamento
62	agent (ageing protective agent)	agent protecteur m contre le vieillissement	Alterungsschutzmittel n	agente m protettivo contro l'invecchiamento
63	agent (anti-scorching agent)	agent m antigrillant	Vulkanisationsverzögerer m	agente m ritardante della prevulcanizzazione
64	agent (anti-sunchecking agent)	agent m antisolaire	Lichtschutzmittel n	agente m protettivo anti luce solare
65	agent (binding agent)	agent m de liaison	Bindemittel n	agente m legante
66	agent (blowing agent)	agent m gonflant	Treibmittel n, Blähmittel n	agente m gonfiante
67	agent (crosslinking agent)	agent m de rétification	Vernetzungsmittel n	agente m reticolante
68	agent (curing agent)	agent m de vulcanisation	Vulkanisationsmittel n	agente m vulcanizzante
69	agent (deodorising agent)	agent m déodorant	Desodorisierungsmittel n	agente m deodorante
70	agent (emulsifying agent)	émulsifiant m	Emulgiermittel n, Emulgator m	agente m emulsionante
71	agent (stabilising agent)	agent m stabilisateur	Stabilisiermittel n	agente m stabilizzante
72	agent (surface active agent)	agent m tensio-actif	oberflächenaktive Substanz f	agente m tensio-attivo
73	agent (swelling agent)	agent m gonflant	Quellmedium n	agente m rigonfiante
74	agent (tackifying agent)	agent m collant, de collage	Klebrigmacher m	agente m promotore di appiccicosità
75	agglomerate	agglomérat m	Agglomerat n	agglomerato
76	aggregate	agrégat m	Aggregat n	aggregato, conglomerato
77	air bomb	bombe f à air	Druckluftflasche f	bomba f ad aria
78	air brake hose	tuyau m de frein pneumatique	Druckluftbremsschlauch m	tubo m per freni ad aria compressa
79	air cure	vulcanisation f en étuve	Raumtemperaturvulkanisation f	vulcanizzazione f in aria
80	air cushion	coussin m pneumatique	Luftkissen n	cuscino m d'aria
81	air (entrapped)	air m emprisonné	Lufteinschluß m	aria f intrappolata
82	air humidity	humidité f atmosphérique	Luftfeuchtigkeit f	umidità f atmosferica
83	air (moisture of the)	humidité f de l'air	Luftfeuchtigkeit f	umidità f dell'aria, (umidità dell'aria)
84	air oven	étuve f à air	Luftofen m	stufa f ad aria calda
85	air oven ageing	vieillissement en étuve	Luftofenalterung f	invecchiamento m in stufa ad aria calda
86	air oven curing	vulcanisation f en étuve	Heißluftvulkanisation f	vulcanizzazione f in stufa ad aria calda
87	air pocket	poche f d'air	Lufteinschluß m	bolla f d'aria
88	air shot (air shot moulding)	injection f d'air (moulage)	Luftblase f	colpo m a vuoto di prova, (nello stampaggio)
89	aircraft hose	tuyau m d'aviation	Flugzeugschlauch m	tubo m per aerei
90	ALCRYN melt-processable rubber	ALCRYN caoutchouc transformable par fusion	ALCRYN thermoplastisches Elastomer	ALCRYN gomma lavorabile allo stato fuso

Nr	ESPAÑOL	PORTUGUÊS	SVENSKA	NEDERLANDS
46	poder m de adhesión	poder m de adesão	vidhäftningsförmåga, -styrka	kleefvermogen, kleefkracht
47	adhesión (aparato de medicion de la)	aderência (aparelho m para ensaio de)	vidhäftningsprovare	hechtingsmeter
48	adhesivo m	cola f, adesivo m	bindemedel	lijm
49	adhesivo	adesivo adj	vidhäftande	klevend
50	tela f adhesiva; teijido m adhesivo	tecido m adesivo	självhäftande väv	zelfklevend weefsel
51	autoadhesivo	autocolante	tryck-känsligt lim	lijm (drukgevoelige)
52	ajuste m, regulación f	ajustagem f, ajustamento m	justering	instelling
53	adsorción f	adsorção	adsorption	adsorptie
54	aparato m de envejecimiento	aparelho m para ensaio de envelhecimenato	apparat för åldringsprov	verouderingsapparaat
55	envejecimiento m dinámico	envelhecimento m dinâmico	dynamisk åldring	dynamische veroudering
56	envejecimiento m del caucho	envelhecimento m da borracha	gummiåldring	veroudering van rubber
57	envejecimiento m por aceite	envelhecimento m pelo óleo	oljeåldring	veroudering in olie
58	estufa f de envejecimiento	estufa f de envelhecimento	åldringskammare	verouderingsoven
59	envejecimiento m en estufa	envelhecimento m em estufa	ugnsåldring	veroudering in een oven
60	resistencia f al envejecimiento	resistência f ao envelhecimento	åldringsbeständighet	weerstand tegen veroudering
61	ensayo m de envejecimiento	ensaio m de envelhecimento	åldringsprov	verouderingsproef
62	agente m protector contra el envejecimiento	protetor m contra o envelhecimento; agente m protector contra o envelhecimento	åldringskyddsmedel	antioxydant
63	agente m retardante de la prevulcanización	agente m retardador da vulcanização prematura	antivulkmedel, retarder	antiscorch-middel
64	agente m antisolar	inibidor m do fendilhamento pela luz solar; agente m inibitório da deterioração pela luz solar	solskyddsmedel	beschermmiddel tegen zon aantasting
65	agente m aglutinante	agente m aglutinante; (Port.) agente m de colagem	bindemedel	bindmiddel
66	agente m hinchante	agente m de expansao, agente esponjante	jäsmedel	blaasmiddel
67	agente de reticulación	agente de reticulação	primärt vulkmedel	verknopingsmiddel, vernettingsmiddel
68	agente m vulcanizante	agente m de vulcanização	vulkmedel	vulcaniseermiddel
69	agente m desodorizante	agente m desodorizante	luktborttagningsmedel	euk verdrijvend middel
70	agente m emulsionante	agente m emulsificante	emulgator	emulgeermiddel, emulgator
71	agente m estabilizante	agente m estabilizador	stabiliseringsmedel	stabilisator
72	surfactivo m, agente tensoactivo	agente m tenso activo	ytaktivt medel	oppervlakte actieve stof
73	agente m hinchante	agente m de inchamento	svällningsmedel	zwelmiddel
74	agente m de mordiente	agente m de adesividade	klibbmedel	kleefmiddel
75	aglomerado	aglomerado	agglomerat	agglomeraat
76	agregado	agregado, aglomerado	aggregat	aggregaat
77	bomba f de aire	bomba f de ar	luftkammare	luchtcylinder
78	manguera f para freno neumático	tubo m para travão pneumático	slang till tryckluftbroms	luchtremslang
79	vulcanización f al aire	vulcanização f ao ar	luftvulkning	hete lucht vulcanisatie
80	cojin de aire	almofada f pneumática	luftkudde	luchtkussen
81	aire ocluído	bôlha f de ar, ampôla f	luftblåsa	ingesloten lucht
82	humedad f atmosférica	humidade f atmosférica	luftfuktighet	luchtvochtigheid
83	humedad f atmosférica	humidade f do ar	luftfuktighet	luchtvochtigheid
84	estufa f de aire caliente	estufa f de ar	varmluftsugn	hete-luchtoven
85	envejecimiento en estufa de aire	envelhecimento m em estufa de ar	åldringsprov i varmlufts ugn	verouderingsproef in hete-luchtoven
86	vulcanización f en estufa a aire caliente	vulcanização f em estufa de ar	varmluftvulkning	vulcanisatie in hete-luchtoven
87	oclusión f de aire, burbuja f de aire, ampolla	bôlha f de ar, bolsa f de ar	luftficka	luchtinsluiting
88	golpe de aire (en el moldeo)	AR (moldação)	luft skott	lucht stoot
89	manguera para aviones	tubo m para aviões	flygplansslang	vliegtuigslang
90	ALCRYN caucho procesable por fusion	ALCRYN borracha processável por fusão	ALCRYN termoelast	ALCRYN thermoplastisch elastomeer

aliphatic hydrocarbon

Nr	ENGLISH	FRANÇAIS	DEUTSCH	ITALIANO
91	aliphatic hydrocarbon	hydrocarbure m aliphatique	Kohlenwasserstoff m (aliphatisch)	idrocarburo m alifatico
92	alkali resisting	résistant aux bases	alkalibeständig	resistente agli alcali
93	alternating current	courant m alternatif	Wechselstrom m	corrente f alternata
94	ammeter	ampèremètre m	Amperemeter n	amperometro m
95	amplitude	amplitude f	Amplitude f	ampiezza f
96	angle (angle shearing)	angle m de cisaillement	Scherwinkel m	angolo m di taglio
97	angle loss	angle m de perte	Verlustwinkel m	angolo di perdita
98	angle of contact	angle m de contact	Berührungswinkel m	angolo m di contatto
99	aniline value	point m d'aniline	Anilinzahl f	punto m di anilina
100	anti-aging agent	agent anti-vieillissement m	Alterungsschutzmittel n	anti-invecchiante m
101	anti-tack agent	agent m anticollant	Klebrigkeitsverminderer m, Antihaftmittel n	agente m anti adesivo
102	antioxidant	antioxygène m	Oxidationsschutz m, Alterungsschutzmittel n	antiossidante
103	antioxidant (staining)	antioxygène m tachant	verfärbendes Alterungsschutzmittel n	antiossidante m macchiante
104	antioxidising agent	agent m antioxygène	Oxidationsschutzmittel n	agente m antiossidante
105	antiozonant (US)	agent m antiozone	Ozonschutzmittel n	antiozonante
106	antiscorcher	antigrillant m	Vulkanisationsverzögerer m	agente m ritardante della prevulcanizzazione
107	apex	apex m	Scheitel m	apice m
108	apparatus	appareillage m	Apparat m, Gerät n	apparecchio m
109	apparent density	densité f apparente, masse volumique apparente	scheinbare Dichte f, Schüttdichte f, Schüttgewicht n	densità f apparente
110	aqueous extract	extrait m aqueux	wäßriger Auszug m	estratto m acquoso
111	aromatic hydrocarbon	hydrocarbure m aromatique	aromatischer Kohlenwasserstoff m	idrocarburo m aromatico
112	aromatic oil	huile f aromatique f	Weichmacher m (aromatisch)	olio m aromatico
113	article (moulded articles)	article m moulé	Formartikel m	articolo m stampato
114	articles (hygienic articles)	articles mpl sanitaires	sanitäre Artikel mpl	articoli m igienici
115	asbestos (rubber-asbestos goods)	articles mpl en caoutchouc-amiante	Gummi-Asbest-Artikel fpl	articoli m di gomma-amianto
116	ash	cendre f	Asche f	ceneri f
117	assembling	assemblage m	Montieren n	montaggio m
118	automotive part	pièce f pour automobile	Kraftfahrzeugteil n	pezzo m per autovettura
119	baking	cuisson f	Erhitzen n, Einbrennen n, Härten n	cottura f
120	balata	balata f	Balata f	balata f
121	ball valve	clapet m à bille	Kugelventil n	valvola f a sfera
122	balloon	ballon m	Ballon m	pallone m
123	band	bande f	Band n	fascia f
124	band (heating band)	manchon m chauffant	Heizband n	fascia f riscaldante
125	barrel	fourreau m	Zylinder m	cilindro
126	base coat	première couche f	Grundstrich m, Grundierung f	mano f di fondo
127	batch dryer	séchoir m	Trockner m	essiccatore m per mescole
128	battery box	boîte f de batterie	Akkumulatorgehäuse n, Batteriegehäuse n	recipiente m per batterie, batteria f
129	battery separator	séparateur m d'accumulateur	Akkumulatorscheidewand f	separatore m per batterie
130	bead breaker	tourniquet m dégage-talon	Reifenhebel m, Wulstlösegerät n	apparecchio m per staccare i talloni dei pneumatici
131	bead wire	tringle f de talon	Wulstdraht m	armatura f dei talloni dei pneumatici
132	bearing	appui m	Lager n	appoggio m (cuscinetto)
133	bearing (absorbing bearing)	appui m amortisseur	Dämpfungslager n	cuscinetto m assorbente
134	bearing plate	plaque f d'appui	Lagerplatte f	piastra f di appoggio
135	beater (in latex foam industry)	batteuse f (dans l'industrie de la mousse de latex)	Schlagmaschine f	frullatore m, nell'industria della schiuma di lattice

Nr	ESPAÑOL	PORTUGUÊS	SVENSKA	NEDERLANDS
91	hidrocarburo m alifático	hidrocarboneto m alifático	alifatiskt kolväte	alifatische koolwaterstof
92	resistente a los álcalis	resistente aos álcalis	alkalibeständig	alkalibestand, loogbestand
93	corriente f alterna	corrente f alternada	växelström	wisselstroom
94	amperímetro m	amperímetro m	amperemätare	ampère meter
95	amplitud f	amplitude f	amplitud	amplitude
96	ángulo m de cizallamiento	ângulo m de cisalhamento, ângulo m de corte	skärvinkel	afschuifhoek
97	angulo de perdida	âgulo de perda	förlust vinkel	verlieshoek
98	ángulo m de contacto	ângulo m de contato	kontaktvinkel	aanrakingshoek
99	índice m de anilina, punto m de anilina	índice m de anilina	anilinpunkt	aniline getal
100	antienvejecedor m	anti-degradante m; agente m contra o envelhecimento	åldringsskyddsmedel	antioxydant
101	agente m antipegajosidad	agente m anti-pegajoso, agente m anti-adesivo	antiklibbmedel	antikleefmiddel
102	antioxidante m	antioxidante m	antioxidant	antioxidant
103	antioxidante m manchante o que mancha	antioxidante m causador de manchas	missfärgande antioxidant	verkleurende antioxidant
104	agente m antioxidante	agente m antioxidante	antioxidationsmedel	antioxidant
105	antiozonante m	antiozonizante m, antiozonante	antiozonant	antiozonant
106	retardante m de la prevulcanización	retardador m da vulcanização prematura	fördröjningsmedel, retarder	antiscorch-middel
107	punta f	ápice m, vértice m	spets	apex strip
108	aparato m	aparelho m	apparatur	apparaat
109	densidad f aparente	densidade f aparente	skenbar densitet	schijnbare soortelijke massa
110	extracto m acuoso	extrato m aquoso	vätske-extrakt	waterig extract
111	hidrocarburo m aromático	hidrocarboneto m aromático	aromatiskt kolväte	aromatische koolwaterstof
112	aceite aromatico	óleo aromático	aromatisk olja	aromatische olie
113	artículo m moldeado	artigo m moldado	formpressade artiklar	gietvormen
114	artículos mpl sanitarios	artigos mpl higiênicos, artigos mpl sanitários	hygieniska artiklar	hygiënische artikelen
115	artículos mpl de goma con amianto	artigos mpl de borracha com asbestos; (Port.) artigos mpl de borracha com amianto	gummi-asbestvaror	artikelen uit rubber/asbest-mengsel
116	ceniza f	cinza f	aska	as
117	montaje m	montagem f, reunião f	montering	assemblage, monteren
118	pieza f para automóvil	peça f para veículo automóvel	bildel	auto onderdeel
119	hornada f	cozimento m, recozimento m	härdning	inbranden
120	balata f	balata f	balata	balata
121	válvula f de bola	válvula f de esfera	kulventil	kogelventiel
122	balón m, globo	balão m	ballong	ballon
123	banda f, cinta f	cinta f, faixa f	fäll	band
124	banda calefactora	banda de aquecimento	värme band	verhittingsband
125	cilindro	cilindro	cylinder	cylinder
126	capa f de base	demão f de base	grundskikt	onderlaag, grondlak
127	secador m por lotes	secador m por lotes	satstork	batch droger
128	caja f de batería	caixa f para bateria	batterikärl	accubak
129	separador m de batería	separador m para bateria	batteriskiljevägg	accuwand
130	máquina destalonadora (neumáticos)	aparelho m para litertar o talão de pneu; (Port.) aparelho m para libertar a vira do pneu	apparat för lossnande av däckfoten från fälgen	hiellosser
131	armadura o cable del talón	armadura f do talão do pneu; (Port.) armadura f da vira do pneu	trådkärna	hielendraad
132	cojinete m, soporte m	apoio m	lager	lager
133	cojinete m absorbente	apoio m amortecedor	dämpande lager	trillingsdempend lager
134	placa f de apoyo	chapa f de apoio	lagerplatta	oplegplaat, lagerplaat
135	batidora	batedor m, batedeira f	blandare, visp	klopper, klopmachine

B

beater (in paper industry)

Nr	ENGLISH	FRANÇAIS	DEUTSCH	ITALIANO
136	beater (in paper industry)	pile à papier f	Holländer m	mescolatore m olandese nell'industria della carta
137	bed sheeting	draps mpl de lit	Betteinlage f	foglio m calandrato di supporto
138	bellow (rack and pinion steering)	soufflet m de direction	Lenkungsmanschette f	soffietto per scatola guida
139	belt fastener	attache f de courroie, agrafe f de courroie	Riemenverbinder m	graffa f per cinghie
140	belt fastener inserting machine and tools	machine f et outillage m pour insérer les attaches de courroie	Gurtverbindereindrückmaschine und -werkzeuge f	macchina f e utensili per applicare le graffe alle cinghie
141	belt (raw-edge V-belt)	courroie f trapézoïdale à bords tranchés (sans enveloppage)	flankenoffener Keilriemen m	cinghia trapezoidale a spigoli grezzi
142	belt (ribbed belt)	courroie striée	Keilrippenriemen m	cinghia scanalata
143	belt stretching gearing	engrenages fpl pour étirer les courroies	Riemenstreckgetriebe n	ingranaggio m per tirare le cinghie
144	belt (timing belt)	courroie synchrone de distribution	Zahnriemen m	cinghia dentata (per trasmissioni sincrone)
145	belt vulcanising press	presse f à vulcaniser les courroies	Riemenvulkanisierpresse f, Förderbandpresse f	pressa f per vulcanizzare le cinghie
146	belting	courroies fpl	Riemenmaterial n	cinghia f
147	belting (grain belting)	courroies fpl transporteuses pour grains	Getreideförderband n	nastri m trasportatori per granuli
148	bend	pliage m	Biegung f	piegamento m (flessione f)
149	bend (to)	plier	biegen	piegare
150	bevelled (adj)	biseauté	abgekantet	smussato, bisellato
151	bias cut	coupe f en biais	Diagonalschnitt m	tagliato secondo un angolo, tagliato diagonalmente
152	binder	aggloménant m, liant m	Binder m, Bindemittel n	agente m agglomerante (legante)
153	bitumen	bitume m	Bitumen n	bitume m
154	black (animal black)	noir m animal	Blutkohle f	nero m animale
155	bladder	vessie f	Blase f, Heizschlauch m	camera f d'aria
156	blade (adjustable blade)	lame f ajustable, réglable	einstellbares Messer n, Klinge f	lama f regolabile
157	blade (doctor blade)	racle f	Rakel f	spatola f per spalmatura
158	bleaching	blanchiment m	Bleichung f	candeggio m
159	bleed (to)	dégorger	ausbluten	trasudare
160	bleeding	dégorgement m (de couleurs)	Ausbluten n (von Farben)	trasudamento m (dei colori)
161	blend (to)	mélanger	mischen, verschneiden	mescolare
162	blister	bulle f d'air	Blase f (Luftblase)	bolla f d'aria
163	blistering	boursouflures fpl	blasenbildend	rigonfiamento m
164	block polymer	polymère m bloc	Block-Polymer n	polimero m a blocchi
165	blooming	efflorescence f	Ausblühen n	efflorescenza f
166	blooming of sulphur, sulphur bloom	efflorescence f de soufre	Schwefelausblühung f	efflorescenza f di zolfo
167	blow	soufflage m	Blasen n, Blähen n	bolla f, soffiatura
168	blow extrusion	extrusion/soufflage f	Extrusionsblasverfahren n	estrusione per soffiaggio
169	blow moulding	injection-soufflage f	Blasformverfahren n	stampaggio per soffiaggio
170	blush (to)	repousser	anlaufen	affiorare
171	blushing	efflorescence f	Anlaufen n	affioramento m
172	boat (collapsible bomb)	canot m pneumatique	Schlauchboot n	canotto m pneumatico
173	bomb (ageing apparatus bomb)	bombe f (appareil de vieillissement)	Bombe f (Alterungsapparat)	bomba f (apparecchio per l'invecchiamento) f
174	bomb (oxygen bomb)	bombe f à oxygène	Sauerstoffbombe f	bomba f ad ossigeno
175	bond	liaison f	Bindung f, Haftung f	legame m
176	bond separation	décollage m, décollement m	Lagentrennung f	distacco m dell'incollaggio
177	bond strength	force f d'adhérence	Haftfestigkeit f	forza f dell'incollaggio, resistenza f dell'adesione
178	bond stress	contrainte f dans la couche de collage	Haftspannung f	sforzo m d'incollaggio
179	bonding agent	agent m d'adhérisation	Haftvermittler m	agente m d'incollaggio
180	bonding of rubber	collage m du caoutchouc	Kautschukbindung f, Kautschukhaftung f	incollaggio m della gomma

Vocabulary of the rubber industry

Nr	ESPAÑOL	PORTUGUÊS	SVENSKA	NEDERLANDS
136	molino m de muelas verticales	moínho m de cilindros	holländare	hollander
137	sábanas fpl	lençois mpl para cama	bädd-duk	kalanderdoek
138	fuelle para la cremallera de la dirección	fole para a cremalheira da direcção	bälg f. kuggstångs styrning	tandheugel aandrijving
139	grapa f para correa	fixador m para correia, grampo m para correia	remlås	transportbandverbinder
140	máquina f y herramientas para aplicar grapas y correas	máquina f e ferramentas para aplicar grampos em correias	remlåsinsättningsmaskin och -verktyg	verbinding van transportband einden
141	correa (correa trapezoidal de cantos rugosos)	correia trapezoidal de arestas rugosas	skuren kilrem	V-snaar
142	correa estriada	correia estrada	kuggrem	getande V-snaar
143	engranajes fpl para estirar correas	equipamento m para estirar correias	remspänningsdrev	drijfwerk voor het strak zetten van de riemen
144	correa de transmisión	correa de sincronizacáo	transmission rem	getande aandrijfsnaar
145	prensa f para vulcanizar correas	prensa f para vulcanizar correias	remvulkpress	transportbanden pers
146	correas fpl	correias fpl, correame m	remmar	transportbanden
147	correas mpl transportadoras de granos	correias fpl transportadoras gara gràos	transportband för säd	transportbanden voor korrels
148	curva f, doblez f	dobramento m, flexão f	böjning	buiging
149	plegar, doblar	dobrar	(att) böja	buigen
150	sesgado	biselado, chanfrado	avfasad	afgeschuind
151	corte m al bies	corte m enviesado	diagonalsnitt	diagonaal gesneden
152	aglomerante m, aglutinante m	aglutinante m, cola f, ligante	bindemedel	bindmiddel
153	betún m, asfalto m	betume m	bitumen, asfalt	bitumen
154	negro animal	negro animal	sot	dierlijk roet
155	vejiga f, vejiga f para balón de fútbol	bexiga f, bexiga de bola	blåsa	blaas (bladder)
156	cuchilla f ajustable	lâmina f ajustável	vridbart blad	verstelbaar mes
157	bisturi cuchilla	racla, lâmina	rakel	strijkmes
158	blanqueo m	branqueamento m	blekning	bleken
159	sangrar	sangrar, desbotar	blöda	migratie van kleurstoffen
160	sangrado m (de colores)	sangria f, escorrimento m de côres	blödning	uitlopen van kleurstoffen
161	mezclar	misturar	blanda	mengen
162	ampolla f	empôla f, bôlha f de ar	luftblåsa	luchtblaas
163	formación de ampollas	aparecimento de bolhas	blås bildande	blaasvorming
164	polímero en bloque	polímero em blocos	block polymer	blok polymeer
165	eflorescencia f	eflorescência f	blomning	uitbloemend
166	eflorescencia f del azufre	eflorescência f do enxôfre	utsvavling	uitbloemen van zwavel
167	soplado	sopro insuflação	blåsa	blazen
168	extrusion soplado	extrusão insuflação	extrusions blåsning	blaas extrusie
169	moldeo por soplado	moldação por sopro, moldação por insuflação	formblåsning	blaasgieten
170	eflorescer	eflorescer	blomma	uitbloemen
171	eflorescencia f	eflorescência f	blomning	uitbloemend
172	canoa f plegable	barco m pneumático	hopfällbar båt	opblaas boot
173	bomba f (aparato para envejecimiento)	bomba f (aparelho para envelhecimento)	åldringsbomb	cylinder (verouderingsapparaat)
174	bomba f de oxígeno	bomba f de oxigênio	syrebomb	zuurstofcylinder
175	ligazón f	ligação f união	bindning	verbinding, hechting
176	despegarse	descolagem f, despegamento m	fogbristning	loslaten van verbinding
177	fuerza de unión, resistencia del pegado	resistência f da colagem, resistência da ligação f	foghållbarhet	hechtkracht
178	tensión f en la capa de adherencia	tensão f na camada de ligação	fogpåkänning	spanning in hechtlaag
179	agente m de unión o pegado	agente m de ligação, agente m de colagem	bindemedel	bindmiddel, hechtmiddel
180	pegado m del caucho	colagem f da borracha, união f da borracha	limning av gummi	hechten van rubber

B

B

boot(constant velocity joint boot)

NR	ENGLISH	FRANÇAIS	DEUTSCH	ITALIANO
181	boot(constant velocity joint boot)	soufflet m de joint homocinétique, de transmission	Achsmanschette f	cuffia per giunti a velocità constante
182	boot (US), bonnet (UK)- (automotive)	capot m (automobile)	Balg m, Stulpe f	soffietto m, cuffia f
183	boots (rubber boots)	bottes fpl en caoutchouc	Gummistiefel mpl	stivali m di gomma
184	braided hose	tuyau m tressé	Schlauch m (umflochten)	tubo m rinforzato con treccia
185	braiding machine	tresseuse f	Klöppelmaschine f, Umflechtmaschine f	macchina f trecciatrice
186	brass	laiton m	Messing n	ottone m
187	break (elongation at break)	allongement m à la rupture	Bruchdehnung f	allungamento m a rottura
188	break (load at break)	charge f à la rupture	Bruchlast f	carico m a rottura
189	breakdown	plastification f, ramollissement	Abbau m	premasticazione f
190	breakdown (electricity)	claquage m, décharge f disruptive	Durchschlag m	scarica f perforante
191	breakdown (mechanics)	dérangement m, panne f	Störung f	guasto m, avaria f
192	breakdown (physics-chemistry)	dégradation f (physique-chimique)	Abbau m	decomposizione f, deterioramento m
193	breaking strength	résistance f à la rupture	Bruchfestigkeit f	resistenza f alla rottura
194	brittle (adj)	fragile, cassant	spröde, brüchig	fragile
195	brittleness	fragilité f	Sprödigkeit, Brüchigkeit f	fragilità f
196	brittleness (cold brittleness)	fragilité f à basse température	Sprödigkeit f in der Kälte	fragilità f a basse temperature
197	brittleness temperature	température de fragilisation	Versprödungstemperatur f	temperatura f di infragilimento
198	brush coated	revêtu au pinceau	beschichtet (mit Pinsel aufgetragen)	rivestito a pennello
199	brush on (to)	revêtir au pinceau	anstreichen, mit dem Pinsel auftragen	pennellare
200	brush-coat (to)	appliquer au pinceau	anstreichen, beschichten (mit Pinsel)	rivestire a pennello
201	buffer (rubber buffer)	tampon m en caoutchouc, butoir m en caoutchouc	Gummipuffer m	respingente m di gomma
202	building	confection f	Konfektionierung f, Zusammenbau m	confezione f, costruzione
203	building tack	collant m de confection	Konfektionsklebrigkeit f	potere m adesivo
204	building up	confection f	Konfektionierung f, Aufbau m, Montage f	confezione f montaggio m
205	bulk density	masse volumique apparente	Schüttdichte f, Schüttgewicht n	densità f apparente
206	bumper	pare-choc m	Puffer m, Stoßstange f	paraurti m
207	buoyancy jacket	gilet m de sauvetage	Schwimmweste f	giubbetto f di salvataggio
208	bushing	manchon m	Buchse f, Lagerbuchse f	boccola f
209	butt joint	soudure f bout à bout	Stumpfstoß m	giunto m testa a testa
210	butt seam	joint m bout à bout	Stumpfnaht f	unione f testa a testa
211	cable	câble m	Kabel n	cavo m
212	cable covering	câble (enveloppe f de)	Kabelmantel m	rivestimento m del cavo
213	cable distribution box	boîte f de dérivation pour câbles	Kabelüberführungskasten m, Kabelverteiler m	cassetta f di derivazione per cavi
214	cable (lead covered cable)	câble m sous plomb	Bleikabel n	cavo m ricoperto in piombo
215	cable making machinery	machine f pour la fabrication de câbles	Kabelherstellungsmaschine f	macchina f per la fabbricazione dei cavi
216	cable (overhead)	câble m aérien	Luftkabel n, Freiluftkabel n	cavo m sospeso
217	cable (power)	câble m de transmission de puissance	Kraftstromkabel n	cavo m di potenza
218	cable sheathing compound	mélange m pour gainage de câble	Kabelmantelmischung f	mescola f per rivestimento di cavi
219	calcined magnesia	magnésie f calcinée	Magnesiumoxid n kalziniert	magnesia calcinata f
220	calender	calandre f	Kalander m	calandra f

B

Nʀ	ESPAÑOL	PORTUGUÊS	SVENSKA	NEDERLANDS
181	junta homocinética	junta homocinética	konstant hastighetsbälg	stofbalg voor een homokinetische koppeling
182	portaequipajes (automovil)	porta bagagem (automovel)	bälg	stofbalg
183	botas fpl de goma	botas fpl de borracha	gummistövlar	rubber laarzen
184	manguera f trenzada	tubo m entrançado, mangueira f entrançada	slang med flätat inlägg	slang met gevlochten wapening
185	máquina de trenzar, trenzadora	máquina f de entrançar, entrançadeira f	flätmaskin	vlechtmachine
186	latón m	latão m	mässing	messing
187	alargamiento a la rotura	alongamento m à roptura	brott-töjning	rek bij breuk
188	carga f a la rotura	carga f na ruptura; (Port.) carga f na rotura	brottbelastning	breukbelasting
189	rotura	mastigação	nedbrytning	afbraak
190	descarga f disruptiva	descarga f disruptiva	genomslag	doorslag
191	falla f, averia f	avaria f	förstörande	storing
192	degradación f	decomposição f	nedbrytning	ontleding
193	resistencia f a la ruptura	resistência f à ruptura	brottshållfasthet	treksterkte
194	frágil, quebradizo	friável, quebradiço, frágil	spröd	bros
195	fragilidad f	friabilidade f, fragilidade f	sprödhet	brosheid
196	fragilidad f a baja temperatura	friabilidade f às temperaturas baixas	köldsprödhet	brosheid lage temperatuur
197	temperatura de fragilidad	temperatura de fragilidade	sprödpunkt	temperatuur waarbij materiaal bros wordt
198	aplicado con pincel	revestido a pincel	penselstruken	een coating aangebracht met kwast
199	pintar	pincelar	bestryka med pensel	met de kwast aanbrengen
200	aplicar con pincel	revestir a pincel	penselstryka	een coating aanbrengen met kwast
201	amortiguador m de caucho	pára-choques m de borracha, amortecedor m de borracha	gummibuffert	rubber buffer, trillingsdemper
202	construccion f, montaje m	confecção f, montagem f	byggning (av däck)	opbouw
203	mordiente o pegajosidad de las mezclas en crudo	pegajosidade f das misturas para confecção	klibbförmåga hos blandningar för konfektionering	kleefkracht van mengsels bij opbouw, tack
204	confección f	confecção f, montagem f	uppbyggning	opbouw
205	densidad aparente	densidade f aparente	volym densitet	volume dichtheid
206	parachoques m	pára-choques m	stötdämpare	bumper
207	chaqueta f salvavidas	colete m salva-vidas	flytväst	zwemvest
208	manguito m, casquillo m	casquilho	bussning	doorvoermanchet
209	unión por testa, junta a testa	junta f de tôpo	stumskarv	stootnaad
210	costura por testa	costura f de tôpo	stumsöm	stuiknaad
211	cable m	cabo m	kabel	kabel
212	recubrimiento de cable	revestimento f de cabo	kabelmantel	kabelmantel
213	caja f de derivación para cables	caixa f de distribuição para cabos	kabelförgrening	kabelverdeelkast
214	cable m con blindaje de plomo	cabo m com blindagem de chumbo	blymantlad kabel	loodkabel
215	maquinaria f para la fabricación de cables	maquinaria f para fabricação de cabos	maskineri för kabeltillverkning	machines voor kabelfabricage
216	cable suspendido	cabo suspenso	luftledning	stroomdraad
217	cable m para transmisión de fuerza	cabo m de potência	kraftkabel	krachtstroomkabel
218	composición para recubrir cables	composição f para revestimento de cabos	kabelmantelblandning	mengsel voor kabelmantel
219	magnesia f calcinada	magnésia f calcinada	rostad magnesiumoxid	gecalcineerde magnesium
220	calandra f	calandra f	kalander	kalander

C

Nr	ENGLISH	FRANÇAIS	DEUTSCH	ITALIANO
221	calender (cross axis, i.e. cross axis calender)	calandre à désaxage	Kalander m mit Schränkung einer Walze f	calandra f con cilindri ad assi incrociati
222	calender grain	grain m de calandrage	Kalandereffekt m	effetto m della calandra
223	calender (to)	calandrer	kalandrieren, kalandern	calandrare
224	calendered friction	friction à la calandre	Friktionierung f mittels Kalander	frizione f della calandra
225	calendering	calandrage m	Kalandern n	calandratura f
226	calorimetry	calorimétrie f	Kalorimetrie f	calorimetria f
227	canvas	canevas m, toile f de chanvre	Segeltuch n	tela f olona
228	capacity	capacité f	Kapazität f, Fassungsvermögen n	capacità f
229	capacity (load carrying capacity)	capacité f de charge	Tragfähigkeit f	capacità f di carico
230	capacity (thermal capacity)	capacité f calorifique	Wärmekapazität f	capacità f termica
231	capping	capsulage m	Abdeckung f	capping m, interruzione f crescita della catena molecolare
232	car mat	tapis m pour auto	Automatte f	tappetino m per autovetture
233	car tire (US), car tyre (GB)	pneu m tourisme	Autoreifen m	pneumatico m per autovetture
234	carbon (activated)	charbon m activé	Aktivkohle f	carbone m attivo
235	carbon (active)	charbon m actif	Aktivkohle f	carbone attivo m
236	carbon black HAF (high abrasion furnace)	noir m de carbone HAF	Ruß m (HAF)	nero fumo m
237	carbon black (activated)	noir m de carbone activé	Ruß m aktiviert	nerofumo m attivo
238	cast (to)	mouler par coulée	gießen	colare, fondere
239	caster (US), castor (GB)	roulette f	Laufrolle, Streubüchse f	rotella f orientabile
240	casting (flow casting)	moulage m par coulée	Gießen n	colata f per scorrimento
241	catalyst	catalyseur m	Katalysator m	catalizzatore m
242	caulk (to)	calfater	abdichten	stuccare
243	caulking	calfatage m	Abdichten n	stuccatura f
244	cavity (mould cavity)	cavité f (de moule)	Formhöhlung f, Formnest n	cavità dello stampo, impronta
245	cellular (adj)	cellulaire, alvéolaire	zellig	cellulare
246	cellular rubber	caoutchouc m cellulaire	Zell-, Moos-, Schwammgummi n	gomma cellulare f
247	celluloid base adhesive	colle f à base de celluloid	Celluloseklebstoff m	adesivo m a base di celluloide
248	cement (glue, adhesive, etc.)	dissolution f, colle f, adhésif m	Kleber m, Klebstoff m	adesivo m, colla f
249	cement (in construction work)	ciment m (dans la construction)	Zement m	cemento m
250	cement (quick setting)	colle f à prise rapide	schnellabbindender Kleber m	adesivo m a presa rapida
251	cement suction hose	tuyau m d'aspiration pour ciment	Zementsaugschlauch m	tubo m d'aspirazione per cemento
252	centipoise	centipoise f	Centipoise n	centipoise m
253	chain structure	structure f de chaîne	Kettenstruktur f	struttura f della catena
254	channel (channel depth)	canal m (profondeur)	Gangtiefe f	profondità del canale
255	channel (channel width)	canal m (largeur)	Gangbreite f	larghezza del canale
256	chart	graphique m	Diagramm n, Aufstellung f	grafico m
257	chemical hose	tuyau m pour produits chimiques	Chemikalienschlauch m	tubo m per prodotti chimici
258	chemically modified	modifié chimiquement	chemisch modifiziert	modificato chimicamente
259	chemicals for the rubber industry	produits mpl chimiques pour l'industrie du caoutchouc	Chemikalien für die Kautschukindustrie	prodotti m chimici per l'industria della gomma
260	chloroprene	chloroprène m	Chloropren n, Chlorbutadien n	cloroprene m
261	chopper	guillotine f, découpeur m	Zerkleinerungsmaschine f	ghigliottina
262	clamp	mâchoire f	Klemme f	staffa f
263	clamping force	force f de fermeture	Schließkraft f	pressione f di chiusura, di stampo
264	clay	kaolin m	Kaolin n	caolino m
265	clay (china clay)	kaolin m	Kaolin n	caolino m
266	clay (hard)	kaolin m dur	Hartkaolin n	caolino m duro
267	clicker die	emporte-pièce m	Stanzmesser n	stampo m per tranciare
268	cloth (double cloth)	tissu m doublé	Doppeltuch n	tessuto m doppio
269	cloth (protective clothing)	vêtement m de protection	Schutzkleidung f	indumento m protettivo
270	clothing (wind and waterproof clothing)	vêtements mpl de protection contre le vent et l'eau	Kleidung f (wind- und wasserfest)	indumenti m impermeabili

Nr	ESPAÑOL	PORTUGUÊS	SVENSKA	NEDERLANDS
221	calandra	calandra	kalander	kalander (met dwarscorrectie)
222	efecto o grano de calandra	grão da calandra f, efeito da calandra	kalandereffekt	kalandereffect
223	calandrar	calandrar	kalandrera	kalanderen
224	calandrado m por fricción	fricção f na calandra	docka för friktionering	frictionering
225	calandrado m	calandragem f	kalandrering	kalanderen
226	calorimetría f	calorimetria f	kalorimetri	calorimetrie
227	lona f, tela f de lona	lona f, pano de lona	kanvas	canvas
228	capacidad f	capacidade f	kapacitet	capaciteit
229	capacidad f de soportar una carga	capacidade f de carga	lastförmåga	draagvermogen
230	capacidad térmica	capacidade térmica	termisk kapacitet	thermische capaciteit
231	encapsulamiento	encapsulamento	kappa	afdekking
232	esterilla f de automóvil	tapete m para automóvel	bilmatta	automat
233	neumático m, cubierta f	pneumático m para automóvel	bildäck	autoband
234	carbón m activado	carvão m activado	aktiverat kol	geactiveerde koolstof
235	carbón m activo	carvão activo	aktivt kol	aktieve koolstof, aktief roet
236	negro m de humo HAF	negro m de fumo	kimrök	roet
237	negro m humo activado	negro m de fumo activado	aktiverad kimrök	aktief roet
238	colar, fundir	vazar para	gjuta	gieten
239	ruedecilla movible, rueda de guía	rodízio m	svänghjul	zwenkwiel
240	colada f por vaciado	moldação f por vazamento	gjutning	gietproces
241	catalizador m	catalizador m	katalysator	katalysator
242	calafatear	calafetar	dikta	kitten, afdichten
243	calafateo	calafetagem f, calafêto m	diktning	kitten, afdichten
244	cavidad del molde, huella	cavidade do molde	formutrymme	matrijsvorm
245	celular	celular, alveolar	cellformig	cellulair
246	caucho m celular	borracha f celular	cellgummi	celrubber
247	cola f a base de celuloide	cola f à base de celulóide	cellulosanitratlim	celluloidlijm
248	cola f, adhesivo m	cola f, adesivo m	lim	lijm, solutie
249	cemento	cimento m	cement	cement
250	cola f de cuajado rapido, cemento	cola f de pega rápida	snabbindande cement	snel hardende lijm
251	tubo m de aspiración para cemento	mangueira f de aspiração para cimento	cementsugslang	aanzuigslang voor cement
252	centipoise m	centipoise m	centipoise	centipoise
253	estructura f de cadena	estrutura f de cadeia	kedjestruktur	ketenstructuur
254	profundidad del canal	profundidade do canal	kanaldjup	kanaaldiepte
255	anchura del canal	largura do canal	kanal bredd	kanaalbreedte
256	cuadro m, gráfica f	quadro m, gráfico m	diagram	diagram, grafiek
257	tubo m para productos químicos	mangueira f para produtos químicos	kemikalieslang	slang voor chemicaliën
258	químicamente modificado	modificado quimicamente	kemiskt modifierad	chemisch gemodificeerd
259	productos mpl químicos para la industria del caucho	produtos mpl químicos para a indústria da borracha	kemikalier för gummi industrin	chemicaliën voor de rubberindustrie
260	cloropreno m	cloropreno m	kloropren	chloropreen
261	cortadora f	guilhotina f, talhadeira f	skärmaskin	guillotine, balensnijder
262	mordaza f, abrazadera f	grampo m, braçadeira f	inspänna	inspanklem
263	fuerza de cierre	força de fecho	ihopklämnings styrka	klemkracht
264	arcilla f, caolín m	argila f caolino	lera	klei, kaolien
265	caolín	caolino argila	kaolin	kaolien
266	arcilla f dura, caolín duro	caulino m duro, argila f dura	hård lera	harde kaolien
267	matriz f de troquelar	matriz f de estampagem, cunho m	stansplatta	stansmes
268	tejido m doble	pano m de textura dupla	dubblerat tyg	(dubbel weefsel) doek
269	traje m protector	roupa f de proteção	skyddskläder	veiligheidskleding
270	prendas fpl a prueba de viento y agua	roupa f impermeável, vestuário m impermeável	vind- och vattentät klädsel	wind- en waterdichte kleding

co-agent

C

NR	ENGLISH	FRANÇAIS	DEUTSCH	ITALIANO
271	co-agent	co-agent m	Coagens n	coagente
272	co-precipitation	coprécipitation f	gemeinsame Ausfällung f	co-precipitazione f
273	coagulation	coagulation f	Gerinnung f, Koagulation f	coagulazione f
274	coat, to	revêtir, gommer, enduire	auskleiden, streichen, beschichten	rivestire
275	coat (to calender coat)	gommer (à la calandre)	beschichten mittels Kalander	rivestire a calandra
276	coat with doctor blade (to)	enduire à la râcle	mit einer Rakel f auftragen	rivestire a spatola
277	coated (adj)	revêtu, gommé, enduit	ausgekleidet, angestrichen, beschichtet	rivestito
278	coated fabric	tissu m enduit	Gewebe n (beschichtet), gummiertes Gewebe n	tessuto m rivestito
279	coating	revêtement m, enduit m	Auskleiden n, Anstreichen n, Beschichten n	rivestimento m
280	coating (dip coating)	enduction f par trempé	Tauchlackierung f, Tauchbeschichtung f	rivestimento f per immersione
281	coating gun	pistolet m à enduire	Spritzpistole f	pistola f per spruzzatura
282	coating machine	machine f à enduire	Beschichtungsmaschine f	macchina f spalmatrice
283	coating on the calender	gommage m à la calandre, enduction f par calandrage	Beschichten n auf dem Kalander	rivestimento in calandra
284	coating pistol	pistolet m à enduire	Spritzpistole f	pistola f per spruzzatura
285	coefficient (friction)	coefficient m de friction	Reibungskoeffizient m	coefficiente m di frizione
286	coefficient of elasticity	coefficient m d'élasticité	Elastizitätskoeffizient m	coefficiente m di elasticità
287	coefficient of expansion	coefficient m de dilatation	Ausdehnungskoeffizient m	coefficiente m di dilatazione
288	coefficient of friction	coefficient m de friction	Reibungskoeffizient m	coefficiente m di attrito
289	coefficient of friction (dynamic) 3 m/min. 0,7 MPa	coefficient m de friction (dynamique) 3 m/min, 0,7 MPa	Reibungszahl f (dynamisch) oder Gleitreibungszahl f 3 m/min, 0,7 MPa	coefficiente m di attrito (dinamico) 3 m/min, 0,7 MPa
290	coefficient of friction (static)	coefficient m de friction (statique)	Reibungszahl f (statisch) oder Haftreibungszahl f	coefficiente m di attrito (statico)
291	coefficient of linear thermal expansion x-y °C	coefficient de dilatation thermique linéaire x-y °C	Wärmeausdehnungs-, Längenausdehnungskoeffizient m (linear) x-y °C	coefficiente m di dilatazione termica lineare x-y °C
292	coefficient of permeability	coefficient m de perméabilité	Durchlässigkeitskoeffizient m	coefficiente m di permeabilità
293	coefficient of thermal conductivity	coefficient m de conductibilité thermique	Wärmeleitzahl f, Wärmeleitungskoeffizient m	coefficiente m di conduttività termica
294	coefficient of viscosity	coefficient m de viscosité	Viskositätskoeffizient m	coefficiente m di viscosità
295	coefficient of vulcanisation	coefficient m de vulcanisation	Vulkanisationskoeffizient m	coefficiente m di vulcanizzazione
296	cohesion	cohésion f	Kohäsion f	coesione f
297	coiled hose	tuyau m à serpentin	Spiralschlauch m	tubo a spirale
298	cold creep	fluage m à froid	Kaltfluß m	creep m a freddo
299	cold feed extruder	extrudeuse f alimentée à froid	Kaltfütterextruder m	estrusore m (alimentato a freddo)
300	cold flow	fluage m à froid	Kaltfluß m	scorrimento m a freddo
301	collar (heating collar)	collier m (chauffant)	Heizkragen m	banda riscaldante
302	colloid	colloïde m	Kolloid n	colloide m
303	colour fastness	stabilité f de couleur	Farbechtheit f	stabilità f del colore
304	colour fastness to light	résistance f de la couleur à la lumière	Lichtechtheit f der Farbe, Lichtbeständigkeit f	stabilità f del colore alla luce
305	colour (GB) color (US)	couleur	Farbe f	colore m
306	colour stability	tenue f de la couleur	Lichtechtheit f, Farbenbeständigkeit f	stabilità f del colore
307	compatibility	compatibilité f	Verträglichkeit f	compatibilità f
308	compound (caulking compound)	mélange m (pour mastic)	Abdichtmasse f	mescola f per stucco
309	compound (chemical compound)	composition f (chimique)	Verbindung f	composto m
310	compound (compound mixture)	mélange m	Mischung f	mescola f

Vocabulary of the rubber industry

Nr	ESPAÑOL	PORTUGUÊS	SVENSKA	NEDERLANDS
271	coagente	coagente	co-agent	co-agent
272	co-precipitación f	co-precipitação f	sammanfällning	co-precipitatie
273	coagulación f	coagulação f	koagulering	coagulatie
274	revestir, engomar	revestir, recobrir, cobrir	belägga	bekleden
275	engomar por calandrado, recubrir, revestir	revestir à calandra, revestir por calandra	belägga i kalander	met rubber bedekken op de kalander
276	engomar a cuchilla, recubrir, revestir	revestir à espátula/por raclagem	bestryka med rakel	opbrengen met een rakel
277	revestido, engomado, recubierto	revestido, coberto	gummerad, belagd	bekleed
278	tejido m engomado, recubierto, revestido	tecido m revestido	gummerad väv	met rubber bedekt weefsel
279	revestimiento m, capa f	revestimento m, capa f	beläggning	coating
280	revestimiento m por inmersión	revestimento m por imersão	doppbeläggning	coating verkregen door dompelen
281	pistola f pulverizadora, pistola f para revestir	pistola f pulverizadora, pistola f para revestir	sprutpistol för beläggning	spuitpistool voor coatings
282	máquina f para engomar, recubrir, máquina f para revestir	máquina f para revestir, máquina f para recobrir	beläggningsmaskin	machine om te coaten
283	engomado m con calandra, recubierto, revestido	revestimento m em calandra	belägga i kalander	coaten op de calander
284	pistola f pulverizadora, pistola f para revestir	pistola f pulverizadora, pistola f para revestir	sprutpistol för beläggning	spuitpistool voor coatings
285	coeficiente m de fricción	coeficiente m de atrito	friktionskoefficient	wrijvingscoëfficiënt
286	coeficiente m de elasticidad	coeficiente m de elasticidade	elasticitetskoefficient	elasticiteits coëfficiënt
287	coeficiente m de dilatación	coeficiente m de expansão, coeficiente m de dilatação	utvidgningskoefficient	uitzettingscoëfficiënt
288	coeficiente m de fricción	coeficiente m de atrito	friktionskoefficient	wrijvingscoëfficiënt
289	coeficiente de rozamiento (dinámico)	coeficiente m de fricção (dinâmico)	dynamisk friktionskoefficient, 3 m/min. 0,7 MPa	dynamische wrijringscoëfficiënt, 3m/min. 0,7 MPa
290	coeficiente de rozamiento (estático)	coeficiente m de fricção (estatico)	statisk friktionskoefficient	statische wrijringscoëfficiënt
291	coeficiente de dilatation térmica lineal x-y °C	coeficiente m de expansão termica linear	linjär termisk utvidgningskoefficient	lineaire thermische uitzettings coëfficiënt x-y °C
292	coeficiente m de permeabilidad	coeficiente m de permeabilidade	permeabilitets koefficient	doorlaatbaarheidscoëfficiënt
293	coeficiente m de conductividad térmica	coeficiente m de condutividade térmica	värmeledningstal	warmtegeleidingscoëfficiënt
294	coeficiente m de viscosidad	coeficiente m de viscosidade	viskositetskoefficient	viscositeitscoëfficiënt
295	coeficiente m de vulcanización	coeficiente m de vulcanização	vulkkoefficient	vulcanisatiecoëfficiënt
296	cohesión f	coesão f	kohesion	cohesie
297	manguera en espiral	tubo em espiral - tubo helicoidal	spiral slang	spiraalslang
298	deslizamiento m en frío	escoamento m a frio	kall krypning	kruip bij lage temperatuur
299	extrusora de alimentación en frio	extrusora f (alimentação a frio)	sträng spruta, kallmatad	spuitmachine (koude voeding)
300	flujo m a frio	fluxo f a frio	kallflytning	bijlage temperatur vloei
301	banda calefactora	banda de aquecimento	värmeband	verwarmingsband, verwarmingsmanchet
302	coloide m	colóide m	kolloid	colloid
303	solidez o de color	estabilidade f da cor	färgbeständighet	kleurechtheid
304	solidez f del color a la luz	estabilidade f da cor na exposição à luz	ljusäkthet hos färg	lichtechtheid
305	color	cor	färg	kleur
306	estabilidad f del color	estabilidade f do colorido	färgstabilitet	kleurechtheid
307	compatibilidad f	compatibilidade f	förenlighet	verdraagzaamheid
308	composición f para calafateo, masilla	composição f para calafetar	tätningsmassa	kit, afdichtmengsel
309	compuesto m químico	composto m	förening	chemische samenstelling, verbinding
310	mezcla f	composição f, mistura	blandning	mengsel

C

compound (gum compound)

C

Nʀ	ENGLISH	FRANÇAIS	DEUTSCH	ITALIANO
311	compound (gum compound)	mélange caoutchouc	füllstofffreie Kautschukmischung f	mescola di gomma
312	compound (raw compound)	mélange m cru	Rohmischung f	mescola f cruda
313	compression	compression f	Kompression f	compressione f
314	compression curve	courbe f de compression	Kompressionskurve f	curva f di compressione
315	compression modulus	module m de compression	Druckmodul m	modulo m a compressione
316	compressive modulus	module m d'élasticité en compression	Druck-E-Modul m	modulo m a compressione
317	compressive strength	résistance f à la compression	Druckfestigkeit f	resistenza f alla compressione
318	compressor hose	tuyau m pour compresseur	Kompressorschlauch m	tubo m per compressore
319	computer controlled	contrôlé par ordinateur	computergesteuert	controllato da computer (calcolatore)
320	condensation (aldehyde aniline condensation product)	produit m de condensation (aldéhyde-aniline)	Aldehyd-Anilin-Kondensationsprodukt n	prodotto m di condensazione a base di aldeide-anilina
321	conduction	conduction f	Übertragung f, Fortleitung f	conduzione f
322	conductivity	conductivité f	Leitfähigkeit f	conduttività
323	conductivity (thermal conductivity)	conductivité thermique	Wärmeleitfähigkeit f	conducibilità f termica
324	conductor (cable conductor)	conducteur m de câble	Kabelleiter f, Kabelader f	conduttore m (cavo conduttore)
325	conduit	conduite f	Leitung f	condotto m
326	conduit (rubber-lined conduit)	conduit m garni en caoutchouc	Leitung f (mit Gummi ausgekleidet)	condotto m rivestito in gomma
327	configuration	configuration f	Struktur f	configurazione f
328	connection (ground connection, earthing)	mise f à la terre	Erdverbindung f	messa f a terra
329	connector box	boîte f de dérivation	Abzweigdose f, Verbindungsstück n	cassetta f di derivazione
330	constant velocity joint	joint m homocinétique	Gleichlaufgelenk n, homokinetisches Gelenk n	giunto m omocinetico
331	consultant	conseiller m (technique)	technischer Berater m	consulente m
332	container	container m, réservoir m	Behälter m	contenitore m
333	contraction	contraction f	Kontraktion f, Zusammenziehung f	contrazione f
334	contraction(measure of contraction)	mesure f de retrait	Schwundmaß n	misura f della contrazione
335	control valve	soupape f de contrôle	Regulierventil n, Steuerventil n	valvola f di controllo
336	conveyor belt	courroie f transporteuse	Förderband n, Transportband n	nastro trasportatore
337	cooling air	refroidissement par air	Luftkühlung f	raffreddamento m (aria di)
338	copolymer	copolymère m	Copolymer(isat) n, Mischpolymer(isat) n	copolimero m
339	copolymerisation	copolymérisation f	Copolimerisation f	copolimerizzazione
340	cord	câble m pour pneumatiques, corde f	Kord (Festigkeitsträger)	corda f, filo m
341	core (cable core)	âme f du câble	Kabelseele f	anima f del cavo
342	cork	liège m	Kork m	sughero m
343	cork gasket	joint m d'étanchéité en liège	Korkdichtung f	guarnizione f di sughero
344	cork sheet (rubberised cork sheet)	feuille f de liège caoutchoutée	gummierte Korkplatte f	foglia f di sughero gommata
345	corona	corona (effet m)	Korona f	effetto m corona, scarica f corona
346	corona discharge	corona (décharge f)	Koronaentladung f	scarica f corona
347	corona effect	corona (effet m)	Koronaeffekt m	effetto m corona
348	corona (internal)	corona interne (effet m)	innere Koronawirkung f	effetto m corona interno
349	corona resistivity	corona (résistivité f)	Koronafestigkeit f	resistenza f a scarica corona
350	corrosion	corrosion f	Korrosion f	corrosione f
351	corrosion resisting	résistant à la corrosion	korrosionsbeständig	resistente alla corrosione
352	corrugated hose	tuyau m circonvoluté, cannelé	Faltenschlauch m	tubo m corrugato
353	cotton linter	bourre f de coton	Baumwoll-Linters pl	linter m di cotone
354	cover (belt cover)	revêtement m de courroie	Transportbanddecke f	rivestimento m esterno delle cinghie
355	cover (cover hose)	robe f extérieure de tuyau	Schlauchdecke f	rivestimento m di un tubo

C

Nᴿ	ESPAÑOL	PORTUGUÊS	SVENSKA	NEDERLANDS
311	compuesto de goma	composto de borracha	ofylld blandning	ongevuld rubbermengsel
312	composición f cruda	composição f crua; mistura f bruta	ovulkad blandning	ongevulcaniseerd mengsel
313	compresión f	compressão f	sammantryckning	compressie, samendrukking
314	curva f de compresión, línea f de compresión	curva f de compressão	kompressionskurva	druk-kromme
315	módulo m de compresión	módulo m de compressão	kompressionsmodul	compressie-modulus
316	módulo de compresión	modulo de compressão	tryckmodul	drukmodulus
317	fuerza de compresión	resistência f à compressão	tryckhållfasthet	druksterkte
318	manguera f para compresor	mangueira f para compressor	kompressorslang	hogedrukslang
319	controlado por ordenador	controlado por computador	dator styrd	computer gestuurd
320	producto m de condensación aldehido-anilina	produto m de condensação de aldeído-anilina	aldehyd-anilin kondensationsprodukter	aldehyde-aniline condensatie produkt
321	conducción f	condução f	överföring	geleiding
322	conductividad f, conductibilidad f	condutividade f, condutibilidade f	ledningsförmåga	geleidingsvermogen, geleidbaarheid
323	conductividad térmica	condutividade térmica	termisk kondutivitet	thermische geleidbaarheid
324	conductor m (cable conductor)	condutor m de cabo	kabelledare	geleider
325	tubo m, conducto m	conduta f	ledare	leiding
326	tubo m con forro interior de caucho	condutor revestido a borracha	gummibeklädd ledning	leiding (rubber)
327	configuración f	configuração f	konfiguration	structuur, configuratie
328	conexión f a tierra	ligação f a terra	jordledning	aardleiding
329	caja f de derivación	caixa f de derivação	förbindelselåda	aansluitdoos
330	junta homocinética	junta homocinética	konstant hastighets koppling	homokinetische koppeling
331	consultor m, asesor m	consultor m tecnico	konsult	adviseur
332	envase m, recipiente m, depósito m	recipiente m, vazilha f	behållare	container, vat
333	contracción f	contracção f	kontraktion	contractie, samentrekking
334	medida f de la contracción	medida f da contracção, medida f da retracção	krympmått	graad van samentrekking, mate van krimp
335	válvula f de control	válvula f de contrôle, válvula f reguladora, válvula f de comando	reglerventil	regelklep
336	cinta transportadora	correia transportadora	transport band	transportband
337	refrigeracion por aire	arrefecimento, ar	luftkylning	luchtkoeling
338	copolímero m	copolímero m	co-polymer, sampolymer	copolymeer
339	copolimerización	copolimerização	copolymerisation	copolymerisatie
340	cuerda f	corda m para pneumáticos	kord	koord
341	alma o ánima del cable	alma f de cabo, núcleo de cabo	kabeltråd	kabelkern
342	corcho m	cortiça f	kork	kurk
343	empaquetadura f de corcho	vedante m de cortiça	korkpackning	kurkpakking
344	lámina f de corcho engomada	fôlha f de cortiça revestida com borracha	gummerad korkplatta	met rubber bedekte kurkplaten
345	efecto m corona	corona f, coroa	korona	corona
346	descarga f del efecto	descarga f de corona	koronaurladdning	corona ontlading
347	efecto m corona	efeito m de corona	koronaeffekt	corona-effect
348	efecto m corona interno	efeito m de corona interna	inre koronaeffekt	inwendige ontlading
349	resistencia al efecto corona	resistividade de corona	korona resistens	corona weerstand
350	corrosión f	corrosão f	korrosion	corrosie
351	resistente a la corrosión	resistente à corrosão	korrosionsbeständig	bestand tegen corrosie
352	manguera f ondulada	mangueira f corrugada; (Port.) mangueira f ondulada	räfflad slang	gespiraliseerde slang
353	linters m, borra de algodón	linter m de algodão	bomullsremsor	katoenpluis
354	recubrimiento m para correas	capa f para correias	remtäckplatta	deklaag van transportband
355	cubierta f de manguera	revestimento m para mangueiras	ytterslang	slangbekleding

covering (cable covering machine)

C

Nʀ	ENGLISH	FRANÇAIS	DEUTSCH	ITALIANO
356	covering (cable covering machine)	machine f à revêtir les câbles	Kabelummantelungsmaschine f	macchina f per rivestire i cavi
357	covering (floor covering)	revêtement m (de sol)	Bodenbelag m	rivestimento m di pavimenti
358	covering (roller covering)	revêtement m (de cylindre), garniture f (de cylindre)	Walzenüberzug m	rivestimento m di cilindro
359	covering (wheel covering)	garnissage m (de roue)	Radüberzug m, Radverkleidung f	rivestimento m per ruote o copriruota
360	CPE - chlorinated polyethylene	CPE - polyéthylène chloré	CPE - chloriertes Polyethylen n	CPE - polietilene clorurato
361	cracking	craquelage m	Rißbildung f	rottura f
362	cracking (flex-cracking)	craquelures fpl par flexion	Biegerisse mpl	screpolature f da flessione
363	crawl	fluage m	Runzelbildung f (Farbe), Kriechen n (Kunststoff)	strisciamento m
364	creamery hose	tuyau m pour crèmeries	Molkereischlauch m	tubo m per prodotti caseari
365	creaming latex	crémage m du latex	Aufrahmen n von Latex	crematura f (del lattice)
366	creep	fluage m, écoulement m	Kriechen n, Fließen n	scorrimento m
367	creep resistance	résistance f au fluage	Kriechbeständigkeit f, Dauerstandfestigkeit f, Druckfestigkeit f	resistenza f allo scorrimento
368	crepe rubber	crêpe m	Crêpe m	gomma f crepe
369	cross machine direction	direction f perpendiculaire à l'appareil	quer zur Verarbeitungsrichtung f	direzione perpendicolare alla macchina
370	cross-linking	réticulation f	Vernetzung f	reticolazione f, cross linking m
371	cross-links	rétifications fpl	Vernetzungsstellen f	reticoli f, legami incrociati
372	crystallinity	degré m de cristallinité	Kristallinität f	cristallinità f
373	crystallisation	cristallisation f	Kristallisation f	cristallizzazione f
374	cubic foot	pied m cubique	Kubikfuß m	piede m cubo
375	cubic inch	pouce f cubique	Kubikzoll m	pollice m cubo
376	curable	vulcanisable	vulkanisierbar	vulcanizzabile
377	curative	curative m, agent de cuisson, agent de vulcanisation	Vernetzungsmittel n	vulcanizzante
378	curative masterbatch	mélange-maître m curatif	Vernetzungsmittel-Masterbatch m	mescola madre a base di agenti vulcanizzanti
379	cure	vulcanisation f, cuisson f	Vulkanisation f, Vernetzung f	vulcanizzazione f
380	cure (uncured)	vulcanisé (non-)	unvulkanisiert	non-vulcanizzato
381	curing	vulcanisation f, cuisson f	Vulkanisieren n, Vernetzen n	vulcanizzazione f
382	cycle	cycle m	Zyklus m	ciclo m
383	dairy hose	tuyau m pour laiterie	Molkereischlauch m	tubo m per latterie
384	DAM, dry ambient	à sec	DAM trocken	a secco
385	damper	amortisseur m	Stoßdämpfer m	ammortizzatore
386	damping	amortissement m	Dämpfung f	ammortizzamento m
387	decrease	réduction, décroissance, diminution	Abnahme f	diminuzione f
388	deflashing	ébarbage m	Entgraten n	sbavatura
389	deformation	déformation f	Deformation f, Verformung f	deformazione f
390	deformation under load ... °C, ... MPa, ... h	déformation f sous charge ... °C, ... MPa, ... h	Deformation f unter Last ... °C, ... MPa, ... h	deformazione f sotto carico ... °C, ... MPa, ... h
391	degasser	dégazeur m	Entgaser m	degasatore m
392	degradation	dégradation f, décomposition f	Abbau m	degradazione f
393	degreasant	dégraissant m	Entfettungsmittel n	sgrassante
394	degree	degré m	Grad m	grado m
395	demoulding	démoulage m	entformen	distacco dallo stampo
396	deodorisation	désodorisation f	Desodorisierung f	deodorizzazione f
397	depolymerisation	dépolymérisation f	Depolymerisierung f	depolimerizzazione f
398	deterioration	détérioration f	Zersetzung f, Zerstörung f	deterioramento m
399	determination of hardness by indentation	détermination de la dureté f par pénétration	Eindruckhärteprüfung f	determinazione f della durezza per indentazione
400	diaphragm	diaphragme m	Membran f	diaframma m

C

Nr	ESPAÑOL	PORTUGUÊS	SVENSKA	NEDERLANDS
356	máquina recubridora de cables	máquina f para revestir cabos	kabelbeklädnadsmaskin	extruderen van kabelmantels
357	revestimiento m para pisos	revestimento m para soalhos	golvbeläggning	vloerbedekking
358	capa f de cilindro, revestimiento m de cilindro	revestmento m de cilindros	valsbeklädnad	walsbekleding
359	recubrimiento m de ruedas	revestimento m para rodas	hjulbeklädnad	wielbekleding
360	CPE - poliétileno clorado	CPE - polietileno clorado	CPE - klorerad polyeten	CPE - gechloreerd PE
361	agrietado m	fendilhamento m	sprickbildning	barstvorming
362	agrietado m por flexión	fendilhamento m por flexâo; (Port.) gretamento m por flexâo	böjsprickbildning	barstvorming door herhaald buigen
363	arrastre m	enrugamento m	krypa	kruipen
364	manguera f para lecherías	mangueira f	mejerislang	slang voor melkproducten
365	desnatado, descremado (latex)	latex cremoso	skumning (latex)	opromen van latex
366	deslizamiento	relaxamento m à traccçâo	krypning	kruip, koude vloei
367	resistencia al deslizamiento	resistência f ao escorrimento	krypresistens	kruipweerstand
368	crep, o crepe	borracha f crepe	kräppgummi	crêpe-rubber
369	en sentido perpendicular a la maquina	sentido perpendicular à máquina	tvärs maskin riktning	dwars op de machine richting
370	reticulación	recticulação	tvärbindning	verknoping, vernetting
371	reticulaciones	ligaçóes de reticulação	tvärbindningar	knopen
372	cristalinidad	cristalinidade	kristallinitet	kristalliniteit
373	cristalización f	cristalização f	kristallisation	kristallisatie
374	pie m cúbico	pé m cúbico	kubikfot	kubieke voet
375	pulgada f cúbica	polegada f cúbica	kubiktum	kubieke inch
376	vulcanizable	vulcanizável	vulkbar	vulkaniseerbaar
377	agente de vulcanización	agente de vulcanizarão	vulkmedel	vulkanisatie middel
378	mezcla madre vulcanizante	masterbatch de agentes de vulcanização	vulkmedel masterbatch	vulkanisatic batch
379	vulcanización f	vulcanização f	vulkning	vulcanisatie
380	no vulcanizado, crudo	não vulcanizado	ovulkad	ongevulcaniseerd
381	vulcanización f	vulcanização f	vulkning	vulcaniseren
382	ciclo m	ciclo m	cykel	cyclus
383	tubo m para lecherías	mangueira f para leiteria	mejerislang	melkslang
384	en seco	DAM, ambiente seco	torrt	DAM, droog
385	amortiguador m	mangueira f para leitaria	dämpare	bumper
386	amortiguación	amortecimento m	dämpande	demping
387	disminución f	diminuição f, redução f, decréscimo m	minska	afname
388	desbarbado	rebarbar	avskäggning	ontbramen
389	deformación f	deformação f	deformation	vervorming, deformatie
390	deformación f bajo carga	deformação f sob carga	deformation onder tryck ..°C, ...MPa, ...tim	vervorming onder druk ..°C, ...MPa, ...h
391	desgasificador m	desgaseificador m	avgasningsanordning	ontluchtingsinrichting
392	degradación f	degradação f	nedbrytning	afbraak, ontleding
393	desengrasante m	desengordurante m	avfettningsmedel	ontvettingsmiddel
394	grado m	grau m	grad	graad
395	desmoldear	desmoldar	urtagning ur form	lossen van de matrijs
396	desodorización f	desodorização f	luktborttagning	desodorisatie
397	despolimerización f	despolimerização	depolymerisering	depolymerisatie, ontleding
398	deterioración f, deterioro m	deterioração f	förstöring	aantasting, ontleding
399	determinación f de dureza por penetración	determinação f da dureza por entalhe	hårdhetsbestämning genom intrycksmätning	hardheidsbepaling door indrukking
400	diafragma m	diagragma m	membran	membraan

D

Nr	ENGLISH	FRANÇAIS	DEUTSCH	ITALIANO
401	die diameter	diamètre m de la filière	Düsendurchmesser m	diametro m dell'ugello
402	die length	longueur f de la filière	Düsenlänge f	lunghezza f dell'ugello
403	die (punching die)	emporte-pièce m	Stanzmesser n	stampo m per punzonare
404	die swell ratio	gonflement m à la sortie de la filière	Mundstückquellung f, Spritzquellung f	rapporto m di rigonfiamento nell'estrusione
405	die-cut specialties	articles mpl découpés	Stanzartikel mpl	articoli m tranciati
406	dieing-out press	presse f à estamper	Stanzpresse f	trancia f
407	dielectric	diélectrique m	Dielektrikum n	dielettrico
408	dielectric breakdown voltage (ISO = disruptive voltage)	tension disruptive	Durchschlagspannung f	tensione f di scarica dielettrica (ISO = tensione disruttiva)
409	dielectric constant	constante f diélectrique	Dielektrizitätskonstante f, Dielektrizitätszahl f ... Hz	costante f dielettrica
410	dielectric fatigue	fatigue f diélectrique	Ermüdung f (dielektrisch)	fatica f dielettrica
411	dielectric loss	perte f diélectrique	Verlust m (dielektrisch)	perdita f dielettrica
412	dielectric strength (short term: ... mm, ISO = electric strength)	rigidité f diélectrique (... mm, ISO = rigidité électrique)	Durchschlagfestigkeit f, ... mm	rigidità f dielettrica
413	dip coat (to)	enduire par trempé	tauchlackieren, tauchbeschichten	rivestire per immersione
414	direct current	courant m continu	Gleichstrom m	corrente f continua
415	disc diaphragm	diaphragme m à disque	Scheibenmembran f	diaframma m a disco
416	discharge hose	tuyau m de vidange	Entladungsschlauch m	tubo m di scarico
417	discolouration (GB) discoloration(US)	décoloration f	Verfärbung f	scoloramento m
418	dispersability	aptitude f à la dispersion	Dispergierbarkeit f	disperdibilità
419	dispersion	dispersion f	Dispersion f	dispersione f
420	dissipation factor (tan ε) (ISO = dielectric dissipation factor)	facteur de pertes diélectriques (tan ε)	dielektrischer Verlustfaktor m (tan ε)	fattore m di perdita (tan ε) (ISO = fattore m di perdita dielettrica)
421	distortion	distorsion f	Verdrehung f, Verformung f	distorsione f
422	domestic rubber goods	articles mpl de ménage en caoutchouc	Haushaltsgummiwaren fpl	articoli m in gomma per la casa
423	double - threaded	à double - filet	zweigängig	a passo doppio (vite)
424	dough	pâte f	Paste f	impasto m
425	dowel	colonne f de guidage	Paßstift m	perno m di centratura, di guida
426	drawing	étirage m, dessin m	Zeichnung f	disegno m
427	drying oil	huile f siccative	Öl n (trocknend)	olio m essicativo
428	duck	toile f de chanvre, coutil m, canevas m	Segeltuch n, Zeltleinwand f	tela olona f
429	duct(air duct)	tuyauterie f d'air	Luftleitung f	condotto m per aria
430	durability	durabilité f	Haltbarkeit f	durata f
431	durometer	duromètre m	Härteprüfer m, Durometer n	durometro m
432	Dutch linen, Holland cloth	toile f de Hollande	ungebleichtes Leinen n	tela olandese f
433	dynamic loading	charge f dynamique	dynamische Belastung f	carico m dinamico
434	dynamometer	dynamomètre m	Dynamometer m	dinamometro m
435	efficiency	efficacité f	Wirkungsgrad m	efficienza f
436	ejector	éjecteur m	Ausstoßkolben m, Auswerfer m	eiettore
437	ejector pin	éjecteur m	Ausdrückstift m	perno m espulsore
438	elastic (adj)	élastique	elastisch	elastico (agg.)
439	elastic fabric	tissu m élastique	Gewebe n (elastisch)	tessuto m elastico
440	elastic limit	limite f d'élasticité	Elastizitätsgrenze f	limite m elastico
441	elastic memory	mémoire f élastique	Elastizität f gespeichert	effetto m elastico residuo (memoria elastica)
442	elasticity	élasticité f	Elastizität f	elasticità f
443	elastomer	élastomère m	Elastomer n	elastomero m
444	Elcometer thickness gauge	jauge f d'épaisseur Elcometer	Elcometer-Dickemesser m	spessimetro m Elcometer
445	electrical measuring instrument	instrument m de mesure électrique	elektrisches Meßinstrument n	strumento m per misure elettriche

D

Nr	ESPAÑOL	PORTUGUÊS	SVENSKA	NEDERLANDS
401	diámetro de boquilla	diâmetro da fieira	munstycks diameter	spuitkop diameter
402	longitud de la boquilla o hilera	comprimento da fieira	matris längd	spuitkop lengte
403	troquel m, matriz f	matriz f, fieira m	stansdynor	stansmes
404	proporción de hinchamiento en la extrusión	taxa de inchamento na fieira	matris svällningsförhållande	zwellingspercentage t.o.v. mondstuk
405	artículos mpl troquelados	artigos mpl estampados	stansspecialiteter	gestansde artikelen
406	prensa f para troquelar	prensa f de estampar	stanspress	stansmachine
407	dieléctrico m	dieléctrico m	dielektrikum	diëlektricum
408	tensión de descarga dielétrica (ISO = tensión disruptiva)	voltagem de roptura dielétrica	dielektrisk genomslagspännig (ISO = genomslag spänning)	diëlectrische doorslag spanning (ISO = doorslag spanning)
409	constante f dieléctrica	constante f dieléctrica	dielektricitetskonstant	diëlektrische constante
410	fatiga f dieléctrica	fadiga f dieléctrica	dielektrisk utmattning	vermindering dielectrische weerstand bij langdurige belasting
411	perdida f dieléctrica	perda f dieléctrica	dielektrisk förlust	diëlektrisch verlies
412	rigidez f dieléctrica	rigidez f dieléctrica	genomslagshållfasthet(f.. ISO; elektrisk)	doorslagsterkte (kortweg: ... mm. ISO)
413	recubrir por inmersión	revestir por imersão	belägga genom doppning	bekleden door dompelen
414	corriente f continua	corrente f contínua	likström	gelijkstroom
415	diafragma m de disco	diafragma m de disco	skivmembran	schijfmembraan
416	manguera f de descarga	mangueira f de descarga	lossningsslang	slang voor lossing
417	decoloración f	descoloração f	missfärgning	verkleuring
418	dispersabilidad	dispersabilidade	dispergerbarhet	dispergeerbaarheid
419	dispersión f	dispersão f	dispersion	dispersie
420	factor de disipacion dielectrica	factor de dissipação	dielektrisk förlust faktor (ISO = dielectrisk ferlust faktor)	spannings verlies factor (tan δ) (ISO = di-elektrische verlies factor)
421	distorsión f	distorção f	distortion	vervorming
422	artículos mpl domesticos de caucho	artigos mpl caseiros de borracha	hushållsartiklar av gummi	rubberartikelen voor huishoudelijk gebruik
423	doble paso	duplo passo	dubbel gängad	dubbelgangige schroef
424	masa f, pasta f	massa f, composição f crua	deg	deeg
425	clavija	cavilha	tapp	paspen voor pen-gat verbinding
426	estirado	trefilação f, estiramento f à fieira	ritning	tekening
427	aceite m secante	óleo m secativo	torkande olja	drogende olie
428	lona f	brim m, brim f de algodão, brim m de linho	duk	canvas
429	conducto m de aire	conduta f de ar, aeroduto m	luftledning	luchtleiding, luchtkanaal
430	duración (o vida en servicio)	durabilidade f	hållbarhet	houdbaarheid
431	durómetro m	durômetro m	hårdhetsmätare	hardheidsmeter, durometer
432	tela f de Holanda	pano m da Holanda	hollandsväv	ongebleekt linnen
433	carga f dinámica	carga f dinâmica	dynamisk belastning	dynamisch belasting
434	dinamómetro	dinamómetro	dynamometer	dynamometer
435	eficacia f	eficiência f, eficácia f	effektivitet	doelmatigheid
436	eyector	ejector	utstötare	uitstoter
437	expulsor	pino de ejecção	utstötar-pinne	uitstoot pin
438	elástico	elástico	elastisk	elastisch
439	tejido m elástico	tecido m elástico	elastisk väv	elastisch weefsel
440	limite m de elasticidad	limite m elástico	elasticitetsgräns	elasticiteitsgrens
441	memoria elastica	memória elástica	elastisk efterverkan	elastisch geheugen
442	elasticidad f	elasticidade f	elasticitet	elasticiteit
443	elastómero m	elastômero m	elast	elastomeer
444	medidor m de espesor Elcometer	medidor m de espessuras Elcometer	Elcometer tjockleksmätare	dikte meter, Elcometer
445	instrumento eléctrico de medida	instrumento m eléctrico de medida	elektriskt mätinstrument	elektrische meetapparatuur

E

Nʀ	ENGLISH	FRANÇAIS	DEUTSCH	ITALIANO
446	electrically	électriquement	elektrisch	elettricamente
447	electrolyte	électrolyte m	Elektrolyt m	elettrolito m
448	electron beam curing EB	vulcanisation f sous électrons	Elektronenstrahlvernetzung f	vulcanizzazione f con fasci di elettroni
449	elongation	allongement m	Dehnung f	allungamento m
450	elongation at yield	limite f d'allongement	Dehnung f bei Streckspannung	allungamento m a snervamento
451	elongation curve	courbe f d'allongement	Dehnungskurve f	curva f di allungamento
452	elongation strain	effort d'allongement m	Zugdehnung f	deformazione f per allungamento
453	elongation(ultimate elongation)	allongement m à la rupture	Bruchdehnung f	allungamento m a rottura
454	ELVAX EVA resins	ELVAX (résines d'EVA)	ELVAX Ethylen-Vinyl-Acetat Copolymer n	ELVAX EVA
455	emulsifier	émulsifiant m	Emulgator m	emulsionante
456	emulsion	émulsion f	Emulsion f	emulsione f
457	enamel	émail m	Emaille f	smalto m
458	encapsulation	encapsulation f	Einbettung f, Kapselung f	incapsulamento
459	energy loss	perte f d'énergie	Energieverlust m	perdita f di energia
460	engine mounting	support m moteur	Motorlagerung f	supporto m per motori
461	EPC carbon black (easy processing channel),	noir EPC m	EPC-Ruß m, Channell-Ruß für leichte Verarbeitbarkeit	nerofumo m tipo channel a facile dispersione
462	epoxy resin	résine f époxy	Epoxidharz n	resina f epossidica
463	ester gum	gomme ester f	Abietinsäureester m	gomma f a base di esteri
464	evolution of gases	dégagement m (de gaz)	Entwicklung f (von Gasen)	sviluppo (di gas) m
465	expansion	expansion f, dilatation f	Ausdehnung f	espansione f
466	expansion joint	joint m de dilatation	Dehnungsfuge f, Kompensator m	giunto m di dilatazione
467	experiment	essai m	Experiment n, Versuch m	esperimento m
468	extender	agent m diluant, d'extension	Füllstoff m, Streckmittel n	estensore m
469	extensibility	extensibilité f	Dehnbarkeit f, Streckbarkeit f	estensibilità f
470	extrapolation	extrapolation f	Extrapolation f	estrapolazione f
471	extrudability	aptitude f à l'extrusion	Extrudierbarkeit f, Spritzbarkeit f	estrudibilità f
472	extrude (to)	boudiner, extruder	extrudieren, spritzen	estrudere
473	extruded goods	articles mpl extrudés	Spritzartikel m, Extrusionsartikel m	articoli m estrusi
474	extruded hose	tuyau m extrudé	Schlauch m (extrudiert)	tubo m estruso
475	extruder	extrudeuse	Extruder m, Spritzmaschine f	estrusore m, trafila f
476	extruder die	filière d'extrudeuse	Düse f, Spritzmundstück n	ugello m dell'estrusore
477	extruding machine	extrudeuse f	Spritzmaschine f, Extruder m	estrusore m, trafila f
478	extrusion	boudinage m, extrusion f	Extrusion f	estrusione f
479	extrusion back pressure	contrepression f d'extrusion	Extrusionsstaudruck m	contropressione f di estrusione
480	exudation	exsudation f	Ausschwitzung f	essudamento m
481	fabric	tissu m	Gewebe n	tessuto m
482	failure	rupture f, échec	Versagen n, Fehler m	rottura f, guasto m
483	fastness to light	solidité f à la lumière, résistance f à la lumière	Lichtbeständigkeit f	stabilità f alla luce
484	fatigue	fatigue f	Ermüdung f	fatica f
485	fatigue failure	rupture f par fatigue	Materialversagen n durch Ermüdung	rottura f per fatica
486	filler	charge f	Füllstoff m	carica f
487	filler loading	charge f	Füllstoffdosierung f	riempimento m con cariche
488	film	film m	Film m, Folie f	film m, pellicola f
489	film build - up	formation f du film	Schichtaufbau m	formazione f di pellicole
490	finish	finissage m	Oberflächenbehandlung f, Oberflächenbeschaffenheit f	finitura f

Nʀ	ESPAÑOL	PORTUGUÊS	SVENSKA	NEDERLANDS
446	elèctricamente	elèctricamente f	elektriskt	elektrisch
447	electrolito m	electrólito	elektrolyt	elektrolyt
448	vulcanización por bombardeo electrónico	vulcanização por feixe electrónico	elektron strålevulkning	EB-vulcanisatie, vulcanisatie d.m.v. electronenstralen
449	alargamiento m	alongamento m	töjning	rek
450	alargamiento al límite elástico	alongamento	töjning vid sträckspänningen	maximale rek
451	curva f de alargamiento	curva f de alongamento	förlängningskurva	rekkromme
452	deformación f por tracción	deformação f ao alongamento	krafttöjning	rekspanning
453	alargamiento m a la rotura	alongamento m á ruptura, alongamento m máximo	brottöjning	rek bij breuk
454	ELVAX resinas de EVA	ELVAX resinas EVA	ELVAX EVA plast	ELVAX EVA harsen
455	emulsionante m	emulsificante m	emulgator	emulgeermiddel, emulgator
456	emulsión f	emulsão f	emulsion	emulsie
457	esmalte m	esmalte m	emalj	emaille
458	encapsulación	encapsulamento	inkapsling	inkapseling
459	pérdida f de energía	perda f de energia	energiförlust	energie verlies
460	soporte de motor	apoio m de motor	motorupphängning	motorophanging
461	negro carbón ·channel· de fácil elaboración, negro EPC	negro m de fumo tipo channel de dispersão fácil EPC	EPC kimrök	EPC roet
462	resina f epoxi	resina f epóxida	epoxyharts	epoxy hars
463	goma-éster	goma f de éster	hartsester	ester gum (glycerol ester van hars)
464	liberación, desprendimiento (de gases)	evolução f (de gases)	utveckling (av gas)	gasontwikkeling
465	expansión f, dilatación f	expansão f, dilatação f	expansion	uitzetting, expansie
466	junta de dilatación	junta de dilatação	expansionsbälg	expansiestuk
467	experimento m	experiência f	experiment	experiment
468	diluyente, extendedor	diluente m	fyllmedel	niet versterkende mengsel toevoegingen
469	extensibilidad f	extensibilidade f	utdrygbarhet	rekbaarheid
470	extrapolación f	extrapolação f	extrapolering	extrapolatie
471	aptitud para la extrusión, extrusionabilidad	extrusibilidade f	sprutbarhet	extrudeerbaarheid
472	extruir, extrusionar	extrudir, trafilar	spruta, att	extruderen
473	artículos mpl por extrusión	artigos m extrudidos	extruderat gods	spuitartikelen
474	manguera f por extrusión	tubo m extrudido	extruderad slang	ge-extrudeerde slang
475	máquina f de extrusión, extrusionadora	extrusora f, trefila f, trafila f	extruder	spuitmachine, extruder
476	boquilla f (de budinadora), extrusora	fieira f	munstycke	spuitmondstuk
477	máquina f de extrusión, extrusionadora	máquina f de extrusão, trefila f, trafila f	extruder	spuitmachine
478	extrusión f	extrusão f	extrudat	spuiten, extruderen
479	contrapresión de extrusión	contra pressão de extrusão	extrusions mot-tryck	extrusie na-druk
480	exudación f	exsudação f	utsvettning	uitzweting
481	tejido m, tela f	tecido m, pano m	väv	weefsel
482	falla f, fallo m, rotura f	falha f, ruptura f	brott	breuk, fout
483	resistencia a la luz, solidez a la luz	solidez f sob acção da luz	ljusäkthet	lichtechtheid
484	fatiga f	fadiga f	utmattning	vermoeiing
485	fallo por fatiga	falha f pela fadiga, ruptura f pela fadiga	utmattningsbrott	breuk door vermoeiing
486	carga f, ingrediente de relleno	carga f, carga f de enchimento	fyllnadsmedel	vulstof
487	carga de relleno	carga f de enchimento	fyllmedelmängd	vulstofdosering, hoeveelheid vulstof
488	película	película - filme	film	folie, dunne laag
489	película f formada, depositada película	formação f da película, espessamento m da película	filmuppbyggnad	laagopbouw
490	acabado m	acabamento m	ytfinish	afwerking

E

flame - proof (adj)

F

Nʀ	ENGLISH	FRANÇAIS	DEUTSCH	ITALIANO
491	flame - proof (adj)	ininflammable	flammbeständig	ininfiammabile (agg.)
492	flame rating	indice d'inflammabilité f	Brennbarkeitsklasse	indice m di infiammabilità
493	flammability	inflammabilité f	Entflammbarkeit f	infiammabilità f
494	flash point	point m éclair	Flammpunkt m	punto di infiammabilità
495	flex	flexion f	Biegung f	flessione f
496	flex life	résistance f aux flexions répétées	Beständigkeit f gegen dynamische Beanspruchung	resistenza f a flessioni ripetute
497	flex modulus	module de flexion	Biegemodul m	modulo a flessione
498	flex-fatigue	fatigue f à la flexion	Biegeermüdung f	fatica f a flessione
499	flexing	flexions fpl répétées	Biegung f, Biegen n	flessioni f successive
500	flexing machine	machine f de flexion	Biegeprüfer m, Biegeprüfgerät n	macchina f per prove di flessione
501	flexion	flexion f	Biegung f	flessione f
502	flexometer	flexomètre m	Biegeprüfer m	flessometro m
503	flexural modulus	module d'élasticité en flexion	Biege-E-Modul m	modulo m a flessione
504	flexural strength	résistance à la flexion	Biegefestigkeit f	resistenza f a flessione
505	flexure	courbure f	Biegung f	flessione f
506	flexure test	essai m de courbure	Biegeprüfung f	prova f di flessibilità
507	flight width	pas de vis m	Stegbreite f	passo della vite
508	flocculation, flocculation	floculation f	Flockung f, Ausflockung f	flocculazione f
509	flooring	revêtement m de sol	Bodenbelag m	rivestimento m per pavimenti
510	flow index	indice de fluidité m	Fließindex f	indice di scorrimento
511	fluorine	fluor m	Fluor n	fluoro
512	foot	pied m	Fuß m	piede m
513	footwear	chaussures fpl	Fußbekleidung f	calzatura f, scarpa f
514	frequency	fréquence f	Frequenz f, Häufigkeit f	frequenza f
515	friable (adj)	friable	leicht zerbröckelnd, leicht zu zerreiben	friabile (agg.)
516	friction	friction f, frottement m	Reibung f	frizione
517	friction fabric	tissu m pour friction	Friktionsgewebe n	tessuto m di frizionatura
518	frictioning	frictionnage m, friction f	Friktionierung f	frizionatura f
519	frothing	écumage m	Schäumen n	schiuma f
520	fuel hose	tuyau m à essence	Treibstoffschlauch m, Kraftstoffschlauch m	tubo m per carburante
521	furnace black FEF (fine extrusion)	noir FEF (extrusion fine)	FEF-Ruß, feiner Extrusions-Furnace-Ruß	nero fumo m fornace FEF (per estrusione fine)
522	furnace black FF (fine)	Noir FF m	Furnaceruß m (fein), FF-Ruß	nero fumo m fine tipo fornace. carbon black FF m
523	furnace black FT (fine thermal)	Noir FT	Thermalruß m (fein), FT-Ruß	nero fumo fine tipo thermal m, carbon black FT m
524	furnace black HAF (high abrasion)	noir HAF (très bonne abrasion)	HAF-Ruß m, Furnace-Ruß für hohe Abriebfestigkeit	nero fumo m fornace HAF (ad alta abrasione)
525	furnace black HEF (high elongation)	Noir HEF m (allongement élevé)	HEF-Ruß m, Furnace-Ruß für hohe Dehnung	nero fumo m tipo fornace ad alto allungamento HEF m
526	furnace black HMF (high modulus)	Noir HMF m (haut module)	HMF-Ruß m, Furnace-Ruß für hohen Modul	nero fumo m tipo fornace ad alto modulo, carbon black HMF
527	fusion point	point m de fusion	Schmelzpunkt m	punto m di fusione
528	gallon	gallon m	Gallone f	gallone m
529	gas hose	tuyau m à gaz	Gasschlauch m	tubo m per gas
530	gas permeability	perméabilité f aux gaz	Gasdurchlässigkeit f	permeabilità f ai gas
531	gas transmission	transmission f des gaz	Gasdurchlässigkeit f	trasmissione f dei gas
532	gasket	garniture f d'étanchéité	Dichtung f	guarnizione f
533	gasoline hose	tuyau m à essence	Benzinschlauch m	tubo m per benzina
534	gastight (adj)	étanche aux gaz	gasdicht	impermeabile ai gas
535	gate (in injection moulding)	point m d'injection (en moulage par injection)	Anschnittkanal m	punto m di iniezione (nello stampaggio a iniezione)

Nr	ESPAÑOL	PORTUGUÊS	SVENSKA	NEDERLANDS
491	ininflamable	ignífugo m	brandhärdig	vlambestand
492	indice de inflamación	indice de flamabilidade	brand klassning	brandbaarheidsklasse
493	inflamabilidad f	flamabilidade f	brännbarhet	ontvlambaarheid
494	punto de ignicion	ponto de ignição	flampunkt	vlampunt
495	flexión f	flexão f	böjning	buiging
496	vida a la flexión, duración en flexión	duração f útil sob flexão repetida	flex utmattningslivslängd	levensduur bij buiging
497	modulo de flexión	modulo de flexão	flex modul	flex modulus
498	fatiga f a la flexión	fadiga f à flexão	böjutmattning	buigvermoeidheid
499	flexionado	flexão f sucessiva	böjning	buigen
500	máquina f para flexionar	máquina f de flexão	böjningsmaskin	buigbeproevingsmachine
501	flexión f	flexão f, dobramento m	böjning	buiging
502	flexómetro m	flexômetro m	flexometer	flexometer
503	modulo de flexion	módulo de flexão	böjningsmodul	buig modulus
504	resistencia f a la flexión	resistência f à flexão	böjhållfasthet	dynamische buigweerstand
505	flexión f	flexão f, dobramento m	böjning	buiging
506	ensayo m de flexión	ensaio m de flexão	böjprovning	buigproef
507	paso de rosca	largura	gäng bredd	breedte schroefgang
508	floculación f	floculação f, coagulação f	flockulering	uitvlokken
509	pavimento m	material m para soalhos - material de pavimentação	golvbeläggning	vloerbedekking
510	indice de fluidez	índice de fluidez	flytindex	vloeiindex
511	fluor	fluor	fluor	fluor
512	pie m	pé m	fot	voet
513	calzado m	calçado m	skodon	schoeisel
514	frecuencia f	frequência f	frekvens	frequentie
515	quebradizo	friável, quebradiço	spröd	bros, verbrokkeld
516	fricción f	fricção f, atrito m	friktion	wrijving, frictie
517	tejido para friccionar	tecido m para tratamento pro fricção	friktionsväv	gefrictioneerd weefsel
518	friccionado m	friccionamento m	friktionering	frictioneren
519	espuma f	espuma f	skum, skummande	schuim
520	tubo m para combustible	tubo m para combustível	bränsleslang	brandstofslang
521	negro de humo de horno FEF (extrusión fina)	negro de fumo FEF (extrusão fina)	FEF kimrök	FEF roet
522	negro carbón ‹furnace› fino, negro FF	negro m de fumo fino de fornalha, negro m de fumo FF	FF kimrök	FF-roet
523	negro carbón térmico fino, negro FT	negro m de fumo fino de decomposição térmica FT	FT kimrök	FT-roet
524	negro de humo ‹furnace› HAF	negro de fumo furnace HAF (alta abrasão)	HAF kimrök	HAF roet
525	negro carbón ‹furnace› de alto alargamiento, negro HEF	negro m de fumo de fornalha de alongamento elevado HEF	HEF kimrök	HEF-roet
526	negro carbón ‹furnace› de alto módulo, negro HMF	negro m de fumo de fornalha de módulo elevado HMF	HMF kimrök	HMF-roet
527	punto m de fusión	ponto m de fusão	smältpunkt	smeltpunt
528	galón m	galão m	gallon	gallon
529	tubo m para gas	mangueira f para gás, tubo m para gás	gasslang	gasslang
530	permeabilidad f a los gases	permeabilidade f aos gases	gaspermeabilitet	gasdoorlaatbaarheid
531	transmisión de gas	transmissão de gás	gas permeabilitet	gasdoorlaatbaarheid
532	empaquetadura f	vedante m	packning	pakking
533	tubo m para gasolina	mangueira f para gasolina	bensinslang	benzineslang
534	impermeable a los gases	impermeável aos gases	gastät	gasdicht
535	punto de inyección	ponto de injecção	ingöt	injectiepunt

F

The Language of Rubber

gauge pressure

G

Nr	ENGLISH	FRANÇAIS	DEUTSCH	ITALIANO
536	gauge pressure	pression f manométrique	manometrischer Druck m	pressione f manometrica
537	general purpose furnace black GPF	noir GPF (usage général)	Ruß m (GPF), Allzweck-Furnace-Ruß	nero fumo m fornace GPF (per uso generale)
538	grain	grain m, granulés	Korn n	granulo m
539	gram	gramme m	Gramm n	grammo m
540	graph	graphique m	Darstellung f (graphisch)	grafico m
541	gravity	gravité f	Schwere f	gravità f
542	grease	graisse f	Fett n	grasso m
543	grip	adhérence f au sol, tenue f de route	Griff m, Griffigkeit f	aderenza f al suolo
544	grommet	rondelle f, grommet m	Dichtungsring m	rondella f
545	ground n (US), earth n (GB)	terre f	Erde f, Masse f	terra f, massa f
546	ground (to)	mettre à la terre	erden	mettere a terra
547	guillotine	guillotine f	Querschneider m	ghigliottina f
548	guillotine cutting machine	guillotine f, machine f à trancher à guillotine	Hackmaschine f	trancia f a ghigliottina
549	halogen	halogène m	Halogen n	alogeno
550	hand truck tire (tyre)	bandage m pour chariot	Handkarrenbereifung f	pneumatici m per carrelli
551	hardness	dureté f	Härte f	durezza f
552	hardness Durometer A	dureté Shore A	Härte f Durometer A	durezza Durometro A
553	hardness IRHD	dureté IRHD f	Härte f IRHD	durezza IRHD
554	hardness Rockwell	dureté f Rockwell	Härte f (Rockwell)	durezza f Rockwell
555	hardness Shore	dureté f Shore	Härte f (Shore)	durezza f Shore
556	hardness tester	duromètre m	Härteprüfer m	durometro m
557	haze	trouble m	Trübung f	velo m
558	heat ageing	vieillissement m thermique	Hitzealterung f	invecchiamento m a caldo
559	heat ageing inhibitor	inhibiteur m de vieillissement thermique	Hitzealterungsschutzmittel n	inibitore di invecchiamento a caldo
560	heat and oil resisting	résistant à la chaleur et aux huiles	hitze- und ölbeständig	resistente al calore ed agli olii
561	heat build-up	échauffement m interne	Wärmeentwicklung f (innere)	sviluppo m di calore interno
562	heat development	dégagement m de chaleur	Wärmeentwicklung f	sviluppo m di calore
563	heat effect	effet m thermique	Wärmewirkung f	effetto m termico
564	heat of combustion	chaleur f de combustion	Verbrennungswärme f	calore m di combustione
565	heat of fusion	chaleur f de fusion	Schmelzwärme f	calore m di fusione
566	heat resistance	résistance f à la chaleur	Hitzebeständigkeit f, Wärmebeständigkeit f	resistenza f al calore
567	heat resistant	résistant à la chaleur	hitzebeständig	resistente al calore
568	heat sensitive	thermosensible	hitzeempfindlich	termosensibile
569	heat stability	stabilité f à la chaleur	Hitzebeständigkeit f, Wärmebeständigkeit f	stabilità f al calore
570	heat transfer	transfert m de chaleur	Wärmeübertragung f	trasmissione f del calore
571	heat welding	soudure f à la chaleur	Hitzeverschweißen n	saldatura a caldo
572	heating hose	tuyau m de chauffage	Heizschlauch m	tubo m per riscaldamento
573	Hegman grind gauge	jauge f de broyage Hegman	Feinheitsprüfer m (Hegman)	misuratore m di granulometria Hegman
574	hexa fluoro propylene	hexafluoropropylène	Hexafluorpropylen n	esafluoro propilene
575	high frequency	haute fréquence f	Hochfrequenz f	alta frequenza f
576	hold up spot	point de retenue	Stauung f	punto m di ristagno
577	hold-up time	temps m de maintien	Verweilzeit f	tempo di permanenza
578	Holland cloth	toile f de Hollande	Leinen n (ungebleicht)	tela olandese f
579	homogeneisation	homogénéisation f	Homogenisierung f	omogeneizzazione f
580	hopper	trémie f	Einfülltrichter m	alimentatore, tramoggia

NR	ESPAÑOL	PORTUGUÊS	SVENSKA	NEDERLANDS
536	presión f manométrica	pressão f manométrica	mättryck	manometerdruk
537	negro de humo de horno GPF (uso general)	negro de fumo GPF (aplicações gerais)	GPF - kimrök	GPF roet
538	grano m	grão m	säd	korrel
539	gramo m	grama m	gram	gram
540	gráfico m	gráfico m	diagram	grafiek
541	gravedad f	gravidade f	tyngdkraft	zwaartekracht
542	grasa f	massa f consistente, massa lubrificante	fett	vet
543	agarre m, adherencia f	aderência f à estrada, adesão f ao solo	handtag, grepp	greep
544	arandela con solapas	anilha f para juntas, ilhós f	krans	sluitring
545	tierra f, masa f	terra f, massa f	jord	grond, aarde
546	conectar a tierra	ligar a terra	jorda	aarden
547	guillotina f	guilhotina f	balskärmaskin	guillotinemes
548	guillotina f, máquina para cortar	guilhotina f, máquina f para cortar	balskärmaskin	guillotine snijmachine
549	halogeno	halogénio	halogen	halogeen
550	llanta f para carros de mano	pneumático m para carro de mão	handkärrehjul	banden voor handwagens
551	dureza f	dureza f	hårdhet	hardheid
552	dureza durómetro A	dureza durómetro A	hårdhet Durometer A	hardheid Durometer A
553	dureza IRHD	dureza IRHD	hårdhet IRHD	hardheid IRHD
554	dureza Rockwell	dureza Rockwell	Rockwell hårdhet	Rockwell hardheid
555	dureza Shore	dureza Shore	shore hårdhet	Shore hardheid
556	medidor m de la dureza	durômetro m	hårdhetsprovare	hardheidsmeter
557	neblina	névoa	disig	wazig
558	envejecimiento m térmico	envelhecimento m térmico	värmeåldring	veroudering bij hoge temperatuur
559	inibidor m del envejecimiento por calor	inibidor m do envelhecimento térmico	antioxidant, värmeåldrings skyddsmedel	antioxydant tegen hoge temperatuur
560	resistente al calor y a los aceites	resistente ao calor e aos óleos	värme- och oljebeständig	hitte- en oliebestand
561	calor generado, calor interno	desenvolvimento m de calor interno	inre värmeutveckling	inwendige warmteontwikkeling
562	desarrollo m de calor	desenvolvimento m de calor	värmeutveckling	hitte ontwikkeling
563	efecto m térmico	efeito m térmico	värmeeffekt	warmte effect
564	calor de combustion	calor de combustão	förbränningsvärme	verbrandingswarmte
565	calor de fusion	calor de fusão	fusionsvärme	smeltwarmte
566	resistencia f al calor	resistência f térmica	värmebeständighet	warmtebestendigheid
567	resistente al calor	resistente ao calor	värmebeständig	bestand tegen warmte
568	termo-sensible	termo-sensível	värmekänslig	hittegevoelig
569	estabilidad f al calor	estabilidade f térmica	värmestabilitet	hittestabiliteit
570	transferencia f de calor	transmissão f de calor	värmeöverföring	warmteoverdracht
571	soldadura por calor	soldadura por calor	värme skarvning	lassen d.m.v. hitte
572	manguera m de calefacción	mangueira f de aquecimento	uppvärmningsslang	verwarmingsslang
573	calibrador m de finura Hegman	calibrador m de granulometria Hegman	Hegmans finhetsmätare	Hegman's fijnheidstester
574	hexafluoropropileno	hexafluoropropileno	hexa-fluor-propen	hexafluor propyleen
575	alta frecuencia f	alta frequência f	högfrekvens	hoge frequentie, hoog-frequent
576	punto de retención	ponto de retenção	död punkt	verstopping
577	tiempo de permanencia	tempo de retenção	stopp tid	verblijftijd
578	tela f de Holanda	pano m da Holanda	hollandsväv	holland linnen
579	homogeneización f	homogenização f	homogenisering	homogenisatie
580	tolva	tremonha	magasin	vultrechter

G

H

Nr	ENGLISH	FRANÇAIS	DEUTSCH	ITALIANO
581	hose lining	robe f intérieure de tuyau	Schlauchseele f	rivestimento m di un tubo
582	hot air cure	vulcanisation f à l'air chaud	Heißluftvulkanisation f	vulcanizzazione f in aria calda
583	hot feed extruder	extrudeuse f alimentée à chaud	Heißfütterextruder m	estrusore m (alimentato a caldo)
584	household gloves	gants mpl de ménage	Haushaltshandschuhe mpl	guanti m per uso casalingo
585	household rubber goods	articles mpl en caoutchouc pour le ménage	Haushaltsgummiwaren fpl	articoli m casalinghi in gomma
586	HPC carbon black (hard processing channel)	noir HPC m	HPC-Ruß m, Channell-Ruß hochverstärkend	nerofumo m tipo channel difficile da disperdere
587	hub cap	enjoliveur m	Radkappe f	copriruota m, coppa f
588	humidity	humidité f	Feuchtigkeit f	umidità
589	hydraulic fluid	fluide m hydraulique	Hydraulikflüssigkeit f	fluido m idraulico
590	hydraulic hose	tuyau m hydraulique	Hydraulikschlauch m	tubo m idraulico
591	hydrolysis	hydrolyse f	Hydrolyse f	idrolisi f
592	hygroscopic (adj)	hygroscopique	hygroskopisch	igroscopico m
593	HYPALON synthetic rubber	caoutchouc synthétique HYPALON	HYPALON Synthesekautschuk, chlorsulfoniertes Polyethylen	HYPALON gomma sintetica
594	hypothesis	hypothèse f	Hypothese f	ipotesi f
595	hysteresis	hystérésis f	Hysterese f	isteresi f
596	hysteresis (loop)	cycle m d'hystérésis	Hystereseschleife f	ciclo m di isteresi
597	hysteresis loss	perte f par hystérésis	Hystereseverlust m	perdita f per isteresi
598	HYTREL engineering thermoplastic elastomer	HYTREL élastomère thermoplastique technique	HYTREL elastischer Konstruktionswerkstoff	HYTREL elastomero termoplastico
599	impact strength	résistance f au choc	Schlagzähigkeit f	resistenza f all'urto
600	impact strength notched, Izod	résistance f au choc avec entaille, Izod	Kerbschlagzähigkeit f Izod (eingekerbt)	resistenza f all'urto con intaglio Izod
601	impedance	impédance f	Impedanz f	impedenza f
602	impeller	impulseur m	Gebläserad n, Pumpenflügelrad n	girante f, rotore
603	impermeability	imperméabilité f	Undurchlässigkeit f	impermeabilità f
604	impregnating with rubber	imprégnation f au caoutchouc	Imprägnieren n mit Kautschuk	impregnazione f con gomma
605	impregnation	imprégnation f	Imprägnierung f	impregnazione f
606	impression cylinder	cylindre m d'impression	Druckzylinder m, Druckwalze f	cilindro m di impressione
607	impression roller	rouleau m d'impression	Druckwalze f	cilindro m di impressione
608	inch	pouce m	Zoll m	pollice m
609	indentation hardness (Rockwell)	dureté f par pénétration (Rockwell)	Kugeldruckhärte f (Rockwell)	durezza f di penetrazione (Rockwell)
610	index (limiting oxygen index)	indice limite d'oxygène m	Sauerstoffindex m (LOI)	indice m limite di ossigeno
611	index number	indice m	Indexzahl f	numero m d'indice
612	inductance	inductance f	Induktanz f	induttanza f
613	industrial gloves	gants mpl pour l'industrie	Industriehandschuhe mpl	guanti m industriali
614	inertia moment	moment m d'inertie	Trägheitsmoment n	momento m di inerzia
615	inflammable (adj)	inflammable	entflammbar	infiammabile
616	inflatable cushion	coussin m gonflable, coussin m pneumatique	Luftkissen n	cuscino m gonfiabile
617	infrared (adj)	infrarouge	infrarot	infrarosso m
618	inhibitor	inhibiteur m	Inhibitor m, Verzögerer m	inibitore m
619	initial creep	fluage m initial	anfängliches Kriechen n	scorrimento m iniziale
620	initial deformation, creep	déformation f initiale, fluage	Anfangsdeformation f (Kriechen)	deformazione f iniziale (per scorrimento)
621	injection moulding	moulage m par injection	Spritzgießen n	stampaggio a iniezione
622	inner tube	chambre f à air	Reifenschlauch m	camera f d'aria
623	inorganic (adj)	inorganique	anorganisch	inorganico
624	insert	insert m	Einsatz m	inserto m, prigioniero m
625	insulate (to)	isoler	isolieren	isolare

Vocabulary of the rubber industry 173

NR	ESPAÑOL	PORTUGUÊS	SVENSKA	NEDERLANDS
581	forro m de manguera	revestimento m interior para mangueiras	slangfoder	slangvoering
582	vulcanización f en aire caliente	vulcanização f com ar aquecido	varmluftsvulkning	hete lucht vulcanisatie
583	extrusora de alimentación en caliente	extrusora f (alimentação a quente)	strängspruta varmmatad	spuitmachine (hete voeding)
584	guantes mpl domesticos	luvas fpl caseiras	hushållshandskar	handschoenen voor huishoudelijk-gebruik
585	artículos mpl domesticos goma	artigos mpl caseiros de borracha	hushållsartiklar av gummi	rubber producten voor huishoudelijk-gebruik
586	negro carbón canal de difícil elaboración, negro HPC	negro m de fumo de canal de dispersão difícil HPC	HPC"kimrök	HPC-roet
587	tapacubos	tampa	navkapsel	naafdop
588	humedad	humidade	luftfuktighet	vochtigheid
589	fluido m hidráulico	óleo m para travões hidráulicos	hydraulisk vätska	hydraulische vloeistof
590	manguera f hidráulica	mangueira f hidráulica	hydraulslang	hydraulische slang
591	hidrólisis f	hidrólise f	hydrolys	hydrolyse
592	higroscópico	higroscópico	hygroskopisk	hygroscopisch, wateraantrekkend
593	HYPALON caucho sintético	HYPALON borracha sintéctica	HYPALON syntetiskt gummi	HYPALON, synthetische rubber
594	hipótesis f	hipótese f	hypotes	hypothese
595	histéresis f	histerese f	hysteresis	hysterese
596	curva f cerrada de histéresis	curva f de histerese	hysteresisslinga	hysterese lus
597	pérdida f por histéresis	perda f por histerese	hysteresisförlust	hysterese verlies
598	HYTREL elastómero termoplástico	HYTREL elastómero termoplástico	HYTREL termoplastisk konstruktionselast	HYTREL, engineering thermoplastisch elastomeer
599	resistencia al impacto	resistência f ao impacto	slagseghet	slagvastheid
600	resistencia al impacto, Izod	resistência f ao impacto	slaghållfasthet markerad, izod	slagvastheid met kerf volgens Izod
601	impedancia f	impedância f	impedans	impedantie
602	impulsor m	impulsor m	pumphjul	rotor, aandrijver
603	impermeabilidad f	impermeabilidade f	ogenomtränglighet	ondoorlaatbaarheid
604	impregnación f con caucho	impregnação f com borracha	gummiimpregnering	geimpregneerd met rubber
605	impregnación f ¼	impregnação f	impregnering	impregnering
606	cilindro m para imprenta, rodillo m impresor	cilindro m de impressão, rôlo m impressor	tryckcylinder	drukcylinder
607	rodillo m impresor, cilindro m para imprenta	rôlo m impressor, cilindro m de impressão	tryckvals	drukwals
608	pulgada f	polegada f	tum	inch
609	dureza por penetración (Rockwell)	dureza f por penetração, (Rockwell)	hårdhet enligt Rockwell	kogeldrukhardheid (Rockwell)
610	indice de oxigeno limitado	indice de oxigénio limitativo	syre index	LOI-waarde, zuurstofindex
611	número m índice	número m índice	indexnummer	indexgetal
612	inductancia f	indutância f	induktans	inductantie
613	guantes mpl industriales	luvas fpl industriais	industrihandskar	handschoenen voor industriële doeleinden
614	momento m de inercia	momento m de inércia	tröghetsmoment	traagheidsmoment
615	inflamable	inflamável	brännbar	ontvlambaar
616	cojín m inflable, cojín neumático	almofada pneumática	uppblåsbar kudde	opblaasbaar kussen, luchtkussen
617	infrarrojo	infravermelho	infraröd	infrarood
618	inhibidor m	inibidor m	skyddsmedel	inhibitor, stabilisator
619	flujo m inicial	relaxamento m inicial à tracção	initial krypning	begin kruip
620	deformación f inicial	deformação f inicial (por relaxamento à tracção)	initial deformering plastisk	begin deformatie
621	moldeo por inyección	moldação por injecção - injecção	formsprutning	spuitgieten
622	cámara f de aire (neumaticos)	câmara f de ar	innerslang	binnenband
623	inorgánico	inorgânico	oorganisk	anorganisch
624	inserción f	inserido	inlägg	inlegstuk
625	aislar	isolar	isolera	isoleren

insulating material

Nr	ENGLISH	FRANÇAIS	DEUTSCH	ITALIANO
626	insulating material	matériau m isolant	Isoliermaterial n	materiale m isolante
627	insulation	isolation f	Isolierung f	isolamento m
628	insulator	isolant m	Isolator m	isolatore m
629	intermediate furnace black ISAF	noir ISAF (intermédiaire)	ISAF-Ruß m, Furnace-Ruß für mittlere Abriebfestigkeit	nero fumo m furnace ISAF (intermedio)
630	internal friction	frottement m interne	Reibung f (innere)	attrito m interno
631	interpolation	interpolation f	Interpolation f	interpolazione f
632	interval of time	intervalle m de temps	Zeitraum m	intervallo m di tempo
633	iodine number	indice m d'iode	Jodzahl f	numero m di iodio
634	jacket	gaine f	Ummantelung f	camicia f, guaina f
635	jaw	mâchoire f d'accouplement, accouplement m à serrage	Einspannklemme f, Klaue f	ganascia f
636	joint	joint m	Gelenk n	giunto m
637	Joule effect	effet m Joule	Joule-Effekt m	effetto Joule m
638	KALREZ perfluoroelastomer parts	KALREZ (pièces fpl en élastomère perfluoré)	KALREZ Teil n aus Perfluorkohlenstoffelastomer	KALREZ articoli finiti in perfluoroelastomero
639	kilogram	kilogramme m	Kilogramm n	chilogrammo m
640	knife spreading	enduction f au couteau	Beschichten n mit Messer oder Rakel, Streichen n	spalmatura f a spatola
641	laboratory equipment	équipement de laboratoire	Laboratoriumsausrüstung f, -ausstattung f	apparecchio m di laboratorio
642	lamination	lamination f	Laminierung f	laminazione f
643	lap joint	joint m à recouvrement	Überlappungsverbindung f	giunto m a sovrapposizione
644	latex	latex m	Latex m	lattice m
645	latex adhesive	adhésif m à base de latex	Latexkleber m	adesivo m a base di lattice
646	latex - bitumen emulsion	émulsion f latex-bitume	Latex-Bitumen-Emulsion f	emulsione f di lattice-bitume
647	latex cement	ciment latex m	Latexkleber m	lattice m cemento
648	latex concentrated by creaming	latex m concentré par crémage	Latex m, durch Aufrahmen konzentriert	lattice m concentrato per crematura
649	latex concentrated by electrical methods	latex m concentré par des méthodes électriques	Latex m, durch elektrische Methoden konzentriert	lattice m concentrato elettricamente
650	latex concentrated by evaporation	latex m concentré par évaporation	Latex m, durch Eindampfen konzentriert	lattice m concentrato per evaporazione
651	latex - dipped goods	articles mpl en latex au trempé	Tauchartikel mpl (Latex)	articoli m da immersione in lattice
652	latex foam	mousse f de latex	Latexschaum m	schiuma di lattice
653	lead extruder for lead-covered cables	boudineuse f à plomb pour câbles sous plomb	Bleiextruder m für Bleikabel	trafila f a piombo per cavi
654	lead press crosslinking	vulcanisation f sous plomb	Bleipressenvulkanisation f	vulcanizzazione f a piombo
655	lining	revêtement m	Auskleidung f	rivestimento m
656	lip gasket	joint m à lèvres	Lippendichtung f	guarnizione f a labbro
657	litharge	litharge f	Bleiglätte f	litargirio m
658	litre	litre m	Liter m	litro m
659	load deflection	déformation f sous charge	Druckverformung f	flessione f sotto carico
660	load deflection curve	courbe f de déformation sous charge	Druckverformungskurve f	curva f carico-flessione
661	loading (static loading)	charge f statique	Belastung f (statisch)	carico m statico
662	long time creep	fluage m à long terme	Kriechen n in langen Zeiträumen	scorrimento m a lungo termine
663	loop (hysteresis)	cycle m d'hystérésis	Hystereseschleife f	ciclo m di isteresi
664	loop (low-speed stress-strain loop)	courbe f tension-déformation à vitesse réduite	Hysterese-Schleife f, für langsame Verformung	ciclo m sforzo-deformazione a bassa velocità
665	loss	perte f	Verlust m	perdita f

Nr	ESPAÑOL	PORTUGUÊS	SVENSKA	NEDERLANDS
626	material m aislante	material m isolante	isoleringsmaterial	isolatie materiaal
627	aislamiento m	isolamento m	isolering	isolatie
628	aislante	isolador m	isolator	isolator
629	negro de humo de horno ISAF (intermedio)	negro de fumo ISAF (intermedio)	ISAF - kimrök	ISAF roet
630	rozamiento m interno, fricción f interna	atrito m interior	inre friktion	inwendige wrijving
631	interpolación f	interpolação f	interpolering	interpolatie
632	intervalo m de tiempo	espaço m de tempo, intervalo m de tempo	tidsintervall	tussenpozen
633	índice m de iodo	índice m de iodo	jodtal	joodgetal
634	envoltura f, camisa f	capa f, camisa f	mantel	mantel
635	mordaza f, abrazadera f	mordente m, grampo m	käft	klemplaat
636	junta f, unión f, empalme m	junta f, união f, costura f	fog	verbinding
637	efecto m Joule	efeito m Joule	Joule-effekt	Joule-effect
638	KALREZ piezas de perfluoro elastómero	KALREZ peças de perfluoro elastómero	KALREZ perfluorgummi delar	KALREZ perfluorelastomere onderdelen
639	kilogramo m	quilograma m	kilogram	kilogram
640	recubrimiento m a cuchilla	aplicação f com espatula	knivbeläggning	bestrijken m.b.v. mes
641	equipo m de laboratorio	equipamento m de laboratório	laboratorieutrustning	laboratorium uitrusting
642	laminación f	laminação f, lâmina f	laminering	laminering
643	junta f con solapa	junta f com sobreposição	överlappsförband	overlapverbinding
644	látex m	látex m, látice m	latex	latex
645	cola f de látex, adhesivo de látex	cola f de látex	latexlim	latexlijm
646	emulsión f de asfalto en latex	emulsão f de látex-betume	latex-bitumen emulsion	latex-bitumen emulsie
647	cola f de látex	cimento m de latex, cola f de latex	latex lim	latex lijm
648	látex m concentrado por descremado o desnatado	látex m concentrado por cremagem	latex koncentrerat genom skumning	door opromen geconcentreerde latex
649	látex m concentrado por métodos eléctricos	látex m concentrado por processos elétricos	elektriskt koncentrerad latex	elektrisch geconcentreerde latex
650	látex m concentrado por evaporación	látex m concentrado por evaporação	latex koncentrerad genom avdunstning	door indamping geconcentreerde latex
651	artículos mpl de látex obtenidos por immersión	artigos mpl obtidos por imersão em látex	latexdoppade artiklar	latex dompel artikelen
652	espuma f de látex	espuma f de latex	latex skum	schuimlatex
653	extrusora f de plomo para cables, recubiertos de plomo	trefiladora f de chumbo para cabos com blindagem de chumbo	blyspruta för blymantlad kabel	lood-extruder voor kabelmantels
654	prensa de plomo	reticulação em prensa de chumbo	bly vulkning	loodpers
655	revestimiento interior m, forro m	revestimento m, fôrro m	beklädnad	bekleding
656	junta de labio	junta de lábio	läpptätning	afdichtring met lip
657	litargirio m	litargírio m	blyglete	loodglit
658	litro m	litro m	liter	liter
659	deformación f por carga	deflexão f sob carga	belastningsintryck	vervorming onder druk
660	curva f de deformación por carga	curva f de carga-deflexão	belastningskurva	drukbelastingskurve
661	carga f estática	carregamento m estático, carga f estática	statisk belastning	statische belasting
662	deformación retardada a largo plazo	relaxamento m à tracção a longo prazo	krypning (långtids)	kruipgedrag over langere periode
663	curva f cerrada de histéresis	curva f de histerese	hysteresisslinga	hysterese lus
664	curva cerrada de tensión-alargamiento a baja velocidad	curva f esfôrço-deformação a baixa velocidade	spännings-töjningskurva vid låg hastighet	spanningskurve bij lage snelheid
665	pérdida f	perda f	förlust	verlies

loss of energy

L

NR	ENGLISH	FRANÇAIS	DEUTSCH	ITALIANO
666	loss of energy	perte f d'énergie	Energieverlust m	perdita f di energia
667	loss of heat	perte f de chaleur	Wärmeverlust m	perdita f di calore
668	loss of soluble materials	perte f des substances solubles	Verlust m an löslichen Bestandteilen	perdita f di materiali solubili
669	low pressure gas hose	tuyau m de gaz basse pression	Niederdruckgasschlauch m	tubo per gas a bassa pressione
670	low-warpage	faible gauchissement m	geringer Verzug m	ad alta planarità f, planare (agg.)
671	lowering	abaissement m, réduction f	Erniedrigung f	riduzione f
672	lubricant	lubrifiant m	Schmiermittel n, Gleitmittel n	lubrificante m
673	machine (coat spreading machine)	métier à gommer m	Streichmaschine f	macchina f spalmatrice a coltello
674	machine direction	direction f de la machine	in Fliessrichtung f	direzione della machina
675	machine (extruding machine)	extrudeuse f	Spritzmaschine f, Extruder m	estrusore m, trafila f
676	machine (flexing machine)	machine f de flexion	Biegeprüfer m, Biegeprüfgerät n	macchina f per prove di flessione
677	machine (guillotine cutting machine)	guillotine f, machine f à trancher à guillotine	Hackmaschine f	trancia f a ghigliottina
678	machine (rubber testing machine)	machine f pour tester le caoutchouc	Gummiprüfmaschine f	macchina f per far prove sulla gomma
679	machine (sizing machine)	métier m à gommer	Appretur (Ausrüstungs)-maschine (für Textilausrüstung) f	macchina f incollatrice
680	machine (spreading machine)	métier m à gommer	Streichmaschine f	spalmatrice f
681	machine (strength testing machine)	machine f pour essais de résistance	Festigkeitsprüfer m	apparecchio m per prove di resistenza
682	machine (taping machine)	machine f à rubanner	Bandwickelmaschine f	nastratrice f
683	machinery (cable making machinery)	machine f pour la fabrication de câbles	Kabelherstellungsmaschine f	macchina f per la fabbricazione dei cavi
684	macrostructure	macrostructure f	Makrostruktur f	macrostruttura
685	magnesia	magnésie f	Magnesiumoxid n	ossido m di magnesio
686	manometer	manomètre m	Manometer n	manometro m
687	masterbatch	mélange m maître	Masterbatch m, Vormischung f	masterbatch m, mescola madre f
688	material (insulating material)	matériau m isolant	Isoliermaterial n	materiale m isolante
689	material (raw material)	matière f première	Rohmaterial n	materia f prima
690	matter (dry matter)	matière f solide	Trockensubstanz f	materia f solida, secca
691	maximum deformation	déformation f maximale	maximale Verformung f	deformazione f massima
692	measurement	mesure f	Messung f, Maß n	misura f
693	measuring instrument	appareil m de mesure	Meßinstrument n	strumento m di misura
694	measuring instrument (electrical measuring instrument)	instrument m de mesure électrique	Meßinstrument n (elektrisch)	strumento m per misure elettriche
695	mechanical goods	articles mpl industriels	Artikel mpl (technische)	articoli m tecnici
696	medium thermal furnace black MT	noir MT	Ruß m (Furnace), MT-Ruß m (mittlerer Teilchengröße)	nero fumo fornace MT (medium thermal)
697	melamine resin	résine f de mélamine	Melaminharz n	resina f melamminica
698	melt flow rate	indice de fluidité à l'état fondu	Schmelzindex m	indice di scorrimento a fusione
699	melting point	point m de fusion	Schmelzpunkt m	punto m di fusione
700	memory (elastic memory)	mémoire f élastique	elastische Nachwirkung f	effetto m elastico residuo (memoria elastica)
701	metal oxide curing system	système m de vulcanisation à base d'oxyde métallique	Metalloxid-Vulkanisations-system n	sistema m di vulcanizzazione con ossidi metallici
702	metallic oxide	oxyde m métallique	Metalloxid n	ossido m metallico
703	meteorological balloon	ballon m météorologique	Wetterballon m	pallone m metereologico
704	meteorological (meteorological balloon)	ballon m météorologique	Wetterballon m	pallone m metereologico
705	meter	mètre m	Meter n	metro m
706	micro-cellular rubber	caoutchouc m micro-cellulaire	feinporiges Zellgummi n	gomma f micro-cellulare
707	micro-porous rubber	caoutchouc m micro-poreux	feinporiges Zellgummi n	gomma f micro-porosa
708	microstructure	microstructure f	Microstruktur	microstruttura
709	migration	migration f	Wanderung f, Migration f	migrazione f
710	mildew inhibitor	inhibiteur de moisissures m	schimmelverhütendes Mittel n	anti-crittogamico m

Nr	ESPAÑOL	PORTUGUÊS	SVENSKA	NEDERLANDS
666	pérdida f de energía	perda f energia	energiförlust	energieverlies
667	pérdida f de calor	perda f de calor	värmeförlust	hitteverlies, warmteverlies
668	pérdida f de substancias solubles	perda f de matérias solúveis	förlust av lösliga beståndsdelar	verlies van oplosbare bestanddelen
669	manguera para gases a baja presión	tubo para gases a baixa pressão	lågtrycksgasslang	gas slang voor lage drukken
670	alabeo reducido	trama reduzida	låg-distorterande	lage afschuiving
671	descenso m, reducción f	abaixamento m, redução f	sänkning	verlaging
672	lubri(fi)cante m	lubrificante m	smörjmedel	smeermiddel
673	máquina f para engomar a cuchilla, máquina engomadora horizontal	máquina f de espalhar o revestimento	strykmaskin	machine om lijm op te brengen
674	dirección de la máquina	direcção da máquina	maskin riktning	machine richting
675	máquina f de extrusión, extrusionadora	máquina f de extrusão, trefila f, trafila f	extruder	spuitmachine
676	máquina f para flexionar	máquina f de flexão	böjningsmaskin	buigbeproevingsmachine
677	guillotina f, máquina para cortar	guilhotina f, máquina f para cortar	balskärmaskin	guillotine, balensnijder
678	máquina f para ensayar el caucho	máquina f para ensaiar borracha	provningsmaskin för gummi	rubber testapparaat
679	máquina f de engomar	máquina f de engomar	justeringsmaskin	machine voor opbrengen van lijm
680	engomadora f horizontal a cuchilla	máquina f de espalhar	strykmaskin	machine om lijm op te brengen
681	máquina f para ensayos de resistencia	máquina f para ensaios de resistência	provapparat för hållfasthet	trekbanken
682	máquina de encintar	máquina f de aplicar fita, cingideira f	bandningsmaskin	bandwikkel machine
683	maquinaria f para la fabricación de cables	maquinaria f para fabricação de cabos	maskin för kabeltillverkning	machines voor kabelfabricage
684	macroestructura	macroestrutura	makrostruktur	macrostruktuur
685	magnesia	magnésia	magnesium	magnesium oxide
686	manómetro m	manómetro m	manometer	manometer
687	mezcla madre	concentrado, masterbatch	masterbatch	masterbatch, concentraat
688	material m aislante	material m isolante	isoleringsmaterial	isolatie materiaal
689	materia f prima	matéria f prima	råvara	grondstof
690	materia f seca	sólidos mpl, matéria f sêca	torrsubstans	indampresidu, droge stof
691	deformación f máxima	deformação f máxima	maximal deformation	maximale vervorming
692	medición f, medida f	medida f, medição f	mått	afmeting, meting
693	instrumento m de medida	instrumento m de medida	mätinstrument	meetapparaat
694	instrumento eléctrico de medida	instrumento m eléctrico de medida	elektriskt mätinstrument	elektrische meetapparatuur
695	artículos mpl mecanicos	artigos mpl industriais	formartiklar	technische rubberartikelen
696	negro de humo de horno MT	negro de fumo MT	MT - kimrök	MT roet
697	resina f de melamina	resina f de melamina	melaminharts	melaminehars
698	índice de fluidez de la masa fundida	quantidade de fluxo por fusão - índice de fluidez por fusão	smält index	smeltindex
699	punto m de fusión	ponto m de fusão	smältpunkt	smeltpunt
700	memoria elastica	efeito m elástico residual, memória elástica	elastisk efterverkan	elastisch geheugen
701	sistema m vulcanizante con óxido metálico	processo m de vulcanização com óxido metálico	vulksystem med metalloxid	vulcanisatie systeem met metaaloxide
702	óxido m metálico	óxido m metálico	metalloxid	metaaloxide
703	globo m meteorológico	balão m meteorológico	meteorologisk ballong	meteorologische ballon
704	globo m meteorológico	balão m meteorológico	meteorologisk ballong	meteorologische ballon
705	metro m	metro m	meter	meter
706	goma f microcelular	borracha f microcelular - borracha microporosa	mikrocellgummi	microcellulaire rubber
707	goma f microporosa	borracha f microporosa	mikro-poröst gummi	microporeuze rubber
708	microestructura	microestrutura	mikrostruktur	microstruktuur
709	migración f	migração f	migrering	migratie
710	inhibidor m del mildiu	inibidor m de bolor	mögelskyddsmedel	schimmelwerend middel

M

mildew resistant

Nʀ	ENGLISH	FRANÇAIS	DEUTSCH	ITALIANO
711	mildew resistant	résistant aux moisissures	schimmelbeständig	resistente alle muffe
712	mill (mixing mill)	mélangeur m ouvert, malaxeur m	Mischwalze f	mescolatore m aperto
713	mill (open mill)	mélangeur m ouvert, mélangeur m à cylindres, malaxeur m	Mischwalze f	mescolatore m aperto
714	milligram	milligramme m	Milligramm n	milligrammo m
715	milling	malaxage m	walzen	impasto m, mescolazione
716	mixer(Banbury mixer)	mélangeur Banbury m	Banburymischer m, Innenmischer m	mescolatore m Banbury
717	mixing	mélangeage m	Mischen n	mescolazione f, mescolamento m
718	mixing mill	mélangeur m ouvert, malaxeur m	Mischwalze f	mescolatore m aperto
719	modulus	module m	Modul m	modulo m
720	modulus (effective dynamic modulus)	module m dynamique effectif	effektiver dynamischer Modul m	modulo m dinamico effettivo
721	modulus (elasticity modulus)	module m d'élasticité	Elastizitätsmodul m	modulo m di elasticità
722	modulus (high frequency modulus)	module m en haute fréquence	Hochfrequenzmodul m	modulo m ad alta frequenza
723	modulus of compression	module m de compression	Druckmodul m	modulo m a compressione
724	modulus of elasticity	module m d'élasticité	Elastizitätsmodul m	modulo m elastico
725	modulus of elasticity in compression	module m d'élasticité en compression	Elastizitätsmodul m bei Druck	modulo m elastico a compressione
726	modulus of elasticity in flexion	module m d'élasticité en flexion	Elastizitätsmodul m bei Biegung	modulo m elastico a flessione
727	modulus of elasticity in shear	module m d'élasticité en cisaillement	Schubmodul m	modulo m elastico a taglio
728	modulus of elasticity in traction	module m d'élasticité en tension	Elastizitätsmodul bei Zug	modulo m elastico a trazione
729	modulus of elongation	module m d'allongement	Dehnungsmodul m	modulo m di allungamento
730	modulus of shearing	module m de cisaillement	Schermodul m	modulo m di taglio
731	modulus of transverse elasticity	module m d'élasticité transversale	Transversalelastizitätsmodul m	modulo m di elasticità trasversale
732	moisture	humidité f	Feuchtigkeit f	umidità
733	moment (inertia moment)	moment m d'inertie	Trägheitsmoment n	momento m di inerzia
734	moment of inertia	moment m d'inertie	Trägheitsmoment n	momento m di inerzia
735	monofilament	monofilament m	Monofilament n	monofilamento m, monobava m
736	mould fouling	encrassement m du moule	Formverschmutzung f	sporcamento stampi
737	mould (multi-cavity mould)	moule m à plusieurs empreintes	Mehrfachform f	stampo a impronte multiple
738	moulded article	article m moulé	Formartikel m	articolo stampato
739	moulded goods	articles mpl moulés	Formartikel mpl	articoli m stampati
740	moulding	moulage m	Formartikelherstellung f, Formen f	stampaggio
741	moulding (blow moulding)	injection-soufflage f	Blasformverfahren n	stampaggio per soffiaggio
742	moulding (injection moulding)	moulage m par injection	Spritzgießen n	stampaggio a iniezione
743	moulding (moulding press)	moulage m sous presse	Preßformen n	stampaggio a compressione
744	moulding (transfer moulding)	moulage par transfert	Spritzpressen n, Transferpressen n	stampaggio in transfer (stampaggio per trasferimento)
745	mount (anti-vibration mount)	support m antivibratoire	Dämpfungslager n	supporto m anti vibrante
746	mounting	support m	Aufhängung f	supporto m
747	MPC carbon black (medium processing channel)	noir MPC m	MPC-Ruß m	tipo channel a lavorabilità media, carbon black MPC
748	MT carbon black(medium thermal)	noir MT m	MT-Ruß m, Thermalruß mittlerer Teilchengröße	nerofumo m tipo thermal a lavorabilità media, carbon black MT
749	multi-cavity mould	moule m à plusieurs empreintes	Mehrfachform f	stampo a impronte multiple
750	natural rubber	caoutchouc m naturel	Naturkautschuk m	gomma f naturale

Nr	ESPAÑOL	PORTUGUÊS	SVENSKA	NEDERLANDS
711	resistente al añublo (mildew)	resistente ao bolor	mögelbeständig	bestand tegen schimmel
712	cilindro m mezclador	misturador m aberto, misturador de cilindros	valsverk	mengwals
713	mezcladora f abierta, mezcladora f de cilindros	misturador m aberto, misturador m de cilindros	öppet valsverk	open mengwals
714	miligramo m	miligrama m	milligram	milligram
715	molido, masticado	mastigação	blandning (på valsverk)	walsen
716	mezclador m Banbury	misturador m Banbury, misturadur interno	Banbury blandningsmaskin	Banbury menger
717	mezclado	misturação	blandning	menging
718	cilindro m mezclador	misturador m aberto, misturador de cilindros	valsverk	mengwals
719	módulo m	módulo m	modul	modulus, stramheid
720	módulo m dinámico efectivo	módulo m dinâmico eficaz	dynamisk modul	effectieve dynamische modulus
721	módulo m de elasticidad	módulo m de elasticidade	elasticitetsmodul	elasticiteitsmodulus
722	módulo m de alta frecuencia	módulo m de alta frequéncia	högfrekvensmodul	hoge frequentie modulus
723	módulo m de compresión	módulo m di compressão	sammantryckningsmodul	drukmodulus, compressiemodulus
724	módulo m de elasticidad	módulo m de elasticidade	kompressionsmodul	elasticiteitsmodulus
725	módulo de elasticidad en compresión	módulo m de elasticidade sob compressão	elasticitetsmodul under tryck	modulus van elasticiteit onder druk
726	módulo de elasticidad en flexión	módulo m de elasticidade sob flexão	elasticitetsmodul under böjning	elasticiteitsmodulus onder buiging
727	módulo de elasticidad en cizalla	módulo m de elasticidade sob cisalhamento	elasticitetsmodul under skjuvning	modulus van elasticiteit onder afschuiving
728	módulo de elasticidad en tensión	módulo m de elasticidade sob tensão	elasticitetsmodul under spänning	modulus van elasticiteit onder spanning
729	módulo m de alargamiento	módulo dm de alongamento	töjningsmodul	rekmodulus
730	módulo m de elasticidad transversal, módulo m de cizalladura	módulo m de cizalhamento, módulo m de corte	skjuvningsmodul	afschuivingsmodulus
731	módulo m de elasticidad transversal, módulo m de cizalladura	módulo m de elasticidade transversal, módulo m de cizalhamento	glidningsmodul	afschuivingsmodulus
732	humedad	humidade	fukt	vocht
733	momento m de inercia	momento m de inércia	tröghetsmoment	traagheidsmoment
734	momento m de inercia	momento m de inércia	tröghetsmoment	traagheidsmoment
735	monofilamento m	monofilamento m, monofio m	enfibertråd	monofilament
736	molde obstruido, sucio	sujidade do molde	form smutsning	malvervuiling, matrijsvervuiling
737	molde de cavidades múltiples	molde com cavidades múltiplas	form med flera urtag	matrijs met meerdere vormen
738	piezas moldeadas	peças moldadas	formgjord detalj	persartikel
739	artículos mpl moldeados	artigos mpl moldados	formpressgods	persartikel
740	moldeo	moldação	formpressning	vormartikel
741	moldeo por soplado	moldação por sopro, moldação por insuflação	formblåsning	blaasextrusie
742	moldeo por inyección	moldação por injecção - injecção	formsprutning	spuitgieten
740	moldeo por compresión	moldação por compressão - compressão	form pressning	compressie-pers
744	moldeo por transferencia	moldação por transferéncia	transfer pressning	transfer persen
745	soporte m antivibratorio	apoio m anti-vibração	fjädringsdämpande gummielement	trillingsdemper ophangingselement
746	soporte m	suporte m, apoio m	upphängning	ondersteuning, ophanging
747	negro carbon ‹canal› de mediana dispersión, negro MPC	negro m de fumo de canal de facilidade média de dispersão MPC	MPC kimrök	MPC-roet
748	negro carbón térmico mediano, negro MT	negro m de fumo de decomposição térmica	MT kimrök	MT-roet
749	molde de cavidades múltiples	molde com cavidades múltiplas	form med flera urtag	matrijs met meerdere vormen
750	caucho m natural	borracha f natural	naturgummi	natuurrubber

Neoprene latex

Nr	ENGLISH	FRANÇAIS	DEUTSCH	ITALIANO
751	Neoprene latex	latex m de Néoprène	Neoprene-Latex m	lattice m di Neoprene
752	Neoprene sponge	Néoprène n spongieux	Neoprene-Moosgummi m	spugna f di Neoprene
753	Neoprene synthetic rubber	caoutchouc synthétique Néoprène m	Neoprene Synthesekautschuk m, Polychloroprenkautschuk m	Neoprene m gomma sintetica
754	nerve	nerf m	Nerv m	nervo m
755	neutralisation	neutralisation f	Neutralisation f	neutralizzazione f
756	neutralise, to	neutraliser	neutralisieren	neutralizzare
757	non woven	non tissé	ungewebt	tessuto m non tessuto
758	non-black reinforcing filler	charge f claire	Verstärkerfüllstoff m (hell)	carica rinforzante non nera (bianca)
759	non-discolouring	non-décolorant	nichtverfärbend	non-discolorante
760	non-staining	non-tachant	nichtfleckend, nicht kontaktverfärbend	non-macchiante
761	non-warp	non-gauchi	ohne Verzug m	non ordito (agg.), senza trama
762	NORDEL hydrocarbon rubber	caoutchouc hydrocarboné NORDEL	NORDEL Kohlenwasserstoff-Ethylen-Propylen-Dien-Kautschuk m	NORDEL gomma idrocarbonica
763	notched impact strength	résistance f au choc avec entaille	Kerbschlagzähigkeit f (eingekerbt)	resistenza f all'urto (con intaglio)
764	nozzle	buse f	Düse f	ugello
765	nozzle block	frette f	Düsenkopf m	porta ugello m
766	oil ageing	vieillissement m dans l'huile	Ölalterung f	invecchiamento m in olio
767	oil and gasoline resisting hose	tuyau m résistant à l'huile et à l'essence	Schlauch m (öl- und benzinbeständig)	tubo m resistente all'olio ed alla benzina
768	oil and petrol resisting hose	tuyau m résistant à l'huile et à l'essence	Schlauch m, öl- und kraftstoffbeständig)	tubo m resistente all'olio ed alla benzina
769	oil (aromatic oil)	huile f aromatique f	Weichmacher m (aromatisch)	olio m aromatico
770	oil (drying oil)	huile f siccative	Öl n (trocknend)	olio m essicativo
771	oil enriched rubber	caoutchouc m étendu à l'huile	Kautschuk m (ölgestreckt)	gomma f estesa con olio
772	oil extended rubber	caoutchouc m étendu à l'huile	Kautschuk m (ölgestreckt)	gomma f estesa con olio
773	oil (heavy fuel oil)	huile f (combustible) lourde	Heizöl n (schwer)	olio m combustibile pesante
774	oil (mineral oil)	huile f minérale	Mineralöl n	olio m minerale
775	oil (naphtenic oil)	huile f naphténique	Weichmacher m (naphthenisch)	olio m naftenico
776	oil (paraffinic oil)	huile f paraffinique	Weichmacher m (paraffinisch)	olio m paraffinico
777	oil resisting	résistant à l'huile	ölbeständig	resistente all'olio
778	oil resisting hose	tuyau m résistant à l'huile	Schlauch m (ölbeständig)	tubo m resistente all'olio
779	oil suction and discharge hose	tuyau m d'aspiration et de dépottage d'huile	Schlauch m, Ölansaug- und -zuleitungsschlauch	tubo m d'aspirazione e di scarico dell'olio
780	open mill	malaxeur, mélangeur m ouvert, mélangeur m à cylindres, m	Mischwalze f	mescolatore m aperto
781	organic curing system	système m de vulcanisation organique	organisches Vulkanisationssystem n	sistema di vulcanizzazione organico
782	oscillogram	oscillogramme m	Oszillogramm n	oscillogramma m
783	oscillograph	oscillographe m	Oszillograph m	oscillografo m
784	osmotic (adj)	osmotique	osmotisch	osmotico
785	ounce	once f	Unze f	oncia f
786	outdoor exposure	exposition f à l'extérieur	Bewitterung f im Freien	esposizione f all'aperto
787	oven ageing	vieillissement m en étuve	Ofenalterung f	invecchiamento m in stufa
788	oven (Geer ageing oven)	étuve f de vieillissement Geer	Geer-Alterungsofen m	stufa f di invecchiamento Geer
789	oven (Geer-Evans oven)	étuve f Geer-Evans	Geer-Evans-Ofen m	stufa f Geer-Evans
790	oven (heated oven)	étuve f	Heißluftofen m	stufa f riscaldata
791	overboot	chaussure f de protection	Schutzstiefel m	soprascarpa f
792	overcured (adj)	survulcanisé	übervulkanisiert	sovravulcanizzato (agg.)
793	overshoe	galoche f	Überschuh m	soprascarpa f
794	oxide (red oxide of iron)	oxyde m de fer rouge	Eisenoxid rot	ossido m di ferro rosso
795	oxygen index	indice d'oxygène m	Sauerstoffindex m	indice d'ossigeno

N

Nr	ESPAÑOL	PORTUGUÊS	SVENSKA	NEDERLANDS
751	látex m de Neopreno	látex m de Neoprene	Neoprene latex	Neoprene latex
752	esponja f de Neopreno	esponja f de Neoprene	Neoprene cell gummi	Neoprene sponsrubber
753	Neopreno m caucho sintético	Neoprene m borracha sintética	Neoprene syntetiskt gummi	Neoprene synth. rubber
754	nervio m	nervo m	nerv	nerf
755	neutralización f	neutralização f	neutralisering	neutralisering
756	neutralizar	neutralizar	neutralisera	neutraliseren
757	no tejido	naõ tecido	fibertyg	niet geweven textiel
758	carga de refuerzo no basada en negro de humo	carga reforçante não negra	ljust, förstärkande fyllnadsmedel	versterkende witte vulstof
759	no decolorante	não-descorante	icke-färgande	niet verkleurend
760	no manchante	que não mancha	icke-missfärgande	niet afgevend
761	sin alabeo	sem trama	icke buktad	zonder vervorming
762	NORDEL caucho hidrocarbonado	NORDEL borracha hidrocarbonada	NORDEL hydro karbon gummi, EPDM-gummi	NORDEL koolwaterstof rubber
763	resistencia al impacto (con entalla)	resistência f ao impacto	slaghållfasthet (markerad)	slagvastheid (met kerf)
764	boquilla	ponteira	munstycke	spuit mondstuk
765	boquilla y adaptador	ponteira	munstycks stopp	spuitblok
766	envejecimiento m por el aceite	envelhecimento m pelo óleo	oljeåldring	veroudering door olie
767	tubo m resistente al aceite y a la gasolina	mangueira f resistente ao óleo e à gasolina	olje- och bensinbeständig slang	olie en benzine bestendige slang
768	tubo m resistente al aceite y a la gasolina	mangueira f resistente ao óleo e à gasolina	olje- och bensinbeständig slang	olie en benzine bestendige slang
769	aceite aromatico	óleo aromático	aromatisk olja	aromatische olie
770	aceite m secante	óleo m secativo	torkande olja	drogende olie
771	caucho m enriquecido con aceite	borracha f diluída com óleo	oljetutdrygat gummi	met olie versneden rubber
772	caucho m extendido al aceite	borracha f estendida com óleo	oljeutdrygat gummi	met olie versneden rubber
773	fuel oil pesado	óleo m (combustível) pesado	tjockolja	zware (brandstof) olie
774	aceite m mineral	óleo m mineral	mineralolja	minerale olie
775	aceite nafténico	óleo nafténico	naftenisk olja	nafta olie
776	aceite parafinico	óleo parafinico	paraffin olja	parafine olie
777	resistente al aceite	que resiste a óleo	oljebeständig	oliebestendig
778	tubo m resistente al aceite	mangueira f resistente ao óleo	oljebeständig slang	oliebestendige slang
779	tubo m de aspiración y descarga de aceite	mangueira f de aspiração e descarga de óleo	olje sug- och lossnings slang	olie aanzuig- en losslang
780	mezcladora f abierta, mezcladora f de cilindros	misturador m aberto, misturador m de cilindros	öppet valsverk	open mengwals
781	sistema m de vulcanización orgánica	processo m de vulcanização orgânico	organiskt vulksystem	organisch vulcanisatie systeem
782	oscilograma m	oscilograma m	oscillogram	oscillogram
783	oscilógrafo m	oscilógrafo m	oscillograf	oscillograaf
784	osmótico	osmótico	osmotisk	osmotisch
785	onza f	onça f	uns	Engelse ounce
786	exposición f al aire libre	exposição f ao ar livre	utomhusexponering (-utsatthet)	blootstelling aan de buitenlucht
787	envejecimiento m en estufa	envelhecimento m em estufa	ugnsåldring	veroudering in de oven
788	estufa f Geer para envejecimiento	estufa f Geer para envelhecimento acelerado	Geer åldringskammare	Geer verouderingsoven
789	estufa f Geer-Evans	estufa f de Geer-Evans	Geer-Evans ugn	Geer-Evans oven
790	estufa f calentada	estufa f aquecida	uppvärmd ugn	verwarmde oven
791	chanclo m	bota-galocha f	ytterkänga	overschoen
792	sobrevulcanizado	sobrevulcanizado	övervulkad	overgevulcaniseerd
793	chanclo m	galocha f	bottin	overschoen
794	óxido m de hierro rojo	óxido m vermelho de ferro	järnoxidrött	ijzeroxide rood
795	indice de oxígeno	índice de oxigénio	syreindex	zuurstof index

oxygen index (limiting oxygen index)

NR	ENGLISH	FRANÇAIS	DEUTSCH	ITALIANO
796	oxygen index (limiting oxygen index)	indice limite d'oxygène	Sauerstoffindex m, LOI	indice m limite di ossigeno
797	ozone ageing	vieillissement m à l'ozone	Ozonalterung f	invecchiamento m all'ozono
798	packing	garniture f	Packung f	imballaggio
799	pad (sponge damper pad)	coussin m amortisseur en caoutchouc spongieux	Moosgummidämpfungspolster n	appoggio m ammortizzante in spugna
800	pad (sponge pad)	coussin m en caoutchouc spongieux	Schwammgummipolster n, Moosgummipolster n	appoggio m di spugna
801	particle	particule f	Partikel fpl	particella
802	pattern (embossed)	dessin m gaufré	Prägemuster n	disegno m goffrato
803	PBTP, polybutylene teraphthalate	PBTP, térephtalate de polybutylène	Polybutylenterephthalat n (PBTP)	polibutilene tereftalato
804	pellet	granulé m	Granulat n	granulo m, pellet m
805	pelletised rubber	caoutchouc m en granulés	Granulatkautschuk m	gomma pellettizzata
806	penetration	pénétration f	Durchdringung f	penetrazione f
807	performance	performance f, comportement m	Einsatzverhalten n, Gebrauchstüchtigkeit f	prestazione
808	period of resting	durée f de repos	Ruheperiode f	periodo m di riposo
809	period (scorch period)	temps m de grillage	Anvulkanisationszeit f	tempo di scottatura
810	permeability	perméabilité f	Durchlässigkeit f	permeabilità f
811	permeability coefficient	coefficient m de perméabilité	Durchlässigkeitskoeffizient f	coefficiente m di permeabilità
812	permeability of water vapour	perméabilité f à la vapeur d'eau	Wasserdampfdurchlässigkeit f	permeabilità f al vapor d'acqua
813	permeability to gas	perméabilité f aux gaz	Gasdurchlässigkeit f	permeabilità f ai gas
814	permeable (adj)	perméable	durchlässig	permeabile (agg.)
815	permitivity, permittivity	constante f diélectrique, permittivité f	Dielektrizitätskonstante f	permittività f
816	permitivity (relative permitivity ... Hz)	permittivité relative ... Hz	Dielektrizitätszahl ... Hz	permittività f relativa ... Hz
817	peroxide cure	vulcanisation f au peroxyde	Peroxidvernetzung f	vulcanizzazione a perossidi
818	PETP	PETP, térephtalate de polyéthylène	Polyethylenterephthalat n	polietilenetereftalato
819	petrol and oil resisting hose (GLB)	tuyau m résistant à l'essence et à l'huile	Schlauch m (benzin- und ölbeständig)	tubo m resistente all'olio ed al petrolio
820	petroleum jelly	petrolatum (vaseline brute)	Vaseline n	petrolato m
821	phthalic acid	acide m phtalique	Phthalsäure f	acido m ftalico
822	pigment	pigment m	Pigment n	pigmento m
823	pigment loading	charger en pigment	Pigmentfüllung f	carica f con pigmento
824	pin (insert)	point m d'injection	Haltestift m	perno m d'inserzione
825	pint	pinte f	Pinte f	pinta f
826	piston rubber	caoutchouc m pour pistons	Kolbengummi n	gomma f per pistoni
827	pitch	pas m	Steigung f	passo
828	plant (degreasing plant)	installation f de dégraissage	Entfettungsanlage f	impianto m per sgrassare
829	plastic	matière f plastique, plastique m	Kunststoff m	(plastica) materia f plastica
830	plasticator (Gordon plasticator)	mélangeur, plastificateur m Gordon	Gordon-Kneter m	mescolatore m Gordon
831	plasticiser	plastifiant m	Weichmacher m	plastificante m
832	plasticiser (ester plasticiser)	plastifiant m ester	Esterweichmacher m	plastificante m a base di esteri
833	plasticiser-time effect	influence f du temps sur la plastification	Weichmacher-Zeiteffekt m	effetto m plastificante-tempo
834	plasticity (recovery in plasticity determination)	reprise f élastique	elastische Erholung f	recupero m elastico (in prove di plasticità)
835	plate (backing plate)	plateau m de fixation	Aufspannkörper m	piastra f di fissaggio, di staffaggio
836	plate (clamping plate)	plateau m de fermeture	Aufspannplatte f	piastra f di chiusura
837	plate (ejector plate)	plateau m d'éjection	Ausdrückplatte f	piastra f di espulsione
838	plate (heating plate)	plaque f chauffante	Heizplatte f	piastra f riscaldante
839	plate (polishing plate)	plateau m de polissage	Polierblech n	piastra f di lucidatura
840	plate (retainer plate)	fixation f	Halteplatte f	piastra f di fissaggio dello stampo

NR	ESPAÑOL	PORTUGUÊS	SVENSKA	NEDERLANDS
796	indice de oxigeno limitado	indice de oxigénio limitativo	syre index	LOI-waarde, zuurstofindex
797	envejecimiento m por ozono	envelhecimento m pelo ozono	ozonåldring	veroudering door ozonaantasting
798	empaquetadura f	guarniçao f, empanque m	packning	pakking
799	cojín m amortiguador de esponja	almofada f amortecedora de esponia	cellgummidämpare	schuimrubber stootkussen
800	cojín m de esponja	almofada f de esponja	cellgummidyna	schiumrubber kussen
801	particula	partícula	partikel	deeltje
802	modelo o dibujo en relieve	desenho m em relêvo	präglat mönster	reliëfpatroon
803	PBTP, polibutlén tereftalato	tereftalato de polibutadieno	polybutenterphtalat	polybutylene tereftalaat
804	granza	grão	granulat	korrel
805	caucho en granza	borracha paletizada	granulerat gummi	gegranuleerde rubber
806	penetración f	penetração	penetrering	penetratie
807	comportamiento	comportamento/prestações	prestanda	gedrag tijdens gebruik
808	período m de reposo	período m de repouso	vilotid	rustduur
809	periodo de prevulcanización	período de scorch	anvulkningstid	scorchtijd, aanvulkanisatietijd
810	permeabilidad f	permeabilidade f	permeabilitet	doorlaatbaarheid, permeabiliteit
811	coeficiente de permeabilidad	coeficiente de permeabilidade	permeabilitetskoefficient	permeabiliteits coefficient
812	permeabilidad al vapor de agua	permeabilidade ao vapor de água	permeabilitet av vattenånga	permeabiliteit van water damp
813	permeabilidad f a los gases	permeabilidade f aos gases	gaspermeabilitet	gasdoorlaatbaarheid
814	permeable	permeável	genomtränglig	doorlaatbaar
815	permitividad	constante f dielétrica, permitividade	permittivitet	dielectrische constante
816	permitividad relativa ... Hz	constante dieléctrica relativa ... Hz	relativ dielektrisk konstant ... Hz	relatieve dielektrische constante ... Hz
817	vulcanisación por peroxido	vulcanização com peróxido	peroxidvulkad	peroxide vulkanisatie
818	polietilén tereftalato	tereftalato de polietileno	polyetenterphtalat	polyethyleen tereftalaat
819	manguera f resistente a la gasolina y al aceite	mangueira f resistente à gasolina e ao óleo	bensin- och oljebeständig slang	benzine en olie bestendige slang
820	vaselina	petrolatum	vaselin	vaseline
821	ácido m ftálico	ácido m ftálico	ftalsyra	ftaalzuur
822	pigmento m	pigmento m	pigment	pigment, kleurstof
823	carga f de pigmento	carga f com pigmento	pigmentfyllnadsgrad	hoeveelheid pigment
824	inserción f	pino de inserção	inläggstapp	insteek pen
825	pinta f	pinto m	pint	pint
826	caucho m para émbolos	borracha f para pistões	kolvgummi	plunjermanchet
827	paso	passo	stigning	spoed van een schroef
828	instalación de desengrasado	instalação f de desengorduramento	avfettningsanläggning	ontvettingsinstallatie
829	plástico m, materia f plástica	plástico m, material plástico m	plast	plastic
830	mezclador m Gordon	misturador m Gordon	Gordon blandningsmaskin	Gordon mixer
831	plastificante m	plastificante m	mjukgörare	weekmaker
832	plastificante de ester	plastificante de ester	estermjukgörare	ester weekmaker
833	efecto m plastificante-tiempo	efeito m plastificante-tempo	verklig plasticeringstid	weekmaker geleertijd
834	recuperación f elástico	recuperação f elástica	återhämtning	elastisch vormherstel
835	placa posterior	placa posterior	uppbackningsplatta	steun plaat
836	placa de cierre	placa de fecho	ihopklämnings platta	klemplaat
837	placa de eyección	placa de ejecção	utstötningsplatta	uitstoot plaat
838	plancha f de calefacción	fogareiro m elétrico	uppvärmningsplatta	verwarmingsplaat
839	pulidora	placa de polimento	poler platta	polijstplaat
840	placa retenedora	placa de retenção	fast pressplatta	borg plaat

plate (stripper plate)

N°	ENGLISH	FRANÇAIS	DEUTSCH	ITALIANO
841	plate (stripper plate)	plateau m intermédiaire de démoulage	Abstreifplatte f	piastra f di estrazione, di smontaggio
842	platen	plateau m	Aufspannplatte f	piano m di pressa
843	platen (heated platen)	plateau m chauffant	erwärmte Platte f	piatto m riscaldato
844	plucking	arrachage m, pincement m	Ausreißen n	distacco m parziale
845	plug (plate dispersion)	cheville f de plateau de distribution	Streuplatte f	piastra f distributrice omogeneizzante
846	plunger	piston m	Kolben m	pistone m
847	ply looseness	séparation f entre plis	Schichtenablösung f	separazione f tra gli strati
848	ply separation	séparation f entre plis	Lagentrennung f	separazione f tra gli strati
849	pneumatic cushion	coussin m pneumatique, coussin m gonflable	Luftkissen n	cuscino m pneumatico
850	point (melting point)	point m de fusion	Schmelzpunkt m	punto m di fusione
851	polymer	polymère m	Polymer n	polimero m
852	polymer (cis-tactic polymer)	polymère m cis-tactique	Polymer n (cis-taktisch)	polimero m cis-tattico
853	polymer (graft polymer)	polymère m greffé	Pfropfpolymer n	polimero graffato
854	polymer (ladder polymer)	polymère m à structure en échelle	Leiterpolymer n	polimero m a scala
855	polymer (raw polymer)	polymère m brut	Rohpolymer n	polimero m greggio
856	polymerisation	polymérisation f	Polymerisation f	polimerizzazione
857	polymerisation (bulk polymerisation)	polymérisation f en masse	Blockpolymerisation f	polimerizzazione f in massa
858	polystyrene	polystyrène m	Polystyrol n	polistirolo m, polistirene
859	polysulphide	polysulfure m	Polysulfid n	polisolfuro m
860	polyurethane	polyuréthanne m	Polyurethan n	poliuretano m
861	porosity	porosité f	Porosität f	porosità
862	post cross-linking	post-réticulation f	Nachvernetzung f	post-reticolazione f
863	post-cure	post-vulcanisation f	nachheizen, tempern	post-vulcanizzazione
864	pot life	vie f en pot	Topfzeit f	stabilità f della soluzione
865	pound	livre f	Pfund n	libbra f
866	pound per square foot	livre f par pied carré	Pfund n pro Quadratfuß	libbra f per piede quadrato
867	pound per square inch	livre f par pouce carré	Pfund n pro Quadratzoll	libbra f per pollice quadrato
868	powder rubber	caoutchouc m en poudre	Pulverkautschuk m	gomma in polvere
869	powder, to	pulvériser	pulverisieren	polverizzare
870	power factor (sin ε)	facteur de puissance (sin ε)	Leistungsfaktor (sin ε)	fattore m di potenza (sen ε)
871	prebake	pré-cuit	einbrennen	preriscaldamento
872	preform pressure	pression f de moulage, de préformage	Vorverformungsdruck m	pressione f di preformatura
873	preheat	préchauffé	vorwärmen	pre-riscaldamento
874	press	presse f	Presse f	pressa
875	press (clicking press (US), clicker press (GB))	presse f à estamper, presse f de découpage	Stanzpresse f	pressa f per tranciare
876	press (cure press)	presse à vulcaniser f	Vulkanisationspresse f	pressa f per vulcanizzare
877	press (double ram press)	presse f à double piston	Doppelkolbenpresse f	pressa f a doppio pistone
878	press (downstroke press)	presse f descendante	Presse f mit obenliegendem Druckzylinder m, Oberkolbenpresse f	pressa f discendente
879	press (fly press)	presse f crocodile	Stanzpresse f	pressa f a bilanciere
880	press (lead covering press for hoses)	presse f à plomb pour tuyaux	Bleiumschlagmaschine f, -presse f für Schläuche	pressa f a piombo per tubi
881	press (lead press)	presse f (à plomb)	Bleipresse f	pressa f a piombo
882	press (lead sheathing press)	presse f (à plomb)	Bleimantelpresse f	pressa f per applicare guaine di piombo
883	press (long stroke press)	presse f (col de cygne)	Presse f mit grosser Hublänge f	pressa f a corsa lunga
884	press (multiplaten)	presse f (à plateaux multiples)	Mehrfach-Etagenpresse f	pressa f a piani multipli
885	press (punching)	presse f (à estamper)	Stanzpresse f	pressa f punzonante

P

Nr	ESPAÑOL	PORTUGUÊS	SVENSKA	NEDERLANDS
841	placa extractora	placa de extracção	ut tagningsplatta	stripper plaat
842	plato de la prensa	prato de prensa	press platta	pers plaat
843	plato m calentado	prato m aquecido	uppvärmt uppspänningsbord	verwarmde platen
844	despegue	extracção	plockning	het uithalen, lossen van de matrijs
845	clavija (plato de dispersión)	ficha de dispersão na placa	dispersions tapp	dispersie plaat
846	pistón	pistão	kolv	plunger
847	separación f de pliegues o capas	separação f das camadas; separação f das lonas	kordlagerglapp	separatie tussen rubberlagen
848	separación f de los pliegues	separação f das camadas; separação f das lonas	kordlagerseparation	loslaten van lagen
849	cojín o almohada neumática	almofada f pneumática	luftkudde	pneumatische vering, pneumatische schokdemping
850	punto m de fusión	ponto m de fusão	smältpunkt	smeltpunkt
851	polímero m	polímero m	polymer	polymeer
852	polímero	polímero cistático	cis polymer	cis-tactisch polymeer
853	polímero injertado	polímero inxertado	ymp-polymer	ent-polymeer
854	polímero (polimero de estructura en escalera)	polimero em escada	steg polymer	ladder polymeer
855	polímero m bruto	polímero m cru	råpolymer	ruwe polymeer
856	polimerización	polimerização	polymerisation	polymerisatie
857	polimerizacion en masa	polimerização em massa	mass polymerisation	bulk polymerisatie
858	poliestireno m	polistireno m	polystyren	polystyreen
859	polisulfuro m	polissulfureto m	polysulfid	polysulfide
860	poliuretano m	poliuretano m	polyuretan	polyurethaan
861	porosidad	porosidade	porositet	poreusheid
862	post reticulado	recticulação posterior	efterhärdning	na-verknoping
863	postvulcanizado	pós-vulcanização	efter härda	na-vulkanisatie
864	vida o duración en envase	duração f na lata	lagringsbeständighet på burk	houdbaarheid
865	libra f	libra f	pund	pond
866	libra por pie cuadrado	libra f por pé quadrado	pound per kvadratfot	pond per vierkante voet
867	libra f por pulgada cuadrada	libra f por polegada quadrada	pound per kvadrattum	pond per vierkante inch
868	caucho en polvo	borracha em pó	pulver gummi	poeder rubber
869	pulverizar	pulverizar	pulverisera	verpoederen
870	factor m de potencia (sen ε)	factor m de potência	effektfaktor (sin ε)	arbeidscoefficiënt (sin ε)
871	precalentado	aquecimento prévio	för värmning	voorbakken
872	presión de preformado	pressão de preformação	förformningstryk	voorvervormingsdruk
873	precalentamiento	preàquecímento	förvärmning	voorverwarmen
874	prensa, prensar	prensa	press	pers
875	balancín m recortador, prensa f troqueladora	prensa f de estampar	stanspress	stanspers
876	prensa de vulcanización	prensa de vulcanizaçao	vulk press	vulkanisatie pers
877	prensa de doble embolo	prensa de duplo embolo	dubbel kolvs press	dubbel plunjer pers
878	prensa descendente	prensa descendente	nedåt gående press	neerwaartse pers
879	prensa	prensa	skruvpress	schroefpers
880	prensa f (recubridora de plomo para mangueras)	prensa f de chumbo para tubos	blyöverdragningspress för slang	loodpers voor slangen
881	prensa f de plomo	prensa f de chumbo	blypress	loodpers
882	prensa f para revestir laminas con plomo	prensa f para aplicar bainhas de chumbo	blymantlingspress	loodmantelpers
883	prensa de carrera larga	prensa de longo curso	lång slag press	pers met een lange slag
884	prensa multiple	prensa multipla	flerplatte press	etage pers
885	prensa f para troquelar	prensa f de estampar	stanspress	stanspers

press (reciprocating screw press)

Nr	ENGLISH	FRANÇAIS	DEUTSCH	ITALIANO
886	press (reciprocating screw press)	presse f (à vis piston)	Schubschneckenpresse f	pressa f a vite punzonante
887	press (single press)	presse f (mono-poste)	Einzelpresse f	pressa f singola
888	press (stamping press)	presse f (à estamper), presse f (à découper)	Stempelpresse f, Gesenkpresse f	pressa f per punzonare
889	press (upstroke press)	presse f (à piston ascendant)	Presse f mit untenliegendem Druckzylinder m	pressa f ascendente
890	pressure (air pressure)	pression f d'air	Luftdruck m	pressione f d'aria
891	pressure (atmospheric pressure)	pression f atmosphérique	Druck m (atmosphärisch)	pressione f atmosferica
892	pressure gauge	manomètre m	Manometer n	manometro m
893	pressure sensitive tape	ruban m adhésif sensible à la pression	Selbstklebeband n	nastro auto-adesivo
894	prevulcanised latex	latex pré-vulcanisé	Latex m (vorvulkanisiert)	lattice pre-vulcanizzato
895	primer	couche primaire f	Grundierung f	mano f di fondo
896	process aid	auxiliaire m de mise en œuvre	Verarbeitungshilfsmittel n	coadiuvante di lavorazione
897	processability	aptitude à la mise f en œuvre	Verarbeitbarkeit f	lavorabilità
898	processing	travail m, usinage m, mise f en œuvre	Behandlung f, Verarbeitung f	lavorazione f
899	profile	profilé m, article m profilé	Profil n	profilo m
900	proofed fabric	tissu m imperméabilisé	Gewebe n (gummiert), imprägniertes Gewebe n	tessuto m impermeabilizzato
901	properties (cryogenic properties)	propriétés fpl (cryogéniques)	Eigenschaften f bei Tiefsttemperaturen f	proprietà f criogeniche
902	properties (dynamic properties)	propriétés fpl (dynamiques)	Eigenschaften fpl (dynamisch)	proprietà f dinamiche
903	properties (mechanical properties)	propriétés fpl (mécaniques)	Eigenschaften fpl (mechanisch)	proprietà f meccaniche
904	properties (principal properties)	caractéristiques fpl (principales)	Haupteigenschaften fpl	proprietà f di base
905	properties (vulcanisate properties)	propriétés fpl (des vulcanisats)	Vulkanisateigenschaften fpl	proprietà f dei vulcanizzati
906	property	caractéristique f, propriété f	Eigenschaft f	proprietà f
907	protective gloves	gants mpl de protection	Schutzhandschuhe mpl	guanti m protettivi
908	pulley (V-belt pulley)	poulie f (pour courroies trapézoïdales)	Keilriemenscheibe f	puleggia f per cinghia trapezoidale
909	pump impeller	impulseur m de pompe	Pumpenflügelrad n, Pumpenrotor m	girante f di una pompa
910	punching machine die	outil m à estamper, à découper	Stanzwerkzeug n	stampo m per tranciare
911	push back	repousser, faire reculer	Auswerfer m (hydraulisch)	ingobbatura f dell'isolante (nei cavi elettrici)
912	radiation (exposure to radiation)	exposition f aux radiations	Strahlung f (einer Strahlung aussetzen)	esposizione f alle radiazioni
913	radiator hose	tuyau m pour radiateur	Kühlerschlauch m	tubo m per radiatore
914	radical (free radical)	radical (libre) m	Radikal n (frei)	radicale (libero)
915	railway hose	tuyau m pour wagons de chemin de fer	Eisenbahnwagenschlauch m	tubo m per carri ferroviari
916	rainshoes	chaussures fpl de pluie	Galoschen fpl	soprascarpe f da pioggia
917	ram	piston m	Stempel m	pistone
918	ram (ejector ram)	piston m (d'éjection)	Auswerferkolben m	pistone d'eiezione
919	ram (injection ram)	piston (d'injection)	Spritzkolben m	pistone m di iniezione
920	range (range of temperatures)	gamme f (de températures)	Temperaturbereich m	gamma f di temperature
921	rate (crystallisation rate)	vitesse f (de cristallisation)	Kristallisationsgeschwindigkeit f	velocità f di cristallizzazione
922	rate (curing rate)	vitesse f (de vulcanisation)	Vulkanisationsgeschwindigkeit f, Vernetzungsgeschwindigkeit f	velocità f di vulcanizzazione
923	rate of crack growth	vitesse f de propagation des craquelures	Rißwachstumsgeschwindigkeit f	velocità f di propagazione delle screpolature
924	rate of crystallisation	vitesse f de cristallisation	Kristallisationsgeschwindigkeit f	velocità f di cristallizzazione
925	rate of cure	vitesse f de vulcanisation	Vulkanisationsgeschwindigkeit f	indice, velocità f di vulcanizzazione

Nr	ESPAÑOL	PORTUGUÊS	SVENSKA	NEDERLANDS
886	prensa de inyección husillo	prensa com fuso reciprocante	reciprocerande skruv press	reciproke schroefpers
887	prensa	prensa	enkel press	enkelvoudige pers
888	prensa f estampadora	prensa f de estampar, prensa f de repuxar	präglingsmaskin	stansmachine
889	prensa ascendente	prensa ascendente	uppåtgående press	pers met opgaande slag
890	presión f de aire	pressão f de ar	lufttryck	luchtdruk
891	presión f atmosférica	pressão f atmosférica	atmosfäriskt tryck	atmosferische druk
892	manómetro m	manômetro m	manometer	drukmeter
893	cinta autoadhesiva	fita autocolante	tryck-känslig tejp	drukgevoelige tape
894	latex prevulcanizado	latex prevulcanizado	förvulkad latex	voorgevulkaniseerde latex
895	imprimación	primário m, primeira f demão	grundfärg, grundlager	grondlaag
896	ayuda de proceso	auxiliar de processo	process hjälpmedel	verwerkingsverbeteraar
897	procesabilidad	processabilidade	processbarhet	verwerkbaarheid
898	trabajo m, elaboración f, procedimiento m	fabricação f, tratamento m, elaboração f	bearbetning	verwerking, behandeling, bewerking
899	perfil m, artículo m perfilado	perfil m, artigo m produzido por extrusão	profil	profiel, loopvlakpatroon
900	tela f impermeabilizada	tecido m impermeabilizado	belagd väv	waterdicht weefsel
901	propriedades criogénicas	propriedades criogénicas	kryogeniska egenskaper	cryogene eigenschappen
902	propiedades fpl dinámicas	propriedades fpl dinâmicas	dynamiska egenskaper	dynamische eigenschappen
903	propiedades fpl mecánicas	propriedades fpl mecânicas	mekaniska egenskaper	mechanische eigenschappen
904	propiedades principales	propriedades de princípio	främsta egenskaper	voornaamste eigenschappen
905	propiedades fpl de los vulcanizados	propriedades fpl dos vulcanizados	vulkanisategenskaper	vulkanisatie eigenschappen
906	propiedad	propriedade	egenskap	eigenschap
907	guantes mpl protectores	luvas fpl de proteção	skyddshandskar	beschermende handschoenen
908	polea f para correa en V o trapezoidales	polia f para correias trapezoidais	kilremsskiva	V-snaarschijf
909	impulsor m de bomba	impulsor m de bomba	pump impeller	pomp rotor
910	cuño para estampar	matriz f, estampo m	stansmatris	stansvorm, stansmes
911	fuerza posterior	empurrar para trás	från stötare	terug slag
912	exposición f a radiaciones	exposição f à radiação	bestrålning	blootstelling aan straling
913	tubo m para radiador	mangueira f para radiador	kylvattenslang	radiatorslang
914	radical (libre)	radical (livre)	fri radikal	vrij radicaal
915	tubo m para vagones de ferrocarril	mangueira f para caminhos de ferro	järnvägsslang	slang voor spoorwegwagon
916	botas de lluvia fpl	galochas fpl	galoscher	regenschoenen
917	émbolo	êmbolo	kolv	stempel
918	embolo de expulsion	embolo de ejecção	utstötnings kolv	uitstootstempel
919	pistón de inyección	pistão de injecção	injektionskolv	injectiestempel
920	margen m de temperaturas	gama f de temperaturas	temperaturomfång	temperatuur bereik
921	velocidad de cristalización	velocidade f de cristalização, regime m de cristalização	kristallisationshastighet	kristallisatie snelheid
922	velocidad f de vulcanización	velocidade f de vulvanização, regime m de vulcanização	vulkhastighet	vulcaniseersnelheid
923	velocidad f de propagación de la grieta	velocidade f de expansão do fendilhamento	sprickbildningshastighet	groeisnelheid van haarscheur
924	velocidad f de cristalización	velocidade f de cristalização	kristalliseringshastighet	kristallisatie snelheid
925	velocidad f de vulcanización	velocidade f de vulcanização, tempo m de vulcanização	vulkhastighet	vulcanisatiesnelheid

Vocabulary of the rubber industry 187

P

rate of flow

Nʀ	ENGLISH	FRANÇAIS	DEUTSCH	ITALIANO
926	rate of flow	vitesse f d'écoulement	Fließgeschwindigkeit f	indice, velocità f di scorrimento
927	rate of shear	vitesse f de cisaillement	Schergeschwindigkeit f	indice, velocità f di taglio
928	rate (price, tariff rate)	prix m, tarif m, taux m	Preis m, Tarif m	prezzo m, tariffa f
929	rate (proportion, ratio)	proportion f, degré m, taux m	Verhältnis n	proporzione f, rapporto m
930	rate (scrap rate)	taux m de déchets	Ausschußquote f	rapporto di scarto
931	rate (velocity rate)	vitesse f, rapidité f, vélocité f	Geschwindigkeit f	velocità f
932	ratio	rapport m	Verhältnis n	rapporto m
933	ratio (compression ratio)	taux m (de compression)	Verdichtungsverhältnis n	rapporto m di compressione
934	ratio (die swell ratio)	gonflement m à la sortie de la filière	Mundstücksquellung f, Spritzquellung f	rapporto m di rigonfiamento nell'estrusione
935	ratio (draw ratio)	rapport m (de tirage)	Reckverhältnis n	rapporto m di stiro
936	ratio (length/diameter (L/D) ratio)	rapport L/D m	L/D-Verhältnis n	rapporto m lunghezza/diametro
937	ratio (reduction of ratio)	taux de réduction	Reduktionsverhältnis n	rapporto m di riduzione
938	raw material	matière f première	Rohmaterial n	materia f prima
939	raw rubber	caoutchouc m brut	Rohkautschuk m	gomma f greggia
940	rayon	rayonne f	Rayon m, Kunstseide f	rayon m
941	reagent	réactif m	Reagenz n	reagente
942	rebound n	rebond m	Rückprall m	rimbalzo m
943	rebound test	test m de rebond	Prüfung der Rückprallelastizität f	prova di resilienza, resa f elastica
944	rebound, to	rebondir	rückprallen	rimbalzare
945	recipe	formule f	Rezept n	ricetta f, formula f
946	reclaim (reclaim acid)	régénéré m (acide)	Säureregenerat n	rigenerazione f con acidi
947	reclaim, to	régénérer	regenerieren	rigenerare
948	reclaimed (reclaimed rubber)	régénéré (caoutchouc m)	Kautschukregenerat n	gomma f rigenerata
949	reclaimed rubber	caoutchouc m régénéré	Kautschuk m (regeneriert), Kautschukregenerat n	gomma f recuperata
950	recovery, recuperation	reprise f	Erholung f, elastische	recupero m
951	red lead	minium m	Mennige f	minio m
952	reduce, to	réduire	reduzieren, vermindern	ridurre
953	reduction	réduction f	Reduzierung f, Verminderung f	riduzione f
954	reel	dévidoir m	Haspel f, Spule f	bobina f, rotolo m di alimentazione
955	refractive index	indice de réfraction	Brechungsindex m	indice m di rifrazione
956	regeneration of rubber	régénération f du caoutchouc	Regenerierung von Kautschuk m	rigenerazione f della gomma
957	reinforcing filler	charge f renforçante	Füllstoff f (verstärkend)	carica f rinforzante
958	relative creep	fluage m relatif	relativer Fluß m	scorrimento m relativo
959	relative humidity	humidité f relative	Feuchtigkeit f (relativ)	umidità f relativa
960	relaxation	relaxation f	Entspannung f	rilassamento m
961	requirement	exigence f	Anforderung f	richiesta
962	resilience	résilience f	Rückprallelastizität f	resilienza f
963	resiliency	résilience f	Elastizität f	resilienza f
964	resilient (adj)	résilient	elastisch	resiliente
965	resin	résine f	Harz n	resina f
966	resin (acrylic resin)	résine f (acrylique)	Acrylharz n	resina f acrilica
967	resin (alkyd resin)	résine f (alkyde)	Alkydharz n	resina f alchidica
968	resin (phenolic resin)	résine f (phénolique)	Phenolharz n	resina f fenolica
969	resin (urea resin)	résine f (d'urée)	Harnstoffharz n	resina f ureica
970	resistance	résistance f	Beständigkeit f	resistenza f
971	resistance (acid resistance)	tenue f (aux acides)	Säurebeständigkeit f	resistenza f agli acidi
972	resistance (arc resistance)	résistance f (à l'arc)	Lichtbogenbeständigkeit f	resistenza f all'arco
973	resistance (chemical resistance)	résistance f (chimique)	Chemikalienbeständigkeit f, chemische Beständigkeit f	resistenza chimica f
974	resistance (collapse resistance)	résistance f (à l'affaissement)	Standfestigkeit f	resistenza f allo schiacciamento
975	resistance (corona resistance)	résistance f (à l'effet corona)	Koronabeständigkeit f	resistenza f a scarica corona

Nr	ESPAÑOL	PORTUGUÊS	SVENSKA	NEDERLANDS
926	velocidad f de flujo	velocidade f do fluxo	flythastighet	vloeisnelheid
927	velocidad f de cizallamiento	velocidade f de cisalhamento	skjuvhastighet	afschuifsnelheid
928	precio m, tasa f, tarifa	taxa f	pris	prijs, tarief
929	proporción m, grado m	proporção f, relação f	förhållande	verhouding
930	proporción de desperdicios	taxa de desperdícios	kassations procent	afvalpercentage
931	rapidez f, velocidad f	rapidez f, velocidade f, regime m	hastighet	snelheid
932	relación f, proporción f	relação f, proporção f, razão f	relation	verhouding
933	relación de compresión	taxa de compressão	kompressionsförhållande	drukverhouding
934	proporción de hinchamiento en la extrusión	proporção de inchamento na fieira	matris svällningsförhållande	zwellingspercentage na extrusie
935	proporción de estirado	taxa de estiramento	ned-dragningsförhållande	trek verhouding
936	proporción entre la longitud y el diámetro	relacção comprimento, diametro	L/D - förhållande	L/D verhouding
937	proporción de reducción	quantidade de redução - taxa de redução	reduktionsförhållande	reductie verhouding
938	materia f prima	matéria f prima	råvara	grondstof
939	caucho m crudo, caucho m bruto	borracha f crua	rågummi	onbewerkte rubber
940	rayón f	rayon m	rayon	rayon, kunstzijde
941	reactivo	reagente	reagens	reagens
942	rebote m	ressalto m, repincho m	studselasticitet	terugvering
943	ensayo de resiliencia	teste de resiliência	studselasticitets	test terugveerelasticiteit
944	rebotar	ressaltar, repinchar	studsa test	terugveren
945	receta f, fórmula f	receita f, fórmula f	recept	recept
946	regeneración ácido	regeneração f pelo processo ácido	syraregenerat	zuur regeneraat
947	regenerar	recuperar, regenerar	regenerera	regenereren
948	caucho m regenerado	borracha f regenerada, borracha f recuperada	regenerat	geregenereerde rubber
949	caucho m regenerado	borracha f recuperada/regenerada	regenerat	geregenereerde rubber
950	recuperación f	recuperação f	återställning	terugwinning
951	minio m, óxido rojo de plomo	óxido m vermelho de chumbo, mínio m, zarcão m	blymönja	loodmenie
952	reducir	reduzir	reducera	reduceren, verminderen
953	reducción f	redução f	reduktion	reductie, vermindering
954	devanadera f	bobina f, carretel m, tambor m	rulle	haspel
955	índice de refracción	índice de refracção	refraktionsindex	brekings index
956	regeneración f del caucho	regeneração f da borracha	gummiregenerering	het regenereren van rubber
957	carga f reforzante, carga f activa	carga reforçadora	förstärkande fyllnadsmedel	versterkende vulstof
958	deformación retardada relativa	relaxamento m relativo à tracção	relativ krypning	relatieve kruip
959	humedad f relativa	humidade f relativa	relativ fuktighet	relatieve vochtigheid
960	relajación f	relaxamento m	relaxering	relaxatie
961	solicitud	requisito	krav	eis
962	resiliencia	resiliência f	studselasticitet	veerkracht
963	resiliencia	resiliência f	studselasticitet	veerkracht
964	resiliente	resiliente	studselastisk	veerkrachtig
965	resina f	resina f	harts	hars
966	resina f acrílica	resina f acrílica	akrylharts	acrylhars
967	resina f alquídica	resina f de alquida, resina alquídica	alkydharts	alkydhars
968	resina f fenólica	resina f fenólica	fenolharts	fenolhars
969	resina f de urea	resina f de ureia	ureaharts	ureumhars
970	resistencia f	resistência f	beständighet	weerstand
971	resistencia ácida	resistência f aos ácidos	syra resistens	zuurbestendigheid
972	resistencia al arco eléctrico	resistência ao arco	krypströmsresistens	kruipstroom weerstand
973	resistencia f química	resistência f química	kemikaliebeständighet	chemische bestendigheid
974	resistencia f al aplastamiento	resistência f ao achatamento	formbeständighet	vormvastheid
975	resistencia f al efecto corona	resistência f ao efeito corona	koronabeständighet	corona-weerstand

R

R

resistance (corrosion resistance)

Nʀ	ENGLISH	FRANÇAIS	DEUTSCH	ITALIANO
976	resistance (corrosion resistance)	résistance f (à la corrosion)	Korrosionsbeständigkeit f	resistenza f alla corrosione
977	resistance (crack resistance)	résistance f (aux craquelures)	Rißbeständigkeit f	resistenza f alla rottura
978	resistance (crystallisation resistance)	résistance f (à la cristallisation)	Kristallisationsbeständigkeit f	resistenza f alla cristallizzazione
979	resistance (cut-through resistance)	résistance f (à la propagation de l'entaille)	Schneidwiderstand m	resistenza f all'incisione (di isolante)
980	resistance (flexing resistance)	résistance f (à la flexion)	Biegebeständigkeit f	resistenza f alle flessioni
981	resistance (freeze resistance)	résistance f (au froid, au gel)	Frostbeständigkeit f	resistenza f al freddo
982	resistance (gasoline/petrol resistance)	résistance f (à l'essence)	Benzinbeständigkeit f, Kraftstoffbeständigkeit f	resistenza f alla benzina
983	resistance (gasoline/petrol resistant)	résistance f (à l'essence)	benzinbeständig	resistente alla benzina
984	resistance (grease resistance)	résistance f (aux matières grasses)	Fettbeständigkeit f	resistenza f ai grassi
985	resistance (green resistance)	résistance f (à cru)	Festigkeit f unvulkanisierter Mischung f	tenacità f a crudo
986	resistance (impact resistance)	résistance f (au choc)	Schlagfestigkeit f	resistenza f all'urto
987	resistance (ozone resistance)	résistance f (à l'ozone)	Ozonbeständigkeit f	resistenza f all'ozono
988	resistance (radiation resistance)	résistance f (aux rayonnements)	Strahlenbeständigkeit f	resistenza f alle radiazioni
989	resistance (resistance to mildew)	résistance f (aux moisissures)	Schimmelbeständigkeit f	resistenza f alle muffe
990	resistance (roll resistance)	résistance f (au roulement)	Rollwiderstand m	resistenza f al rotolamento
991	resistance (rolling resistance)	résistance f (au roulement)	Rollwiderstand m	resistenza f al rotolamento
992	resistance (specific resistance)	résistance f spécifique, résistivité f	spezifischer Widerstand m	resistenza f specifica, resistività f
993	resistance to acids	résistance f aux acides	Säurebeständigkeit f	resistenza f agli acidi
994	resistance to alkalis	résistance f aux bases	Alkalienbeständigkeit f	resistenza f agli alcali
995	resistance to checking	résistance f aux craquelures	Beständigkeit gegen Oberflächenrißbildung	resistenza f alle screpolature
996	resistance to chemicals	résistance f aux produits chimiques	Chemikalienbeständigkeit f	resistenza f agli agenti chimici
997	resistance to cracking	résistance f aux craquelures	Rißbeständigkeit f	resistenza f alle rotture
998	resistance to cutting	résistance f à la coupure, à l'entaille	Schneidbeständigkeit f	resistenza f ai tagli
999	resistance to exposure	résistance f à l'exposition	Witterungsbeständigkeit f	resistenza f all'esposizione
1000	resistance to flame	résistance f à la flamme	Flammbeständigkeit f	resistenza f alla fiamma
1001	resistance to flexing	résistance f aux flexions	Biegebeständigkeit f	resistenza f alle flessioni
1002	resistance to heat softening	résistance f au ramollissement par la chaleur	Hitzeerweichungsbeständigkeit f	resistenza f al rammollimento a caldo
1003	resistance to light	résistance f à la lumière	Lichtbeständigkeit f	resistenza f alla luce
1004	resistance to oil	résistance f à l'huile	Ölbeständigkeit f	resistenza f all'olio
1005	resistance to oxidation	résistance f à l'oxydation	Oxidationsbeständigkeit f	resistenza f all'ossidazione
1006	resistance to ozone	résistance f à l'ozone	Ozonbeständigkeit f	resistenza f all'ozono
1007	resistance to ply separation	résistance f à la séparation des plis	Beständigkeit f gegen Lagentrennung f	resistenza f al distacco degli strati
1008	resistance to rolling	résistance f au roulement	Rollwiderstand m	resistenza f al rotolamento
1009	resistance to scraping	résistance f au grattage, aux éraflures	Kratzbeständigkeit f	resistenza f alla raspatura
1010	resistance to shear	résistance f au cisaillement	Scherbeständigkeit f	resistenza f al taglio
1011	resistance to steam	résistance f à la vapeur	Dampfbeständigkeit f	resistenza f al vapore
1012	resistance to sunlight	résistance f au soleil	Sonnenlichtbeständigkeit f	resistenza f alla luce solare
1013	resistance to swelling	résistance f au gonflement	Quellbeständigkeit f	resistenza f al rigonfiamento
1014	resistance to tearing	résistance f au déchirement	Weiterreißfestigkeit f, Kerbzähigkeit f	resistenza f alla lacerazione
1015	resistance to water	résistance f à l'eau	Wasserbeständigkeit f	resistenza f all'acqua

R

Nr	ESPAÑOL	PORTUGUÊS	SVENSKA	NEDERLANDS
976	resistencia f a la corrosión	resistência f à corrosão	korrosionsbeständighet	corrosie-weerstand
977	resistencia al agrietado	resistência f ao fendilhamento; (Port.) resistência f ao gretamento	beständighet mot sprickbildning	weerstand tegen barstvorming
978	resistencia f a la cristalización	resistência f à cristlização	kristallisationsbeständighet	kristallisatie weerstand
979	resistencia al corte	resistência ao corte	genomskärnings beständighet	weerstand tegen doorscheuren
980	resistencia f a la flexión	resistência f à flexão	böjhållfasthet	dynamische buigweerstand
981	resistencia a la congelación	resistência f ao congelamento	köldbeständighet	bestand tegen vorst
982	resistencia f a la gasolina	resistência à gasolina	bensinbeständighet	bestand tegen benzine
983	resistente a la gasolina	resistente à gasolina	bensinbeständig	bestand tegen benzine
984	resistencia f a las grasas	resistência f à massas lubrificantes	fettbeständighet	vetbestandheid
985	resistencia antes de la vulcanización	resistencia em cru	ovulkad resistens	green-strength, sterkte ongevulkaniseerde mengsel
986	resistencia f al impacto	resistência f ao choque	slaghållfasthet	slagvastheid
987	resistencia f al ozono	resistência f ao ozono	ozonbeständighet	bestandheid tegen ozon
988	resistencia a la radiación	resistência à radiação	strålningsbeständighet	bestandheid tegen straling
989	resistencia al moho (mildew)	resistência f ao bolor	mögelbeständighet	schimmelbestendigheid
990	resistencia f al rodamiento	resistência f ao rolamento	rullmotstånd	rolweerstand
991	resistencia f al rodamiento	resistência f ao rolamento	rullmotstånd	rolweerstand
992	resistencia f específica, resistividad f	resistência f específica, resistividade f	specifik beständighet	soortelijke weerstand
993	resistencia f a los ácidos	resistência f aos ácidos	syrabeständighet	bestandheid tegen zuren
994	resistencia f a los álcalis	resistência f aos álcalis	alkalibeständighet	bestandheid tegen alkaliën
995	resistencia f al agrietamiento	resistência f ao fendilhamento	beständighet mot ytsprickter	weerstand tegen oppervlakte scheurtjes
996	resistencia f a los productos químicos	resistência f aos produtos químicos	kemikaliebeständighet	bestandheid tegen chemicaliën
997	resistancia f al agrietamiento	resistência f ao fendilhamento	sprickbeständighet	weerstand tegen barstvorming
998	resistencia f al corte	resistência f ao corte	skärskadebeständighet	weerstand tegen inscheuren
999	resistencia f a la exposición	resistência f à exposição	beständighet mot exponering	bestand tegen het blootstellen aan
1000	resistencia f a la llama	resistência f à chama	flambeständighet	vlamvast, vlambestendig
1001	resistencia f a la flexión	resistência f à flexão	böjbeständighet	buigvast
1002	resistencia f al ablandamiento por el calor	resistência f ao amolecimento pelo calor	värmemjukningsbeständighet	bestand tegen thermische verweking
1003	resistencia f a la luz	resistência f à luz	ljusbeständighet	lichtbestandheid
1004	resistencia f al aceite	resistência f ao óleo	oljebeständighet	oliebestand
1005	resistencia f a la oxidación	resistência f à oxidação	oxideringsbeständighet	bestand tegen oxidatie
1006	resistencia f al ozono	resistência f ao ozono	ozonbeständighet	bestand tegen ozon
1007	resistencia f a la separación de pliegues	resistência f à separação das camadas; (Port.) resistência f à separação das lonas	vidhäftningskraft mellan kordlager	separatieweerstand (tussen koordlagen)
1008	resistencia f al rodamiento	resistência f ao rolamento	rullmotstånd	rolweerstand
1009	resistencia f al desconchado, al raspado	resistência f à raspagem	skavningsbeständighet	bestand tegen krassen
1010	resistencia f al cizallamiento	resistência f ao cisalhamento; (Port.) resistência f ao corte	skärbeständighet	weerstand tegen afschuiving
1011	resistencia f al vapor	resistência f ao vapor	ångbeständighet	bestand tegen stoom
1012	resistencia f a la luz del sol	resistência f à luz solar	solljusbeständighet	bestand tegen zonlicht
1013	resistencia f al hinchamiento	resistência f ao inchamento	svällningsbeständighet	bestand tegen zwelling
1014	resistencia al desgarro	resistência f ao rasgamento	rivhållfasthet	scheurvastheid
1015	resistencia f al agua	resistência f à água	vattenbeständighet	waterbestendigheid

R

resistance to wear and tear

NR	ENGLISH	FRANÇAIS	DEUTSCH	ITALIANO
1016	resistance to wear and tear	résistance f à l'usure et au déchirement	Abnutzungsbeständigkeit f	resistenza f all'usura e alla lacerazione
1017	resistant (adj)	résistant	beständig, widerstandsfähig	resistente (agg.)
1018	resistant (grease resistant)	résistant (aux matières grasses)	fettbeständig	resistente ai grassi
1019	resistant hose (gasoline/petrol[UK] resistant)	tuyau m résistant (à l'essence)	benzinbeständiger Schlauch m	tubo m resistente alla benzina
1020	resistant (oil resistant)	résistant à l'huile	ölbeständig	resistente all'olio
1021	resistant (ozone resistant)	résistant (à l'ozone)	ozonbeständig	resistente all'ozono
1022	resistant to liquid fuel	résistant aux combustibles liquides	beständig gegen flüssigen Treibstoff	resistente ai combustibili liquidi
1023	resistant to salt solutions	résistant aux solutions salines	beständig gegen Salzlösungen	resistente alle soluzioni saline
1024	resistant to solvents	résistant aux solvants	beständig gegen Lösungsmittel	resistente ai solventi
1025	resistivity	résistivité f	spezifischer Widerstand m	resistività f
1026	resistivity (direct current resistivity)	résistivité f en courant continu	Gleichstromwiderstand m	resistività f alla corrente continua
1027	resistivity (electrical resistivity)	résistivité f électrique	spezifischer Widerstand m	resistività f elettrica
1028	resistivity (surface resistivity)	résistivité f superficielle	spezifischer Oberflächenwiderstand m	resistività f superficiale
1029	resistor1100	résistance f (électricité)	Widerstand m	resistore
1030	resting period	durée f de repos	Ruheperiode f	periodo m di riposo
1031	retarder	retardateur m	Verzögerer m	ritardante m
1032	retention of colour	tenue f de la couleur	Farbechtheit f	ritenzione f del colore
1033	retraction	rétraction f	Retraktion f	contrazione f
1034	reversion	réversion f	Reversion f	reversione m (ad un altro stato)
1035	revolutions per minute (RPM)	tours mpl par minute, r.p.m., t/min.	Umdrehungen pro Minute, 1/min	giri m al minuto
1036	rheometer	rhéomètre m	Rheometer n	reometro m
1037	rheometer (oscillating disc rheometer), ODR	rhéomètre à disque oscillant - ODR	Schwingelastometer n, Schwingscheibenrheometer n	reometro a disco oscillante - ODR
1038	rheometer (piston rheometer)	rhéomètre m à piston	Kolbenrheometer n	reometro a pistone
1039	ring	anneau m	Ring m	anello m
1040	ring (cylinder ring)	anneau m de cylindre	Zylinderring m	segmento m del pistone
1041	ring (O-ring)	joint m torique	O-Ring m, O-Ringdichtung f	O-ring m
1042	ring (sealing ring)	anneau m d'étanchéité	Abdichtungsring m	anello m di tenuta
1043	rise	montée f, hausse f, élévation f	Anstieg m	salita f, aumento m
1044	roll for spinning mill	cylindre m de filature	Spinnstuhlwalze f	cilindro m per filatura
1045	roll (milling roll)	cylindre m de malaxage	Walze f	rullo m di mescolazione
1046	roll (printer's roll)	cylindre m d'imprimerie, rouleau m d'imprimerie	Druckwalze f	rullo m per stampanti
1047	roll (printing roll)	cylindre m d'imprimerie, rouleau m d'imprimerie	Druckwalze f	rullo m stampatore
1048	roll (rubber covered roll)	cylindre m recouvert de caoutchouc	Gummiwalze f	cilindro m rivestito di gomma
1049	rollers (lining of rollers)	cylindres (revêtement m de)	Walzenbezug m	rivestimento m di cilindri
1050	rolling resistance	résistance f au roulement	Rollwiderstand m	resistenza f al rotolamento
1051	room - temperature curing	vulcanisation f à température ambiante	Vernetzung f bei Raumtemperatur	vulcanizzazione a temperatura ambiente
1052	rotor	rotor m	Rotor m, Läufer m	rotore m
1053	rubber	caoutchouc m	Gummi m, Kautschuk m	gomma f
1054	rubber (acrylo-nitrile rubber)	caoutchouc m (nitrile acrylique)	Nitrilkautschuk m	gomma f acrilo-nitrilica
1055	rubber and cork compound	mélange m caoutchouc-liège	Kautschuk-Korkmischung f	mescola f di gomma e sughero
1056	rubber (butyl rubber)	caoutchouc-butyle m	Butylkautschuk m	gomma f butilica
1057	rubber cement	colle f à base de caoutchouc	Kautschukkleber m	adesivo m a base di gomma
1058	rubber (chemical resisting rubber)	caoutchouc m (résistant aux produits chimiques)	Kautschuk m (chemisch beständiger Kautschuk)	gomma f resistente ai prodotti chimici
1059	rubber (chlorinated rubber)	caoutchouc m chloré	chlorinierter Kautschuk m	gomma clorurata
1060	rubber clothing	vêtements mpl en caoutchouc	Gummibekleidung f	indumento m di gomma

NR	ESPAÑOL	PORTUGUÊS	SVENSKA	NEDERLANDS
1016	resistencia f al desgaste y a los cortes	resistência f ao desgaste	nöt- och rivhållfasthet	weerstand tegen slijtage en scheuren
1017	resistente	resistente	beständig	bestendig (bestand tegen)
1018	resistente a las grasas	resistente às massas lubrificantes	fettbeständig	bestand tegen vet
1019	tubo m resistente a la gasolina	mangueira f resistente à gasolina	bensinbeständig slang	benzine bestendige slang
1020	resistente al aceite	resistente ae óleo	oljebeständig	oliebestendig
1021	resistente al ozono	resistente m ao ozono	ozonbeständig	ozonbestendig
1022	resistente a los combustibles líquidos	resistente aos combustíveis líquidos	beständig mot flytande bränsle	benzine- en oliebestendig
1023	resistente a las soluciones salinas	resistente às soluçoes salinas	beständig mot saltlösningar	bestand tegen zoutoplossingen
1024	resistente a los disolventes	resistente aos solventes	beständig mot lösningsmedel	bestand tegen oplosmiddelen
1025	resistividad f	resistividade f	resistivitet	soortelijke weerstand
1026	resistividad a corriente continua	resistividade f à corrente contínua	likströmsresistivitet	gelijkstroomweerstand
1027	resistividad f eléctrica	resistividade f eléctrica	elektrisk resistivitet	electrische weerstand
1028	resistividad superficial	resistividade superficial	yt resistivitet	oppervlakte weerstand
1029	resistencia f	resistência f	motstånd	weerstand
1030	periodo m de reposo	período m de repouso	viloperiod	rustperiode
1031	retardante m	retardador m	fördröjare	vertrager
1032	retención o conservación de color	retenção do colorido	färgbeständighet	kleurechtheid
1033	retracción f	retração f	återgång	retractie
1034	reversible	reversão f	reversion	reversie
1035	revoluciones fpl por minuto, r.p.m.	rotaçoes fpl por minuto, r.p.m.	varv per minut	omwentelingen per minuut. RPM
1036	reómetro	reómetro	reometer	rheometer
1037	reómetro de disco oscilante	reómetro de disco oscilante	ODR - reometer	oscillerende schijf rheometer - ODR
1038	reómetro de pistón	reómetro de pistão	kolv reometer	piston rheometer
1039	anillo m, aro m	anel m, aro m, arruela f, anilha f	ring	ring
1040	anillo m de cilindro	anel m de segmento, anel m de pistão	cylinderring	cylinderring
1041	junta tórica	junta f tórica	o-ring	O-ring
1042	anillo tórico	anel m de vedação	tätningsring	afdichtingsring
1043	subida f, elevación f	subida f, elevação f	uppgång	stijging
1044	rodillo m para hilatura	cilindro m para fiação	spinnerivals	spinrol
1045	cilindro mezclador	moínho de rolos	valsverks rulle	walsrol
1046	cilindro m para imprenta, rodillo m impresor	cilindro m de impressão, cilindro m impressor	tryckvals	drukrol
1047	cilindro m para imprenta, rodillo m impresor	cilindro m de impressão, cilindro m impressor	tryckvals	drukwals
1048	cilindro m revestido de goma	cilindro m com revestimento de borracha	gummibeklädd vals	met rubber beklede rol
1049	revestimiento m de cilindros	revestimento m de cilindros	valsbeläggning	walsbekleding
1050	resistencia f al arrastre	resistência f à tracção	draghållfasthet	rolweerstand
1051	vulcanización a temperatura ambiente	vulcanização à temperatura ambiente	vulkning vid rumstemperatur	vulkanisatie bij kamer temperatuur
1052	rotor m	rotor m	rotor	rotor
1053	caucho m	borracha f	gummi	rubber
1054	caucho m acrilonitrílico	borracha f de acrilo nitrilo, borracha f nitrílica	nitrilgummi	nitrilrubber
1055	mezcla f caucho y corcho	composição f de borracha e cortiça	gummi- och korkblandning	kurkrubbermengsel
1056	caucho m butílico	borracha f de butilo, borracha butílica	butylgummi	butyl rubber
1057	adhesivo de caucho	cola f de borracha	gummilim	rubber lijm
1058	goma f resistente a los productos químicos	borracha f resistente aos produtos químicos	kemikaliebeständigt gummi	chemisch bestendige rubber
1059	caucho m clorado	borracha f clorada	klorgummi	chloorrubber
1060	prendas de vestir de caucho	vestuário m de borracha	gummiklädsel	rubber kleding

R

R

rubber content(dry)

Nʀ	ENGLISH	FRANÇAIS	DEUTSCH	ITALIANO
1061	rubber content(dry)	teneur f en caoutchouc sec	Kautschuktrockengehalt m	contenuto m in gomma solida
1062	rubber (crude rubber)	caoutchouc m brut	Rohkautschuk m	gomma f greggia
1063	rubber (dry rubber)	caoutchouc m sec	Festkautschuk m	gomma solida
1064	rubber flooring	revêtement m de sol en caoutchouc	Gummibodenbelag m	rivestimento in gomma di pavimenti
1065	rubber (foam rubber)	caoutchouc m mousse	Schaumgummi m	gomma-spugna (gommapiuma)
1066	rubber garment	vêtement m en caoutchouc	Gummibekleidung f	indumento m di gomma
1067	rubber (hard rubber battery box)	ébonite (bac m d'accumulateur en)	Hartgummi-Batteriekasten m	involucro m d'ebanite per batterie
1068	rubber heel	talon m en caoutchouc	Gummiabsatz m	tacco m di gomma
1069	rubber (oil resisting rubber)	caoutchouc m résistant à l'huile	Kautschuk m (ölbeständig)	gomma f resistente all'olio
1070	rubber (oil softened rubber)	caoutchouc m plastifié à l'huile	Kautschukmischung f (ölplastizierte)	gomma f estesa con olio
1071	rubber (oxidised rubber)	caoutchouc m oxydé	Kautschuk m (oxidiert)	gomma f ossidata
1072	rubber (regenerated /reclaimed rubber)	caoutchouc m régénéré	Kautschuk m (regeneriert), Regenerat n	gomma f rigenerata
1073	rubber (rubber bumper)	absorbeur m de choc en caoutchouc	Gummistoßfänger m	paraurti m di gomma
1074	rubber (rubber spring)	ressort m en caoutchouc	Gummifeder f	molla f di gomma
1075	rubber sheet(solid)	feuille f de caoutchouc	Vollgummiplatte f	foglio m di gomma solida
1076	rubber (snappy rubber)	caoutchouc m nerveux	Gummi n mit hoher Elastizität	gomma f con molto nervo
1077	rubber solution	dissolution f de caoutchouc	Gummilösung f	soluzione f di gomma
1078	rubber solvent	solvant m pour caoutchouc	Gummilösungsmittel n	solvente m per gomma
1079	rubber sponge	caoutchouc m spongieux	Gummischwamm m	spugna f di gomma
1080	rubber (sponge rubber)	caoutchouc m spongieux	Schwammgummi n, Moosgummi n	gomma f spugnosa
1081	rubber stop	butée f en caoutchouc	Gummipuffer m	tampone m di gomma
1082	rubber (surgical cellular rubber)	caoutchouc m (cellulaire pour usage chirurgical)	Zellgummi n für chirurgische Zwecke	gomma f cellulare per chirurgia
1083	rubber testing machine	machine f pour tester le caoutchouc	Gummiprüfmaschine f	macchina f per far prove sulla gomma
1084	rubber thread(extruded)	fil m élastique extrudé	extrudierter Gummifaden m	filo m di gomma estruso
1085	rubber to fabric bond	adhérence f caoutchouc-tissu	Gummigewebebindung f	adesione f tra gomma e tessuto
1086	rubber wall covering	revêtement m de paroi en caoutchouc	Gummiwandverkleidung f	rivestimento m in gomma per pareti
1087	rubber wheel	roue f en caoutchouc	Gummirad n	ruota f di gomma
1088	rubberise, to	gommer, caoutchouter	gummieren	gommare
1089	rubberised fabric	tissu m gommé, tissu caoutchouté	Gewebe n (gummiert)	tessuto m gommato
1090	rubberised goods	articles mpl caoutchoutés	Artikel mpl (gummierte)	articoli m gommati
1091	rubberising	caoutchoutage m, gommage m	Gummieren n	gommatura f
1092	runnerless	canaux mpl régulés	angußlos	senza canali
1093	rupture	rupture f	Bruch m	rottura f
1094	safety factor	facteur m de sécurité	Sicherheitsfaktor m	fattore m di sicurezza
1095	safety margin	marge f (de sécurité)	Sicherheitsspielraum m	margine m di sicurezza
1096	safety (safety processing)	sécurité f (de mise en œuvre)	Verarbeitungssicherheit f	sicurezza f di lavorazione
1097	sample	échantillon m	Materialprobe f, Muster n	campione m
1098	sand and gravel hose	tuyau m pour sable et gravier	Sand- und Kiesschlauch m	tubo m per sabbia e ghiaia
1099	sand blast	sablage m	Sandstrahlen n	sabbiatura f
1100	sand blast, to	sabler	sandstrahlen	sabbiare
1101	scorch	grillage m	Anvulkanisation f	scottabilità f
1102	Scorch (Mooney)	grillage m Mooney	Mooney-Anvulkanisation f	scottabilità f Mooney
1103	scorch time (TS2)	temps de grillage (TS2)	Anvulkanisationszeit f im Rheometer f	tempo di scottabilità (TS2)
1104	scorching tendency	tendance f au grillage	Anvulkanisationsneigung f	tendenza f alla scottabilità
1105	scorchy	précoce, grillant	leicht anvulkanisierend	scottabile

R

Nᴿ	ESPAÑOL	PORTUGUÊS	SVENSKA	NEDERLANDS
1061	contenido m en caucho (seco)	conteúdo m de borracha (seco)	torrhaltgummi	droge rubber gehalte
1062	caucho m bruto, caucho m crudo	borracha f crua	rågummi	ruwe rubber
1063	caucho seco	borracha	gummi	droge rubber
1064	pavimento m, recubrimiento de caucho para pisos	revestimento m de borracha para soalhos	gummigolv	rubber vloerbedekking
1065	caucho-espuma	borracha f esponjosa	skumgummi	schuimrubber
1066	prenda f de vestir de caucho	artigo m de vestuário de borracha	gummikläder	rubber kleding
1067	caja f de ebonita para batería	caixa f de ebonite para bateria	batterikärl av hårdgummi	eboniet accubak
1068	tacón m de caucho	tacão m de borracha	gummiklack	rubber hak
1069	caucho m resistente al aceite	borracha f resistente ao óleo	oljebeständigt gummi	oliebestendige rubber
1070	caucho m ablandado al aceite	borracha f amolecida por óleo	mjukgjort gummi	met olie weekgemaakte rubber
1071	caucho m oxidado	borracha f oxidada	oxiderat gummi	geoxydeerde rubber
1072	regenerado m de goma, caucho m regenerado	borracha f regenerada	regenererat gummit	geregenereerde rubber, regeneraat
1073	parachoques mpl de caucho	pára-choques m de borracha	gummi stötfångare	rubber bumper
1074	muelle m de caucho	mola f de borracha	gummifjäder	rubber veer
1075	lámina maciza de caucho	fôlha f de borracha maciça	ark av massivt gummi	rubbervel
1076	caucho m con mucho nervio	borracha com muito nervo	mycket elastiskt gummi	rubber die snel terugveert
1077	solución f de caucho	solução f de borracha	gummilösning	rubberoplossing (solutie)
1078	disolvente m de caucho	solvente m da borracha	lösningsmedel för gummi	rubber oplosmiddel
1079	esponja f de caucho	esponja f de borracha	gummisvamp	rubberspons
1080	caucho m esponjoso	borracha f esponjosa	cellgummi	sponsrubber
1081	tope m de caucho	batente m de borracha	gummistopp	rubber stop
1082	goma f celular para cirugia	borracha f celular para artigos cirúrgicos	medicinskt cellgummi	cellulaire rubber voor chirurgie
1083	máquina f para ensayar el caucho	máquina f para ensaiar borracha	provningsmaskin för gummi	rubber testapparaat
1084	hilo de goma extruído	fio m de borracha extrudido	sprutad gummitråd	gespoten rubber draad
1085	unión f goma-tela	união f da borracha ao tecido	gummi till väv vidhäftning	rubber-weefsel hechting
1086	revestimiento m de caucho para paredes	revestimento m de borracha para paredes	väggbeklädnad av gummi	rubber muurbekleding
1087	rueda f de caucho	roda f de borracha	gummihjul	rubber wiel
1088	engomar, cauchutar	revestir com borracha	gummera	met rubber bedekken
1089	tejido m cauchutado, tela f engomada	tecido m revestido com borracha	gummerad väv	met rubber bedekte doek
1090	artículos mpl cauchutados	artigos mpl revestidos com borracha	gummerade varor	met rubber bedekte artikelen
1091	engomado m, revestimiento m con goma	revestimento m com borracha, cauchutagem f	gummering	met rubber bedekken
1092	sin canales	sem canais	kanalfri	zonder aanspuitkanalen
1093	rotura f	ruptura f, rotura f	brott	breuk
1094	factor m de seguridad	fator m de segurança	säkerhetsfaktor	veiligheidsfactor
1095	margen m de seguridad	margem m de segurança	säkerhetsmarginal	veiligheidsmarge
1096	seguridad f de elaboración o trabajo	segurança f da fabricação	bearbetningssäkerhet	veiligheid bij verwerking
1097	muestra f, prueba f	amostra f, corpo de prova	prov	monster
1098	manguera f para arena y gravilla	mangueira f para areia e saibro	slang för sand och grus	slang voor zand en grind
1099	chorro m de arena	jato m de areia	sandblästring	zandstraler
1100	chorrear con arena, arenar	limpar a jato de areia	sandblästra	zandstralen
1101	prevulcanización f, vulcanización f prematura	vulcanização f prematura, ‹scorch› m	anvulkning	aanvulkaniseren
1102	prevulcanización f Mooney	vulcanização f prematura Mooney, ‹scorch› Mooney	Mooney anvulkning	Mooney Scorch
1103	tiempo de prevulcanización	tempo de scorch	anvulkningstid (TS2)	aanvulkanisatie, scorchtijd (TS2)
1104	tendencia f a la prevulcanización	tendência f à vulcanização prematura	anvulkningstendens	neiging tot voortijdig vulkaniseren
1105	prevulcanizado	susceptível de vulcanizar prematuramente	lätt anvulkad	snel aanvulkaniserend

Nr	ENGLISH	FRANÇAIS	DEUTSCH	ITALIANO
1106	scrap	déchets mpl	Abfall m, Ausschuß m	scarto
1107	screw	vis f	Schnecke f	vite
1108	screw diameter	diamètre m de la vis	Schneckendurchmesser m	diametro della vite
1109	screw geometry	géométrie f de la vis	Schneckengeometrie f	geometria della vite
1110	seal (glazing seal)	joint m (de vitrage)	Scheibendichtung f	guarnizione f per vetrate
1111	seal (oil seal)	joint m (étanche à l'huile)	Ölabdichtung f	guarnizione di tenuta per olio
1112	seal (shaft seal)	joint m (d'arbre)	Wellendichtung f	anello paraolio
1113	seal (valve stem seal)	joint m (de queue de soupape)	Ventilschaftdichtung f	guidavalvola m, guarnizione f guidavalvola f
1114	sealing force	force de jointoiement f	Dichtkraft f	forza di tenuta
1115	seam (butt seam)	joint m bout à bout	Stumpfnaht f	unione f testa a testa
1116	seamless dipped article	article m trempé sans soudure	nahtloser Tauchartikel m	articolo m per immersione, senza giuntura
1117	section (feeding section)	section f d'alimentation	Fütterzone f	zona di alimentazione
1118	section (metering section)	section f de dosage	Kompressionszone f	zona di misurazione
1119	self curing	autovulcanisant	selbstvulkanisierend	auto-vulcanizzante
1120	self-healing	auto-cicatrisant	selbstheilend	autocicatrizzante (agg.)
1121	self-ignition temperature (ISO = spontaneous ignition temp.)	température d'auto-allumage (ISO = température d'allumage spontané)	Selbstentzündungstemperatur f	temperatura f di autoaccensione (ISO = temp. di accensione spontanea)
1122	semi-reinforcing furnace black SRF	Noir SRF m (semi-renforçant)	SRF-Ruß m (SRF) (halbverstärkend)	nero fumo m semi-rinforzente tipo fornace, carbon black SRF
1123	separation of a bond	décollage m	Trennung f einer Klebverbindung	scollaggio m
1124	separation of an emulsion	rupture f d'une émulsion	Instabilwerden n einer Emulsion	separazione di un'emulsione
1125	serrated (adj)	dentelé	gezahnt, geriffelt	dentellato (agg.)
1126	service life	durée f de service	Gebrauchsdauer f	durata f di esercizio
1127	service test	essai m en service	Gebrauchsprüfung f	prova f d'esercizio
1128	set (assortment)	assortiment m, jeu m	Satz m zusammengehörender Stücke	assortimento m
1129	set (permanent)	déformation f permanente	Verformung f (bleibend)	deformazione f permanente
1130	set (residual set)	déformation f résiduelle, déformation f rémanente	Verformung f (bleibend)	deformazione f residua
1131	set (solidification, hardening)	solidification f, durcissement m	Erstarrung f, Erhärtung f	solidificazione f, indurimento m
1132	shaft	arbre m	Welle f, Achse f	albero m
1133	shallow-flighted	pas m de vis peu profond	flachgängig	a luce poco profonda
1134	shape	forme f, profil m	Form f, Gestalt f	forma f
1135	shear modulus	module m de cisaillement, module d'élasticité en cisaillement	Schermodul, Schubmodul m	modulo m di taglio
1136	shear resistance	résistance f au cisaillement	Scherfestigkeit f	resistenza f al taglio
1137	shear strength	résistance au cisaillement	Scherfestigkeit f	forza di taglio
1138	shear-test, to	soumettre à l'essai de cisaillement	einem Scherversuch m unterziehen	eseguire una prova di taglio
1139	shearing rate	cisaillement (vitesse f de)	Schergeschwindigkeit f	velocità f di taglio
1140	sheathing (cable sheathing)	gaînage m de câble	Kabelummantelung f	rivestimento m del cavo
1141	sheet (calendered sheet)	feuille f (calandrée)	kalandrierte Platte f	lastra f calandrata
1142	sheet (extruded sheet)	feuille f (extrudée)	Folie f (extrudiert)	lastra f estrusa
1143	sheet (hospital sheet)	drap m (d'hôpital, alèse f)	Betteinlage f	teloni m per ospedali
1144	sheeting (rubber and cork compound sheeting)	feuilles fpl (en caoutchouc-liège)	Platten fpl aus einer Gummi-Korkmischung	lastra f di miscela di gomma e sughero
1145	shelf life	durée f de stockage	Lagerbeständigkeit f	durata f di magazzinaggio
1146	shock absorber	amortisseur m, tampon m de choc, pare-chocs m	Stoßdämpfer m	ammortizzatore m
1147	shock resistance	résistance f au choc	Stoßbeständigkeit f	resistenza f agli urti
1148	Shore durometer	Shore (duromètre) m	Shore-Härteprüfer m	Shore durometro m
1149	shrinkage	retrait m	Schrumpfung f, Schwindung f, Schwund m	ritiro m
1150	shrinkage (diameter shrinkage)	retrait m au niveau du diamètre	Durchmesserschwindung f	ritiro m del diametro

S

Nr	ESPAÑOL	PORTUGUÊS	SVENSKA	NEDERLANDS
1106	desperdicios	desperdícios	avfall	uitval
1107	husillo	fuso	skruv	schroef
1108	diámetro del husillo	diámetro do fuso	skruv diameter	schroef diameter
1109	geometría del husillo	geometria do fuso	skruv geometri	schroef geometrie
1110	junta para cristales	junta para vidrados	fönster tätning	glas afdichting
1111	retén m	vedante m para óleo	oljetätning	olie pakking
1112	junta del eje	junta lateral	axel tätning	asafdichting, manchet
1113	junta para eje de valvula	junta para valvula de vapor	ventil tätning	klepsteelafdichting
1114	fuerza de sellado	força de selagem	tätningskraft	afdichtkracht
1115	costura por testa	costura f de tôpo	stumsöm	stuiknaad
1116	artículo m sin costuras por immersión	artigo m sem costura revestido por imersão	sömlös doppad artikel	naadloos dompelartikel
1117	sección de alimentación	secção de alimentação	matnings sektion	voedingszone
1118	sección de medición	secção de medição	uppmätnings sektion	compressiezône
1119	autovulcanizante	autovulcanizante	självvulkande	zelfvulcaniserend
1120	autocicatrizante	autocicatrizzante	självläkande	zelfherstellend
1121	temperatura de autoignición (ISO:temp. de ignición espontanea)	temperatura de ignição expontânea	själv antändningstemperatur (ISO = självantändnings tem)	temperatuur van zelfontbranding (ISO = spontane ontbrandings temperatuur)
1122	negro carbón de horno semi-reforzante, negro SRF	negro m de fumo de fornalha semi-reforçador SRF	SRF kimrök	semi-versterkend roet, SRF roet
1123	separación f de una unión	descolagem f, despegamento m	bristning av en fog	het loslaten van een hechting
1124	separación f de una emulsión	separação f da emulsão	sönderfall av en emulsion	breken van een emulsie, coaguleren
1125	dentado	serrilhado, dentado	sågtandad	getand, zaagvormig
1126	duración f útil, vida de servicio	duração f útil, duração f no serviço	livslängd	levensduur
1127	ensayo en servicio	ensaio m de utilização real	driftprov	gebruiksproef, praktijkproef
1128	juego m, serie f	jôgo m, sortimento m	sats	stel, serie (bij elkaar behorende voorwerpen)
1129	deformación f permanente	deformação f permanente	kvarstående sättning	blijvende vervorming
1130	deformación f residual	deformação f residual	kvarstående sättning	blijvende vervorming
1131	solidificación f, endurecimiento m	solidificação f, endurecimento m	stelnande, hårdnande	stolling (hard worden)
1132	eje m, árbol m	eixo m, veio m	axel	drijfas – schacht
1133	canal del husillo poco profundo	rosca pouco profunda	skruv med grund gängning	vlakke helling
1134	forma f, perfil m	forma f, perfil m	form	vorm
1135	módulo m de cizallamiento	módulo m de cisalhamento	skjuvmodul	afschuifmodulus
1136	resistencia f al cizallamiento	resistência f ao cisalhamento; (Port.) resistência f ao corte	skjuvhållfasthet	afschuifweerstand
1137	resistencia a la cizalla	resistência ao cisalhamento	skjuvbeständighet	afschuifsterkte
1138	someter a prueba de cizallamiento	submeter ao ensaio m de cisalhamento, submeter ao ensaio m de corte	skjuvprova	afschuifproef doen
1139	velocidad f de cizallamiento	velocidade f de cisalhamento; (Port.) velocidade f de corte	skjuvhastighet	afschuifsnelheid
1140	vaina o cobertura del cable	bainha f de cabo	kabelmantel	aanbrengen van kabelmantel
1141	hoja f calandrada	fôlha f calandrada	kalandrerad duk	uitgewalst vel
1142	lámina extrusionada	folha f extrudida	extruderad duk	geëxtrudeerd vel
1143	tela f hospital	lençol m para hospital	sjukhuslakan	hospitaaldoek
1144	plancha f de goma y corcho	fôlhas fpl de composição de borracha e cortiça	duk av gummi- och korkblandning	kurkrubber plaat
1145	vida f en almacenaje	duração f na armazenagem	lagringsbeständighet	houdbaarheid in magazijn
1146	amortiguador m de choques	amortecedor m de choques, amortecedor m	stötdämpare	schokbreker
1147	resistencia f al choque	resistência f aos choques	stötbeständighet	stootvastheid
1148	Shore durómetro m	Shore durómetro m	Shore durometer	Shore hardheidsmeter
1149	encogimiento m, contracción f	contracção f, encolhimento m	krympning	krimp
1150	contracción del diametro	encolhimento do diâmetro	diameter krymp	diameter krimp

shrinkage (mould shrinkage)

Nʀ	ENGLISH	FRANÇAIS	DEUTSCH	ITALIANO
1151	shrinkage (mould shrinkage)	retrait m (au moulage)	Formenschwund m	ritiro m dopo lo stampaggio
1152	silica (fumed)	silice f sublimée	Kieselsäure f, hochaktiv/pyrogen	silice calcinata
1153	silica (precipitated silica)	silice f précipitée	Kieselsäure, gefällt	silice f precipitata
1154	single-threaded	à un filet	eingängig	mono-passo
1155	sintering	frittage m	Sintern n	sinterizzazione
1156	site (cross link site)	site m (de rétification)	Vernetzungsstelle f	punto di reticolazione
1157	size (agglomerate size, average diameter)	granulométrie	Teilchengröße f, durchschnittlicher Durchmesser m	dimensione f agglomerati, diametro m medio
1158	size (particle size, average diameter)	granulométrie f	Teilchengröße f, durchschnittlicher Durchmesser m	granulometria (dimensione della particella)
1159	size (standardised size)	dimensions fpl normalisées	Größe f (genormt)	dimensione f normalizzata
1160	sizing machine	métier m à gommer	Appretur (Ausrüstungs)-maschine (für Textilausrüstung) f	macchina f incollatrice
1161	skid, to	déraper, patiner	rutschen	slittare, frenare
1162	slab	plaque f	Platte f, Prüfklappe f	lastra f
1163	slaughterhouse hose	tuyau m pour abattoirs	Schlachthausschlauch m	tubo m per macelli
1164	slip, to	déraper, patiner, glisser	gleiten, laufen	slittare
1165	smooth bore hose	boyau m à alésage lisse	Schlauch m mit glatter Innenwand	tubo m con interno liscio
1166	snap-fitting	obturateur m	Schnappverschluß m, Klemmverschluß m	chiusura a scatto
1167	soften, to	ramollir	erweichen	rammollire
1168	softener	plastifiant m	Weichmacher m	plastificante m
1169	softness	souplesse f	Weichheit f	morbidezza f
1170	sole	semelle f	Sohle f	suola f
1171	sole (sponge rubber sole)	semelle f (en caoutchouc spongieux)	Moosgummisohle f	suola f di gomma spugnosa
1172	solid n	matière f solide	Festkörper m	solido m
1173	solids content	extrait m sec	Festkörpergehalt m	contenuto in solidi
1174	solids (non-volatile solids)	solides mpl non-volatiles	Festkörpergehalt m	solidi non volatili
1175	solids(total), TS	matières fpl solides totales	Trockensubstanz f	solidi totali
1176	soling	semelles fpl	Sohlen fpl	suola f
1177	solubilise, to	solubiliser	lösen, in Lösung bringen	sciogliere
1178	solubility	solubilité f	Löslichkeit f	solubilità f
1179	soluble (adj)	soluble	löslich	solubile
1180	solution mixer	mélangeur m à dissolution	Lösungsmischer m	mescolatore m per soluzioni
1181	solution polymerisation	polymérisation en solution f	Lösungspolymerisation f	polimerizzazione f in soluzione
1182	solvent (adj)	dissolvant	lösend	solvente m
1183	solvent (aromatic solvent)	solvant m aromatique	aromatisches Lösungsmittel n	solvente m aromatico
1184	solvent n	solvant m, dissolvant m	Lösemittel n, Lösungsmittel n	solvente m
1185	solvent resisting hose	tuyau m résistant aux solvants	Schlauch m (lösungsmittelbeständig)	tubo m resistente ai solventi
1186	solvent welding	soudure f en phase solvant	Lösungsmittelverschweißen n	saldatura con solventi
1187	sound absorbing	insonorisant	geräuschabsorbierend	fonoassorbente (agg.)
1188	sound damping	amortissement m acoustique	geräuschdämpfend	smorzamento del suono
1189	sound deadening	amortissement m du bruit	Geräuschdämpfung f	smorzante del rumore
1190	specialties (extruded specialties)	articles mpl extrudés	extrudierte Artikel mpl	articoli m estrusi
1191	specific gravity	poids m spécifique	Gewicht n (spezifisch), Dichte f	peso m specifico
1192	specific heat	chaleur f spécifique	Wärme f (spezifisch)	calore m specifico
1193	specification	spécification f	Spezifikation f	norma f, specifica f
1194	specimen	spécimen m, échantillon m	Materialprobe f, Muster f, Prüfkörper m	campione m, provino m
1195	speed (even speed of the rolls)	vitesse f égale (des cylindres)	gleichmäßige Geschwindigkeit f (der Walzen)	velocità f eguale (dei cilindri)

Nr	ESPAÑOL	PORTUGUÊS	SVENSKA	NEDERLANDS
1151	contracción f de moldeo, contracción de molde	contracção f no molde	formkrympning	krimp t.o.v. maat matrijs t.g.v. afkoeling
1152	silice anhídrida	silica pirogenada	kiselsyra	silica uit gasfase
1153	silice precipitada	silica precipitada	fälld kiselsyra	geprecipiteerde silica
1154	paso único	passo único	enkel gängad	enkelschroefdraad
1155	sinterizado	sinterização	sintring	sinteren
1156	punto de reticulación	ponto de retiçulação	tvärbindningspunkt	vulkanisatie kant, verknopingspunt
1157	tipo de aglomerado, diametro medio	dimensão do aglomerado, diâmetro médio	agglomerat storlek, genomsnittsdiameter	deeltjesgrootte, gemiddelde diameter
1158	dimension de particula, diametro medio	tamanho da partícula, diâmetro médio	partikel storlek, genomsnittlig storlek	deeltjes grootte, gemiddelde doorsnede
1159	tamaño o medida normalizada	tamanho m padronizado	standardiserad storlek	standaard grootte
1160	máquina f de engomar	máquina f de engomar	justeringsmaskin	machine voor het opbrengen van lijm op textiel
1161	derrapar, patinar, deslizar	patinar, deslizar, escorregar	halka	slippen
1162	plancha f, lámina f gruesa	prancha f/placa	platta	vel
1163	manguera f para mataderos	mangueira f para matadouros	slakterislang	slang voor abatoirs
1164	derrapar, patinar, deslizar	patinar, deslizar, escorregar	halka	slippen
1165	manguera f de pared interior lisa o calibrada	mangueira f de tubo liso	slang med glatt insida	slang met gladde binnenwand
1166	enclavamiento	encravamento	kläm passning	klem fitting
1167	ablandar	amolecer, abrandar, amaciar	mjukna	verweken (week worden)
1168	ablandante m, plastificante m	emoliente m	mjukgörare	weekmaker
1169	blandura f	brandura f, moleza f	mjukhet	zachtheid
1170	suela f	sola f, solado m	sula	zool
1171	suela f de goma esponjosa	sola f de borracha esponjosa	cellgummisula	sponsrubber zool
1172	materia f sólida	matéria f sólida, sólido m	massiv	vaste stof
1173	contenido en sólidos	contendo de sólidos	torrhalt	gehalte vaste stoffen
1174	sólidos no volátiles	sólidos não voláteis	icke flyktiga ämnen	niet-vluchtige bestanddelen
1175	sólidos mpl totales	sólidos mpl totais	torrsubstans (TS)	vaste stof gehalte
1176	suelas fpl	solado m, solas fpl	sulor	zolen
1177	solubilizar	solubilizar	göra löslig	oplosbaar maken
1178	solubilidad f	solubilidade f	löslighet	oplosbaarheid
1179	soluble	solúvel	löslig	oplosbaar
1180	mezcladora f para soluciones	misturador m de soluções	lösningsblandare	menger voor het maken van oplossingen
1181	polimerización de la solución	polimerização em solução	lösnings polymerisation	polymerisatie in oplossing
1182	disolvente	solvente	löslig	oplosbaar
1183	disolvente m aromático	solvente m aromático	aromatiskt lösningsmedel	aromatisch oplosmiddel
1184	disolvente m	solvente m, dissolvente m	lösningsmedel	oplosmiddel
1185	manguera f resistente a los disolventes	mangueira f resistente aos solventes	lösningsmedelsbeständig slang	slang bestand tegen oplosmiddelen
1186	soldadura por disolvente	soladura por solvente	lösningsmedels skarvning	lassen door middel van oplosmiddel
1187	insonorizante	amortecedor acústico	ljudabsorberande	geluid absorberend
1188	amortiguador m del sonido	amortecedor do som	ljuddämpande	geluiddempend
1189	amortiguación f del ruido	amortecimento m do ruído	ljuddämpning	geluiddemping
1190	artículos por extrusión	artigos mpl extrudidos	sprutade specialiteter	spuitartikelen
1191	peso m específico	pêso m específico, densidade f	specifik vikt	soortelijk gewicht, soortelijke massa
1192	calor m específico	calor m específico	specifik värme	soortelijke warmte
1193	especificación f	especificação f	specifikation	specificatie, voorschrift
1194	muestra f, prueba f	espécime m, amostra f	prov	monster, voorbeeld
1195	velocidad f uniforme (de los cilindros)	velocidade f igual (dos cilindros)	lika hastighet (på valsarna)	gelijke snelheid (van de walsrollen)

S

speed (extrusion speed)

Nʀ	ENGLISH	FRANÇAIS	DEUTSCH	ITALIANO
1196	speed (extrusion speed)	vitesse (d'extrusion)	Extrusionsgeschwindigkeit f, Spritzgeschwindigkeit f	velocità di estrusione
1197	speed (variable speed drive)	commande f (à vitesse variable)	Regelantrieb m	comando m a velocità variabile
1198	spider (die spider)	porte-mandrin m	Dornhalter m mit Stegen	porta mandrino m
1199	spiral hose	tuyau m spiralé	Spiralschlauch m	tubo m spiralato
1200	splice	jointure f, soudure f	Spleißstelle f	giunzione f, giunta f
1201	sponge	spongieux m	Schwamm m	spugna f
1202	sponge (blown sponge)	spongieux m	Moosgummi/Schwammgummi n	spugna f gonfiata
1203	sponge (closed cell sponge)	spongieux m à cellules fermées	Zellgummi n (mit geschlossener Zellstruktur)	spugna f a celle chiuse
1204	sponge damper pad	coussin m amortisseur en caoutchouc spongieux	Moosgummidämpfungspolster n	appoggio m ammortizzante in spugna
1205	sponge rubber sheeting	feuilles fpl en caoutchouc spongieux	Schwammgummiplatte f, Moosgummiplatte f	foglio m di gomma spugnosa
1206	spray gun	pistolet m à pulvériser	Spritzpistole f	pistola f spruzzatrice
1207	spray, to	pulvériser, arroser	sprühen, spritzen	spruzzare
1208	spraying	pulvérisation f	Sprühen n, Spritzen n	spruzzatura f
1209	spraying gun	pistolet m à pulvériser	Spritzpistole f	pistola f spruzzatrice
1210	spread by brush (to)	enduire à la brosse	streichen mit Pinsel	spalmare a pennello
1211	spreadcoat (to)	enduire au métier à gommer	beschichten im Streichverfahren n (mit Rakel)	rivestire con spalmatura
1212	spreader	métier m à gommer	Streichmaschine f	spalmatrice f
1213	spreading	enduction f	Streichen, Beschichten, Auftragen n	spalmatura f
1214	spreading machine	métier m à gommer	Streichmaschine f	spalmatrice f
1215	spring	ressort m	Feder f	molla f
1216	sprue	carotte f	Anguß m	matarozza
1217	spun bonded	fil m aggloméré	Spinnvlies n	filato m agglomerato
1218	square inch	pouce m carré	Quadratzoll m	pollice m quadrato
1219	squeeze, to	presser, essorer	quetschen	strizzare
1220	stabilisation	stabilisation f	Stabilisierung f	stabilizzazione f
1221	stabiliser	stabilizzante m	Stabilisator m	stabilizzatore m
1222	stability	stabilité f	Stabilität f	stabilità f
1223	stability (chemical stability)	stabilité f (chimique)	Stabilität f (chemisch)	stabilità chimica f
1224	stability (dimensional stability)	stabilité f (dimensionnelle)	Formbeständigkeit f	stabilità dimensionale
1225	stability (mechanical stability)	stabilité f (mécanique)	Stabilität f (mechanisch)	stabilità meccanica
1226	stability of colour	stabilité f de couleur	Farbbeständigkeit f	stabilità f del colore
1227	stability (storage stability)	stabilité f (au stockage)	Lagerbeständigkeit f	stabilità f al magazzinaggio
1228	stability (thermal stability)	stabilité f (thermique)	Stabilität f (thermisch)	stabilità f termica
1229	stamping foil	feuille f métallique	Prägefolie f	foglio m metallizzato
1230	stamping (hot stamping)	estampage m à chaud	Heißprägen n	impressione f a caldo
1231	standard (adj)	normal, standard	genormt	standard
1232	standard deviation	écart-type m, écart m standard	standard Abweichung f	deviazione f standard
1233	standard error (square root of the variance)	erreur f standard (racine carrée de la variance)	normaler Fehler m (Quadratwurzel der Abweichungen)	errore m standard (radice quadrata della varianza)
1234	standard n	norme f	Norm f	norma f
1235	standardised	normalisé	genormt	normalizzato
1236	static ageing	vieillissement m statique	Alterung f (statisch)	invecchiamento m statico
1237	static friction	frottement m statique	Reibung f (statisch)	attrito m statico
1238	stationer's rubber goods	articles mpl en caoutchouc de papeteries	Gummiwaren fpl des Schreibwarenhandels	articoli m di gomma per cancelleria
1239	stator (pump stator)	stator m de pompe	Pumpenstator m	statore di pompa
1240	steam (live steam)	vapeur f vive	Freidampf m	vapore m vivo

Nr	ESPAÑOL	PORTUGUÊS	SVENSKA	NEDERLANDS
1196	velucidad de extrusion	velocidade de extrusão	sprut hastighet	spuitsnelheid
1197	accionamento m a velocidad variable	acionamento m de velocidade variável, transmissão f de velocidade variável	variator drift	(traploze) toerenregelaar
1198	soporte del troquel	suporte do mandril	munstycks spindel	doornhouder aan spuitmondstuk
1199	manguera f en espiral	mangueira f em espiral, mangueira f helicoidal	spiralslang	spiraalslang
1200	empalme m, junta f	emenda f, união f	skarva	las
1201	esponja f	esponja f	svamp	spons
1202	esponja f hinchada	esponja f inflada	jäst svamp	geblazen spons
1203	esponja de celula cerrada	esponja f de células fechadas	gummisvamp med slutna celler	schuimrubber met gesloten cellen
1204	cojín m amortiguador de esponja	almofada f amortecedora de esponja	svampgummidämpare	schuimrubber stootkussen
1205	láminas fpl de caucho esponjoso	fôlhas fpl de borracha esponjosa	svampgummi duk	schuimrubber platen
1206	pistola f aerográfica	pistola f de pulverização	sprutpistol	spuitpistool
1207	pulverizar, rociar	pulverizar	spruta	verstuiven, spuiten
1208	pulverización f	pulverização f	sprutning	Verstuiving, sproeien
1209	pistola f pulverizadora	pistola f de pulverização	sprutpistol	spuitpistool
1210	extender con pincel	espalhar a pincel, aplicar a pincel	bestrykning med pensel	met een kwast uitstrijken
1211	engomar con máquina horizontal	revestir por espalhamento	sprutbelägga	strijken
1212	engomadora f horizontal a cuchilla	máquina f de espalhar	spridare	strijkmachine
1213	recubrimiento m a cuchilla	espalhamento m, espalhação f	spridning	het bestrijken
1214	engomadora f horizontal a cuchilla	máquina f de espalhar	strykmaskin	strijkmachine
1215	resorte m, muelle m	mola f	fjäder	veer
1216	mazarota	gito	ingöt	aanspuiting
1217	aglomerado de fibra	fiado	samman spunnen	los gesponnen
1218	pulgada f cuadrada	polegada f quadrada	kvadrattum	vierkante inch
1219	apretar, exprimir	apertar, espremer	klämma	uitpersen
1220	estabilización f	estabilização f	stabilisering	stabilisatie
1221	estabilizante m	estabilizador m	stabilisator	stabilisator
1222	estabilidad f	estabilidade f	stabilitet	stabiliteit
1223	estabilidad f química	estabilidade f química	kemisk stabilitet	chemische stabiliteit
1224	estabilidad f dimensional	estabilidade f dimensional	formbeständighet	vormvastheid
1225	estabilidad f mecánica	estabilidade f mecânica	mekanisk stabilitet	mechanische stabiliteit
1226	estabilidad f de color	estabilidade f de colorido	färgstabilitet	kleurstabiliteit
1227	estabilidad f en almacén	estabilidade f na armazenagem	lagringsbestandighet	stabiliteit gedurende opslag
1228	estabilidad f térmica	estabilidade f térmica	termisk stabilitet	thermische stabiliteit
1229	hoja para estampado	chapa de estampar	präglingsfolie	metaal folie
1230	estampación en caliente	estampacem a quente	varmprägling	heet stempelen
1231	normal, patrón, standard	normal, padrão	standard	normaal, standaard
1232	desviación f normal, standard	desvio m normal	standard-avvikelse	standaard deviatie
1233	error f normal o standard (raíz cuadrada de la variación)	êrro m normal, êrro m padrão (raiz quadrada da variação)	medelfel	standaard afwijking
1234	norma f, patrón m, standard m	norma f, padrão m	standard	norm
1235	normalizado, sistematizado	normalizado m, padronizado m	standardiserad	genormaliseerd
1236	envejecimiento m estático	envelhecimento m estático	statisk åldring	statische veroudering
1237	fricción f estática	atrito m estático	statisk friktion	statische frictie
1238	artículos mpl de goma para papelerías	artigos mpl de borracha para papelaria	gummivaror för pappershandeln	rubber artikelen voor de boekhandel
1239	estator de bomba	estator da bomba	pump stator	pomp stator
1240	vapor m vivo	vapor m vivo	direkt ånga	open stoom

S

steam resistant

NR	ENGLISH	FRANÇAIS	DEUTSCH	ITALIANO
1241	steam resistant	résistant à la vapeur	dampfbeständig	resistente al vapore
1242	steelcord	fil m métallique (pneus)	Stahlkord m	filo m metallico per pneumatici
1243	stick (quick stick)	prise f rapide	schnelle Haftung f	potere m di presa rapida
1244	stickiness	adhésivité f, poisseux m	Klebrigkeit f	adesività f, appiccicosità f
1245	sticking	poisseux	anklebend	appiccicoso
1246	sticky	poisseux	klebrig	appiccicoso
1247	stiffening (simple stiffening)	rigidification f simple	einfache Versteifung f	irrigidimento m semplice
1248	stir, to	agiter, remuer	rühren	agitare
1249	stitcher	roulette f lisse, machine f à brocher	Heftmaschine f	cucitrice f
1250	stock	matière f première	Vorrat m, Lagerbestand m	materia f prima, mescola f, materiale
1251	strain	déformation f, contrainte	Dehnung f, Beanspruchung f, Längenänderung f	deformazione f (sotto carico)
1252	strain (elongation strain)	allongement m par traction	Zugdehnung f	allungamento m a trazione
1253	strain gauge	jauge f de contrainte	Zugspannungsmesser m	entità f della deformazione
1254	strain (initial strain)	contrainte f initiale	Anfangsdehnung f	allungamento f iniziale
1255	strain (shearing strain)	contrainte f de cisaillement	Scherbeanspruchung f	deformazione f di taglio
1256	strain tester	appareil m pour la mesure de la contrainte	Zugspannungsprüfer m	apparecchio m per misurare la deformazione
1257	strap (brake strap)	ruban m de frein	Bremsband n	nastro m per freni
1258	strength	résistance f	Kraft f, Stärke f, Festigkeit f	resistenza f, forza
1259	strength (adhesive strength)	pouvoir m adhésif	Haftvermögen n	potere m adesivo
1260	strength (breaking strength)	résistance f à la rupture	Bruchfestigkeit f	resistenza f alla rottura
1261	strength (compressive strength)	résistance f (à la compression)	Druckfestigkeit f	resistenza f alla compressione
1262	strength (flexural strength)	résistance f (à la flexion)	Biegefestigkeit f	resistenza f a flessione
1263	strength (green strength)	résistance f (à cru)	Rohfestigkeit f	resistenza a trazione di mescola non vulcanizzata
1264	strength (peel strength)	résistance f (au pelage)	Schälfestigkeit f	resistenza allo spellamento
1265	strength (pull strength)	résistance f (à la traction)	Haftfestigkeit f	resistenza f allo strappo
1266	strength (shearing strength)	force f de cisaillement	Scherfestigkeit f	resistenza f al taglio
1267	strength (tensile impact strength)	résistance f (au choc en traction)	Zugschlagfestigkeit f	resistenza f all'urto con trazione
1268	strength (tensile strength at break)	résistance f (à la rupture par traction)	Zerreißfestigkeit f	carico m di rottura a trazione
1269	strength (tensile strength at yield)	résistance f (à la traction à la limite élastique)	Streckspannung f	resistenza f allo snervamento
1270	strength testing machine	machine f pour essais de résistance	Festigkeitsprüfer m	apparecchio m per prove di resistenza
1271	strength (wet strength)	résistance f (à l'état humide)	Naßfestigkeit f	resistenza f in condizioni di umidità
1272	stress	contrainte f	Spannung f, Beanspruchung f	tensione f, sforzo m
1273	stress at break	tension, contrainte f de rupture	Bruchspannung f	tensione f a rottura
1274	stress at failure	tension f de contrainte f à la rupture	Bruchspannung f	tensione f a rottura
1275	stress (bending stress)	contrainte f de flexion	Biegebeanspruchung f	sforzo m di flessione
1276	stress (compression stress)	contrainte f de compression	Druckbelastung f	sforzo m a compressione
1277	stress crack	fissure sous contrainte	Spannungsriß m	incrinatura f per sforzo
1278	stress in bending	contrainte f de flexion	Biegespannung f	tensione f per flessione
1279	stress in torsion	contrainte f de torsion	Torsionsspannung f	tensione f per torsione
1280	stress relaxation	relaxation f de tension, de contrainte	Spannungsrelaxation f	rilassamento m dello sforzo
1281	stress (relaxation of stress)	relaxation f de contrainte	Spannung f, Spannungsrelaxation f	rilassamento m dello sforzo
1282	stress (shearing stress)	contrainte f de cisaillement	Scherspannung f	sforzo m di taglio
1283	stress under compression	contrainte f sous compression	Spannung f unter Druck	sforzo m sotto compressione
1284	stress-strain compression curve	courbe f de déformation par compression	Spannungsdeformationskurve f	curva f sforzo deformazione per compressione
1285	stress-strain curve	courbe f contrainte-déformation	Zug-Dehnungskurve f	curva f sforzo deformazione

Nr	ESPAÑOL	PORTUGUÊS	SVENSKA	NEDERLANDS
1241	resistente al vapor	resistente ao vapor	ångbeständig	bestand tegen stoom
1242	hilo m de acero para neumáticos	fio m de aço para pneus	stålkord	staaldraad
1243	adhesion (pegado m rápido)	adesão rápida	snabb vidhäftning	snelle hechting, snelklevend
1244	pegajosidad f, adhesividad f	pegajosidade f, adesividade f	klibbighet	kleverigheid
1245	pegajoso	pegajoso, que pega	klibbning	klevend
1246	pegajoso	pegajoso	klibbig	kleverig
1247	endurecimiento m simple	enrijecimento m simples	enkel förstyvning	eenvoudige verstijving
1248	agitar, remover	agitar, revolver	röra om	roeren
1249	moleta f	rolete m liso, máquina f de pontear	stickare	stikmachine
1250	materia prima	stock	lager	voorraad
1251	deformación f bajo carga	deformação f sob carga	töjning	deformatie door belasting
1252	deformación f por tracción	deformação f ao alongamento	krafttöjning	rekspanning
1253	medidor m del alargamiento	medidor m de deformação	töjprovare	trekspanningsmeter
1254	deformación f inicial	deformação f inicial	initial töjning	begin rek
1255	deformación f por cizallamiento	deformação f por cisalhamento; (Port.) deformação f por corte	skjuvpåkänning	vervorming door afschuiving
1256	aparato m para pruebas de deformación	aparelho m para ensaio de deformação	töjprovare	apparaat voor onderzoek trekspanning
1257	cinta f de freno	cinta de travão	bromsband	rembanden
1258	resistencia f, fuerza f, poder m	resistência f, fôrça f	styrka	sterkte
1259	poder m adhesivo o fuerza adhesiva	poder m adesivo	vidhäftningskraft	kleefvermogen, hechtvermogen
1260	resistencia f a la ruptura	resistência f à ruptura	brotthållfasthet	trekvastheid
1261	resistencia a la compresión	resistência f à compressão	tryckhållfasthet	druksterkte
1262	resistencia f a la flexión	resistência f à flexão	böjhållfasthet	dynamische buigweerstand
1263	resistencia sin vulcanizar	crua/não vulcanizada	grön styrka	sterkte ongevulcaniseerd mengsel
1264	fuerza de pelado	forca de delaminação	skalningshållfasthet	afstripsterkte
1265	resistencia al estiramiento	resistência ao estiramento	drag styrka	trek sterkte
1266	resistencia f al cizallamiento	resistência f ao cisalhamento; (Port.) resistência f ao corte	skjuvhållfasthet	afschuifsterkte
1267	resistencia al impacto por tracción	resistência ao impacto por tracção	slaghållfasthet under spänning	slagvastheid onder trekbelasting
1268	resistencia f a la tracción, carga f de rotura	carga f de ruptura à tracção	brottdraghållfasthet	treksterkte bij breuk
1269	resistencia a la tracción al límite elástico	resistência à tensão no limite elástico	brott spänning	treksterkte bij vloei
1270	máquina f para ensayos de resistencia	máquina f para ensaios de resistência	provapparat för hållfasthet	trekbanken
1271	resistencia f en húmedo	resistência f a húmido, resistência f à humidade	våtstyrka	natte sterkte
1272	tensión f, carga f	tensão f, solicitação f, esfôrço m	spänning	spanning
1273	tensión f a la rotura	tensão f de ruptura	brottspänning	breukspanning
1274	tensión f a la rotura	tensão f de ruptura	brottspänning	breukspanning
1275	tensión f de plegado	esfôrço m de flexão, solicitação f em flexão	böjspänning	buigspanning
1276	tensión f de compresión	tensão f de compressão	tryckspänning	drukspanning
1277	fisura por tensiones	fendilhamento por tensão	spännings sprickning	barstvorming t.g.v. spanning
1278	tensión f por flexión	tensão f por flexão	böjkraftvärde	buigspanning
1279	tensión f por torsión	tensão f de torção	torsionsbelastning	torsiespanning
1280	relajación f de tensión	tensão de relaxamento f	påkänningsrelaxering	spanningsrelaxatie
1281	relajación f de tensión	relaxamento m de esfôrço	spänningsmattning	spänningsrelaxatie
1282	tensión f de cizallamiento	tensão f de cisalhamento; (Port.) tensão f de corte	skjuvspänning	afschuifspanning
1283	tensión f de compresión	tensão f de compressão, solicitação f em compressão	tryckspänning	drukspanning
1284	curva de tensión-deformación por compresión	curva f tensão-deformação à compressão	deformationskurva för tryckbelastning	deformatiekromme bij drukbelasting
1285	curva de tensión-deformación	curva f de deformação sob tensão	spännings-töjningskurva	Trek-rek-diagram

S

stress (compressive stress)

Nr	ENGLISH	FRANÇAIS	DEUTSCH	ITALIANO
1286	stress (compressive stress)	contrainte f de compression	Druckbelastung f, Druckspannung f	sforzo m a compressione
1287	stretch (elongation)	allongement m	Dehnung f	allungamento m
1288	stretch (pre-stretch)	allongement m préalable	Vorspannung f (z.B. von Förderbändern während der Vulkanisation)	allungamento m preliminare
1289	strip	bande f	Streifen m, Leiste f	striscia f
1290	strip force	force f de dénudage	Abstreifkraft f	forza f di spellamento, di strippaggio
1291	structure	structure f	Struktur f	struttura f
1292	structure (black structure)	structure f (du noir)	Rußstruktur f	struttura del carbon black (del nero)
1293	sucker	ventouse f	Saugnapf m	ventosa f
1294	suction and delivery hose	tuyau m d'aspiration et de dépottage	Saug- und Zuleitungsschlauch m	tubo m d'aspirazione e di scarico
1295	suction cup	ventouse f	Saugnapf m	ventosa f
1296	suction device	dispositif m d'aspiration	Saugeinrichtung f	dispositivo m d'aspirazione
1297	sulphur (blooming of), sulphur bloom	efflorescence f de soufre	Schwefelausblühung f	efflorescenza f di zolfo
1298	sulphur cure	vulcanisation f au soufre	Schwefelvulkanisation f	vulcanizzazione a zolfo
1299	sulphur donator	donneur m de soufre	Schwefelspender m	donatore di zolfo
1300	sun-checking	craquelure f par la lumière solaire	Lichtriß m	screpolatura f causata dalla luce solare
1301	sun-checking inhibitor	inhibiteur m de craquelure par la lumière solaire	Lichtrißschutzmittel n	inibitore m di screpolature causate dalla luce solare
1302	sun-cracking	craquelure f due à la lumière solaire	Lichtrißbildung f	screpolatura f causata dalla luce solare
1303	sunlight discolouration	décoloration f par la lumière solaire	Verfärbung f im Sonnenlicht	scolorimento m causato dalla luce solare
1304	sunlight resistance	résistance f à la lumière solaire	Beständigkeit f gegen Sonnenlicht	resistenza f alla luce solare
1305	super furnace black SAF	noir SAF (super)	Ruß m (SAF), Furnace-Ruß mit sehr hoher Abriebbeständigkeit	nero fumo m fornace SAF (super)
1306	surface activity	activité f de surface	Oberflächenaktivität f	attività f superficiale
1307	surface area	zone f superficielle	Oberfläche f (spezifisch)	superficie f, area superficiale
1308	surface tension	tension f superficielle	Oberflächenspannung f	tensione f superficiale
1309	surface treatment product	produit m pour le traitement des surfaces	Oberflächenbehandlungsprodukt n	prodotto m per il trattamento superficiale
1310	surgical goods	articles mpl chirurgicaux	Artikel mpl (chirurgische), Chirurgiebedarf m	articoli m chirurgici
1311	surgical rubber	caoutchouc m à usage chirurgical	chirurgische Gummiwaren fpl	gomma f chirurgica
1312	suspension (colloidal suspension)	suspension colloïdale	Suspension f (kolloidale)	sospensione f (colloidale)
1313	sweating	exsudation f	Ausschwitzen n	essudazione f
1314	swell, to	gonfler	quellen	rigonfiare
1315	swell up, to	gonfler (se)	aufquellen	rigonfiare
1316	swelling	gonflement m	Quellung f	rigonfiamento m
1317	swelling after extrusion	gonflement m après boudinage	Spritzquellung f	rigonfiamento m dopo l'estrusione
1318	swelling power	pouvoir m gonflant	Quellkraft f, Quellstärke f	potere m rigonfiante
1319	swollen	gonflé	gequollen	rigonfiato
1320	synthetic elastomer	élastomère m synthétique	synthetisches Elastomer n	elastomero m sintetico
1321	synthetic latex	latex m synthétique	synthetischer Latex m	lattice m sintetico
1322	synthetic rubber	caoutchouc m synthétique	Synthesekautschuk m	gomma f sintetica
1323	system (curing)	système m de vulcanisation	Vulkanisationssystem n	sistema m di vulcanizzazione
1324	tab (flow tab)	canal m d'écoulement	Fließnase f	canaletto
1325	tack	adhésivité f, collant m	Klebrigkeit f	adesività f

Nr	ESPAÑOL	PORTUGUÊS	SVENSKA	NEDERLANDS
1286	tensión f de compresión	tensão f de compressão	tryckspänning	drukspanning
1287	alargamiento m	estiramento m	töjning	uitrekking
1288	alargamiento m preliminar en la vulcanisación de correas	pré-estiramento m	förspänning	voorrek
1289	tira f	tira f	list	strip
1290	fuerza de desmoldeo	força de desmoldaoem	avskalningskraft	stripkracht
1291	estructura f	estrutura f	struktur	structuur
1292	estructura del negro de humo	estrutura	struktur	roet struktuur
1293	ventosa f	ventosa f	sugapparat	zuiger
1294	manguera f de aspiración y descarga	mangueira f de aspiração e de descarga	sug- och tryckslang	zuig- en laadslang
1295	ventosa f	taça f de aspiração, ·cup· m de aspiração	sugkopp	zuignap
1296	dispositivo m de aspiración	dispositivo m de aspiração	sugapparat	aanzuigapparaat
1297	eflorescencia f del azufre	eflorescência f do enxôfre	utsvavling	uitbloemen van zwavel
1298	vulcanizante a base de azufre	vulcanização com enxofre	svavelvulkad	zwavel vulkanisatie
1299	donante de azufre	dador de enxofre	svavel avgivare	zwavel donor
1300	agrietamiento m por luz solar	fendilhamento m pela luz solar	sprickbildning genom solljus	barstvorming door inwerking zonlicht
1301	inhibidor m de agrietamiento por luz solar	inibidor m do fendilhamento pela luz solar; (Port.) inibidor m do gretamento pela luz solar	solskyddsmedel	inhibitor tegen de zoninwerking
1302	agrietamiento m por luz solar	rachadura f pela luz solar	solsprickning	oppervlakte aantasting door inwerking zonlicht (·olifantenhuid·)
1303	decoloración f por luz solar	descoloração f pela luz solar	missfärgning genom solljus	verkleuring door zonlicht
1304	resistencia f a la luz solar	resistência f à luz solar	solljusbeständighet	bestendig tegen zonlicht
1305	negro de humo de horno SAF (super)	negro de fumo SAF (super)	SAF kimrök	SAF roet
1306	actividad f superficial	actividade f superficial	ytaktivitet	oppervlakte activiteit
1307	area superficial	área superficial	specifik yta	oppervlakte
1308	tensión f superficial	tensão f superficial	ytspänning	oppervlakte spanning
1309	producto m para tratamiento de superficies	produto m para tratamento de superficies	ytbehandlingsprodukt	oppervlakte behandelingsmiddel
1310	artículos mpl quirúrgicos	artigos mpl cirúrgicos	medicinska artiklar	chirurgische artikelen
1311	caucho m quirúrgico	borracha f para artigos cirúrgicos	medicinskt gummi	chirurgisch rubber
1312	suspensoide m	suspensão m	suspensions	colloidale suspensie
1313	exudación f, transpiración f	exsudação f	svettning	uitzweting
1314	hinchar, inflar	inchar	svälla	zwellen
1315	hincharse, inflarse	inchar	svälla upp	(op)zwellen
1316	hinchamiento m	inchamento m, dilatação f	svällning	zwelling
1317	hinchamiento m a la salida de la hilera (extrusion)	inchamento m após a extrusão	svällning efter sprutning	zwelling na spuiten (diameter toename)
1318	poder m hinchante	poder de inchamento	svällningsförmåga	zwelvermogen
1319	hinchado, inflado	inchado, dilatado	svullen	gezwollen
1320	elastómero m sintético	elastômero m sintético	syntetisk elast	synthetische elastomeer
1321	látex m sintético	látex m sintético	syntetisk latex	synthetische latex
1322	caucho m sintético	borracha f sintética	syntetiskt gummi	synthetische rubber
1323	sistema m vulcanizante	processo m de vulcanização	vulksystem	vulcaniseersysteem
1324	rebosadero	aba (aba de circulação)	kanal	vloeinaad
1325	pegajosidad f, adhesividad f	pegajosidade f, adesividade f	klibbförmåga	kleefkracht

S

T

Nr	ENGLISH	FRANÇAIS	DEUTSCH	ITALIANO
1326	tack (dry tack)	pouvoir m collant initial	Trockenklebrigkeit f	appiccicosità f a secco
1327	tack gauge	appareil m pour mesurer l'adhésivité	Klebrigkeitsmesser m	misuratore m di adesività
1328	tack inhibitor	inhibiteur m d'adhésivité	Mittel n (klebrigkeitsverhindernd)	agente m anti-adesivo
1329	tackifier	tackifiant m	Klebrigmacher m	agente di appiccicosità
1330	tackiness	adhésivité f	Haftfähigkeit f, Klebrigkeit f	adesività f
1331	tacky	poisseux	klebrig	appiccicoso
1332	tank (circulating tank)	réservoir m, citerne f	Umwälztank m	serbatoio m di circolazione
1333	tank lining	revêtement m de cuve	Tankauskleidung f, Behälterauskleidung f	rivestimento m per serbatoi
1334	tape	ruban m, bande f	Band n	nastro m
1335	tape (adhesive tape)	ruban m adhésif	Klebeband n	nastro m adesivo
1336	tape (cable tape)	ruban m pour câbles	Kabelband n	nastro m per cavi
1337	tape (friction tape)	ruban m pour friction	Friktionsband n	nastro m per frizionatura
1338	tape (insulating tape)	ruban m isolant	Isolierband n	nastro m isolante
1339	tape (pressure sensitive tape)	ruban m adhésif sensible à la pression	Selbstklebeband n	nastro m autoadesivo
1340	tapered	conique	konisch	conico m
1341	taping machine	machine f à rubanner	Bandwickelmaschine f	nastratrice f
1342	tar	goudron m	Teer m	catrame m
1343	tarnish, to	ternir	anlaufen	opacizzare
1344	tarpaulin	bâche f	Plane f, Persenning f	telone m
1345	tear (Groves tear)	déchirement m Groves	Groves-Weiterreißprüfung f	lacerazione f Groves
1346	tear initiation	amorçage m du déchirement	Rißeinleitung f	inizio m della lacerazione
1347	tear n	déchirure f, déchirement m	Riß m	lacerazione f
1348	tear propagation	propagation f du déchirement	Rißwachstum f	propagazione f della lacerazione
1349	tear resistance	résistance f au déchirement	Weiterreißwiderstand m	resistenza f alla lacerazione
1350	tear resistance determination with a nicked test piece	détermination f de la résistance f au déchirement amorcé	Weiterreißwiderstandsprüfung f an einem eingeschnittenen Prüfkörper	determinazione f della resistenza alla lacerazione su provino intagliato
1351	tear strength	résistance f au déchirement	Weiterreißwiderstand m	resistenza f alla lacerazione
1352	tear, to	déchirer	reißen	lacerare
1353	tearability	aptitude f au déchirement	Zerreißbarkeit f	lacerabilità f
1354	tearing	déchirement m	Reißen n	strappo m
1355	tearing point	point m de déchirement	Einreißpunkt m	punto m di lacerazione
1356	technique	technique f	Technik f	tecnica f
1357	technologist	technicien m	Techniker m, Technologe m	tecnologo m
1358	tellurium	tellure m	Tellur n	tellurio m
1359	temperature	température f	Temperatur f	temperatura f
1360	temperature (ambient temperature)	température f ambiante	Umgebungstemperatur f	temperatura f ambiente
1361	temperature (barrel temperature)	température f du fourreau	Zylindertemperatur f	temperatura del cilindro
1362	temperature (brittleness temperature)	température de fragilisation	Versprödungstemperatur f	temperatura f di infragilimento
1363	temperature (continuous service temperature)	température f de service en continu	Einsatztemperatur f (kontinuierlich)	temperatura f di esercizio continuo
1364	temperature (deflection temperature)	température f de fléchissement sous charge	Formbeständigkeit f in der Wärme	temperatura f di deflessione
1365	temperature drop	chute f de température	Temperaturabfall m	calo m di temperatura

T

Nr	ESPAÑOL	PORTUGUÊS	SVENSKA	NEDERLANDS
1326	pegajosidad o mordiente en seco o crudo	pegajosidade f inicial	torr klibb	droogkleverigheid
1327	medidor m de mordiente	medidor m de adesividade	klibbmätare	kleefkrachtmeter
1328	inhibidor m de mordiente	inibidor m de pegajosidade, inibidor m de adesividade	klibbmotmedel	kleefkracht reducerend middel
1329	agente de pegajosidad	agente de pegajosidade	klibbmedel	kleverigmaker, kleefmiddel
1330	adhesividad f, adherencia f	adesividade f	vidhäftningsförmåga	kleefvermogen, kleverigheid
1331	pegajoso, con mordiente	pegajoso	klibbig	kleverig
1332	tanque o cuba con circulación	tanque m de circulação, cuba f de circulação	cirkulationstank	circulatie tank
1333	revestimiento m de tanques	revestimento m de tanque, fôrro m de depósito	tankbeklädnad	tankbekleding
1334	cinta f	fita f, cinta f	band	band, lint
1335	cinta f adhesiva	fita f adesiva	tejp	kleefband, zelfklevend band
1336	cinta f para cables	fita f para cabos	kabel tape	kabellint
1337	cinta f calandrada por fricción	fita f de fricção	isolerband	gefrictioneerd band
1338	cinta f aislante	fita f isolante, fita f isoladora	isolerband	isolatieband
1339	cinta autoadhesiva	fita autocolante/fita f adesiva	tryckkänslig tape	zelfklevend band
1340	cónico, ahusado	cônico, afusado	konisk	kegelvormig, taps
1341	máquina de encintar	máquina f de aplicar fita, cingideira f	bandningsmaskin	bandwikkelmachine
1342	alquitrán m, brea f	alcatrão m, pez m	tjära	teer
1343	deslustrar, matar el brillo	deslustrar, embaçar	mattera	dof worden, mat worden
1344	lona embreada o alquitranada	encerado m	presenning	dekkleed
1345	desgarro m Groves	rasgamento m Groves	rivhållfasthet enligt Groves	scheursterkte volgens Groves
1346	inicio m del desgarro	início m do rasgamento, rasgo m inicial	rivinitiering	begin van inscheuren
1347	desgarradura f	rasgão m, rasgo m	rivning	scheur
1348	propagación f del desgarro o corte	propagação f do rasgamento, ampliação f do rasgamento	rivfortplantning	groter worden van scheur
1349	resistencia al desgarro	resistência f ao rasgamento	rivhållfasthet	scheurweerstand
1350	ensayo de resistencia al desgarro con probeta con corte inicial	determinação f da resistência ao rasgamento em amostra com entalhe	skärfasthet	meting scheurweerstand bij insnijding vlg ASTM D 624
1351	resistencia f al desgarramiento	resistência f ao rasgamento	rivhållfasthet	scheursterkte
1352	desgarrar	rasgar	riva	scheuren
1353	capacidad de desgarro	aptidão f para ser rasgado, rasgabilidade f	rivbarhet	scheurgevoeligheid
1354	desgarramiento m	rasgamento m	rivning	het scheuren
1355	punto m de desgarramiento	ponto m de rasgamento	rivpunkt	scheurpunt
1356	técnica f	técnica f	teknik	techniek
1357	técnico m	técnico m	tekniker	technicus, technoloog
1358	telurio m	telúrio m	tellur	tellurium
1359	temperatura f	temperature f	temperatur	temperatuur
1360	temperatura f ambiente	temperatura f ambiente	temperatur hos omgivningen	omgevingstemperatuur
1361	temperatura del cilindro	temperatura do cilindro	cylinder temperatur	cylinder temperatuur van spuitmachine
1362	temperatura de fragilidad	temperatura de fragilidade	sprödpunkt	temperatuur waarbij materiaal bros wordt
1363	temperatura de servicio continuo	temperatura de serviço contínua	kontinuerlig service temperatur	continue bedrijfstemperatuur
1364	temperatura de deflexión	temperatura de deflecção	deflektionstemperatur	doorhang temperatuur
1365	caída f de temperatura	queda f de temperatura	temperaturfall	daling van temperatuur

temperature (flash-ignition temperature)

Nr	ENGLISH	FRANÇAIS	DEUTSCH	ITALIANO
1366	temperature (flash-ignition temperature)	température f (d'auto-inflammation)	Flammpunkt m, Entzündungstemperatur f	temperatura f di infiammabilità
1367	temperature index	indice m de température	Temperaturindex	indice m di temperatura
1368	temperature (melt temperature)	température f (de fusion)	Schmelztemperatur f	temperatura di fusione
1369	temperature range	gamme f de température	Temperaturbereich m	gamma f di temperatura
1370	temperature resistance (low temperature resistance)	résistance f à basse température	Tieftemperaturbeständigkeit f	resistenza f alle basse temperature
1371	temperature rise	hausse f de température	Temperaturanstieg m	aumento m di temperatura
1372	temperature (room temperature)	température f ambiante	Raumtemperatur f	temperatura f ambiente
1373	temperature (transition temperature)	température de transition	Umwandlungstemperatur f	temperatura f di transizione
1374	temperature (Vicat softening temperature)	température f de ramollissement Vicat, point Vicat	Vicat Erweichungspunkt m	temperatura f di rammollimento Vicat
1375	tenacity	ténacité f	Zähigkeit f	tenacità f
1376	tensile modulus	module m d'élasticité en traction	Zugspannungswert m, Modul m	modulo m a trazione
1377	tensile strength	résistance f à la traction à la rupture	Zugfestigkeit	resistenza f alla trazione
1378	tensile stress	effort m de traction, contrainte de traction	Zugspannung f	sforzo m di trazione
1379	tension cable (high frequency tension cable)	câble m conducteur pour haute fréquence	Hochfrequenzspannungskabel n	cavo m per alta frequenza
1380	tension control stand for takeup	régulateur m de tension de tirage	Spannungsanzeiger m für Aufwickelmaschine f	regolatore m di tensione per il riavvolgimento
1381	tension (critical surface tension)	tension f de surface critique	kritische Oberflächenspannung f	tensione f superficiale critica
1382	tension-deflection	tension - déformation f	Spannungsbiegung f	tensione f deflessione f
1383	terpolymer	terpolymère m	Terpolymer n	terpolimero
1384	test	essai m, test m	Versuch m, Prüfung f	prova f
1385	test (accelerated ageing test)	essai m (de vieillissement accéléré)	Alterungsprüfung f (beschleunigt)	prova f d'invecchiamento accelerato
1386	test (accelerated light ageing test)	essai m (de vieillissement accéléré à la lumière)	Alterungsprüfung f (beschleunigt) durch Lichteinwirkung	prova d'invecchiamento accelerata dalla luce
1387	test (accelerated weathering test)	essai m (de vieillissement accéléré aux intempéries)	Bewitterungsprüfung f (beschleunigt)	prova d'invecchiamento accelerata agli agenti atmosferici
1388	test (compression cutting test)	essai m (de résistance à la coupure par compression)	Kerbdruckversuch m	prova f di resistenza al taglio per compressione
1389	test (compression deflection test)	essai m (de déformation par compression)	Druckverformungsprüfung f	prova f di deformazione per compressione
1390	test (corrosion test)	essai m (de corrosion)	Korrosionsprüfung f	prova f di corrosione
1391	test (crescent tear test)	essai m de déchirement (méthode du haricot)	Kerbzähigkeitstest m mittels nierenförmigen Prüfkörpers	prova f di propagazione della lacerazione
1392	test (endurance flexing test)	essai m de résistance aux flexions répétées	Dauerbiegeprüfung f	prova f di resistenza alle flessioni ripetute
1393	test (flex test)	essai m de flexibilité	Biegeprüfung f	prova f di flessione
1394	test (flexibility test)	essai m de flexibilité	Biegeprüfung f	prova f di flessibilità
1395	test fluid	liquide m d'essai	Testflüssigkeit f, Prüfflüssigkeit f	liquido m di prova
1396	test (hold test)	essai m (de maintien)	Spannprüfung f	prova f di tenuta
1397	test (laboratory test)	essai m (de laboratoire)	Laboratoriumsprüfung f	prova f di laboratorio
1398	test (mechanical test)	essai m (mécanique)	Prüfung f (mechanisch)	prova f meccanica
1399	test method	méthode d'essai	Prüfmethode f	metodo m di prova
1400	test (notch bending test)	essai m (de flexion sur éprouvette entaillée)	Kerbbiegeprüfung f	prova f di flessione con provini incisi
1401	test (performance test)	essai m (d'endurance)	Gebrauchsfähigkeitsprüfung f, Einsatzprüfung f	prova f di prestazione
1402	test (permanent compression test)	essai m (de compression permanente)	Prüfung f der bleibenden Druckverformung f	prova f di deformazione permanente a compressione
1403	test (shearing test)	essai m (de cisaillement)	Scherversuch m	prova f di taglio
1404	test (simulated service test)	essai m (de service simulé)	Prüfverfahren n unter nachgeahmten Praxisbedingungen	prova f di esercizio simulato
1405	test (swell test)	essai m (de gonflement)	Quellprüfung f	prova di rigonfiamento

Nr	ESPAÑOL	PORTUGUÊS	SVENSKA	NEDERLANDS
1366	temperatura de inflamabilidad	temperatura de inflamação expontânea	flam punkt	ontvlamtemperatuur
1367	indice de temperatura	indice de temperatura	temperaturindex	temperatuur index
1368	temperatura de fusión	temperatura de fusão	smält-temperatur	smelt temperatuur
1369	margen m de temperaturas	amplitude f de temperaturas, gama f de temperaturas	temperaturområde	temperatuurbereik
1370	resistencia f a bajas temperaturas	resistência f às temperaturas baixas	köldbeständighet	bestendigheid tegen lage temperatuur
1371	subida f de temperatura, elevación f de temperatura	subida f de temperatura	temperaturuppgång	stijging van temperatuur
1372	temperatura f ambiente	temperatura f ambiente	rumstemperatur	kamertemperatuur
1373	temperatura de transicion	temperatura de transição	övergångstemperatur	overgangstemperatuur
1374	temperatura de ablandamiento Vicat	temperatura de amolecimento Vicat	vicat mjukningspunkt	vicat verwekingstemperatuur
1375	tenacidad f	tenacidade f	hållfasthet	taaiheid
1376	modulo de tracción	módulo de tensão	spännings modul	modulus, stramheid
1377	resistencia f a la tracción	resistencia à tracção	draghållfasthet	treksterkte
1378	esfuerzo m de tensión	esforço m de tracção, tensão f de tracção	dragpåkänning	trekspanning
1379	cable m de tensión de alta frecuencia	cabo m condutor para alta frequência	högspänningskabel	hoge frequentie spanningskabel
1380	dispositivo de control de la tensión de enrollado o rebobinado	regulador m da tensão para rebobinagem	dragregleringsanordning för upptagning	regelaar van de spanning op wikkelmachine
1381	tensión superficial crítica	tensão superficial crítica	kritisk ytspänning	kritische oppervlakte spanning
1382	tensión-deflexión f	tensão-deflexão f	spänningsinböjning	afbuiging van de spanning
1383	terpolímero	terpolímero	ter polymer	terpolymeer
1384	ensayo m, prueba f	ensaio m, teste m	prov	test, beproeving
1385	ensayo m de envejecimiento acelerado	prova f de envelhecimento acelerado	accelererat åldringsprov	versnelde verouderingsproef
1386	ensayo m acelerado de envejecimiento por acción de la luz	prova f de envelhecimento acelerado pela luz	accelererat ljusåldringsprov	versnelde verouderingsproef door lichtinwerking
1387	ensayo m de intemperización acelerada	prova f acelerada de envelhecimento por intempérie	accelererad väderbeständighetsprovning	versnelde verweringsproef
1388	ensayo m de cortado por compresión	ensaio m ao corte por compressão	tryckrivprovning	drukbestandheidproef
1389	ensayo de compresión deflexión	ensaio m de deformação por compressão	prov avseende formförändring genom tryckbelastning	drukdeformatieproef
1390	ensayo m de corrosión	ensaio m de corrosão	korrosionsprov	corrosieproef
1391	ensayo de desgarro (probeta en media luna o croissant)	ensaio m de aumento de rasgamento	rivhållfasthet med njurformad provstav	crescentscheurproef (scheurweerstandsbepaling met niervormig proefstuk)
1392	ensayo m de resistencia a la flexión repetida	prova f de resistência à flexão repetida	böjuthållighetsprov	testapparaat voor buigbelasting
1393	ensayo m de flexión	ensaio m de flexão	böjprovning	buigproef
1394	ensayo m de flexibilidad	prova f de flexibilidade	böjprovning	buigzaamheidstest
1395	liquido m de ensayo	liquido m de ensaio	provvätska	testvloeistof
1396	prueba de colgado	teste de fixação	håll test	klemproef
1397	ensayo m de laboratorio	ensaio m de laboratório	laboratorieprov	laboratorium proef
1398	ensayo m mecánico	prova f mecânica	mekanisk provning	mechanische beproeving
1399	metodo de ensayo	procedimento de teste	test metod	testmethode
1400	ensayo de doblado con corte inicial	ensaio m de dobramento com entalhe	insnittsböjprov	kerfbuigproef
1401	ensayo m de comportamiento	ensaio m de comportamento prático	användbarhetsprov	onderzoek onder gebruiksomstandigheden
1402	ensayo de deformación por compresión	ensaio m de deformação permanente por compressão	permanent formförändring genom tryck	proef voor bepaling van blijvende vervorming onder druk
1403	prueba f de cizallamiento	ensaio m de cisalhamento, ensaio m de corte	skjuvprov	afschuifproef
1404	prueba f der servicio simulado	ensaio m de serviço simulado	simulerat driftprov	nagebootste bedrijfsproef
1405	ensayo de hinchamiento	ensaio de inchamento, teste de inchamento	svällningstest	zwelproef

T

test (tear propagation test)

Nʀ	ENGLISH	FRANÇAIS	DEUTSCH	ITALIANO
1406	test (tear propagation test)	essai m (de propagation du déchirement)	Weiterreißprüfung f	prova f di propagazione della lacerazione
1407	test (tear test)	essai m (de déchirement)	Weiterreißwiderstandsprüfung f	prova f di lacerazione
1408	test (tearing test)	essai m (de déchirement)	Reißversuch m	prova f di lacerazione
1409	test (water permeability test)	essai m (de perméabilité à l'eau)	Wasserdurchlässigkeitsprüfung f	prova di permeabilità all'acqua
1410	tester (air permeability tester)	perméamètre m à air	Luftdurchlässigkeitsprüfer m	apparecchio m per misurare la permeabilità all'aria
1411	tester (bending tester)	machine f pour essais de flexions	Biegeprüfer m	macchina f per prove di flessione
1412	tester (brittleness tester)	appareil m pour mesures de fragilité	Sprödigkeitsprüfer m	apparecchio m per misurare la fragilità
1413	tester (bursting strength tester)	éclatomètre m	Berstfestigkeitsprüfer m, Falzfestigkeitsprüfer m (Papier)	dinamometro m per prove di scoppio
1414	tester (cold brittleness tester)	appareil m (pour la mesure de la fragilité à basse température)	Kältesprödigkeitsprüfgerät n	apparecchio m per prove di fragilità a bassa temperatura
1415	tester (flex tester)	machine f (pour essais de flexion)	Biegeprüfgerät n	macchina f per prove di flessione (flessometro)
1416	tester (flexing strength tester)	appareil m (de mesure m de résistance à la flexion)	Biegefestigkeitsprüfer m	macchina f per prove di resistenza a flessione
1417	tester (folding tester)	pliographe m	Falzzahlprüfgerät n	macchina f per prove di piegamento
1418	tester (gas permeability tester)	porosimètre m pour gaz	Gasdurchlässigkeitsprüfer m	porosimetro m per gas
1419	tester (Hoffman scratch-hardness tester)	appareil m Hoffman pour la mesure de la dureté par usure	Kratzhärteprüfer m (Hoffman)	apparecchio m Hoffman per misurare la durezza mediante raschiatura
1420	tester (initial wet strength tester)	appareil m pour la mesure de la résistance initiale à l'état humide	Anfangsnaßfestigkeitsprüfer m	apparecchio per la misura della resistenza iniziale ad umido
1421	tester (Schopper machine, tensile tester)	machine f Schopper (pour essais de résistance à la traction)	Zerreißfestigkeitsprüfer m (Schopper)	macchina f Schopper (dinamometro)
1422	tester (Scott tensile tester)	machine f (Scott pour essais de traction)	Zerreißfestigkeitsprüfer m (Scott)	dinamometro m Scott
1423	tester (stiffness tester)	appareil m (pour la mesure de la rigidité)	Steifheitsprüfer m	apparecchio m per prove di rigidità
1424	tester (tearing tester)	machine f (pour essais de déchirement)	Reißfestigkeitsprüfer m	apparecchio m per prove di lacerazione
1425	testing equipment	appareillage m d'essai	Prüfgeräte npl	apparecchio m di prova
1426	testing instrument (electrical)	instrument d'essai m électrique	elektrisches Prüfinstrument n	strumento m per prove elettriche
1427	tetrafluoroethylene - TFE	éthylène tétrafluoré ou tétrafluoroéthylène	Tetrafluorethylen n -TFE	tetrafluoroetilene
1428	textile	tissu m, textile m, toile f	Textil n	tessuto
1429	texture	structure f	Gefüge n, Struktur f	struttura f
1430	thermal decomposition	décomposition f thermique	thermische Zersetzung f	decomposizione f termica
1431	thermal deformation	déformation f thermique	thermische Verformung f	deformazione f termica
1432	thermal expansion	expansion f thermique, dilatation f thermique	Wärmeausdehnung f	dilatazione f termica
1433	thermal instability	indice de stabilité thermique	thermische Instabilität f	instabilità f termica
1434	thermoplastic (adj)	thermoplastique	thermoplastisch	termoplastico (agg.)
1435	thermoplasticity	thermoplasticité f	Thermoplastizität f	termoplasticità f
1436	thermosetting	durcissant à chaud, thermodurcissable	hitzehärtend	termoindurente (agg.)
1437	thick-walled goods	articles mpl à parois épaisses	Artikel mpl (starkwandig), dickwandige Artikel mpl	articoli m a pareti spesse
1438	thixotropic (adj)	thixotrope	thixotrop	tissotropico (agg.)
1439	tile (floor tile)	dalle f de sol	Bodenplatte f, Fußbodenbelag m	mattonella f per pavimenti
1440	tile (rubber tile)	dalle f en caoutchouc	Gummibodenplatte f	piastrella f di gomma
1441	time (curing time)	temps m de vulcanisation	Vulkanisationsdauer f	tempo m di vulcanizzazione
1442	time (hold-up time)	temps m de maintien	Verweilzeit f	tempo di permanenza
1443	time of cure	temps f de vulcanisation	Vulkanisationszeit f	tempo m di vulcanizzazione
1444	time (optimum cure) (TC90)	temps optimum de vulcanisation (TC90)	Vernetzungszeit f, Zeit für 90% Vernetzung im Rheometer (T90)	tempo di vulcanizzazione ottimale (TC90)
1445	time (set-up time)	temps m de solidification	Einrichtungszeit f	tempo m di solidificazione

Nr	ESPAÑOL	PORTUGUÊS	SVENSKA	NEDERLANDS
1406	ensayo m de propagación del desgarramiento	ensaio m da propagação do rasgamento	rivfortplantningsprov	scheurproef (uitgaande van een inscheuring)
1407	ensayo m de desgarramiento	ensaio m de rasgamento	rivprov	scheurproef
1408	ensayo m de desgarramiento	ensaio m de rasgamento	rivprov	scheurproef
1409	prueba f de permeabilidad al agua	ensaio de permeabilidade à agua	prov av vattengenomsläpplighet	waterdoorlaatbaarheidsproef
1410	aparato m para ensayos de permeabilidad al aire	aparelho m para ensaio de permeabilidade	luft permeabilitets testare	luchtporositeits-tester
1411	aparato para ensayos de doblado	máquina f para ensaios de flexão	böjprovare	buigproefmachine
1412	aparato de medicion de fragilidad	aparelhos m para ensaios de friabilidade	sprödhetsprovare	brosheidsmeter
1413	aparato para ensayar la resistencia a reventar	equipamento m para ensaio de resistência ao rebentamento m	sprängtryckprovare	barstdrukmeter
1414	medidor m de fragilidad a temperaturas bajas	aparelho m de ensaio de friabilidade a baixas temperaturas	provningsapparat för köldsprödhet	brosheidsmeter bij een lage temperatuur
1415	flexómetro m	equipamento m para ensaio de flexão	böjprovningsmaskin	buigapparaat
1416	máquina f para ensayos de resistencia a la flexión	máquina f para ensaios de resistência à flexão	provapparat för böjhållfasthet	apparaat voor de bepaling van de buigvastheid
1417	aparato m para ensayar la resistencia al plegado	máquina f para ensaio de dobramento repetido	vikningsprovare	apparaat ter bepaling van de vouwweerstand
1418	comprobador m permeabilidad a los gases	porosímetro m para gases	gas permeabilitetstestare	gasdoorlaatbaarheidstester
1419	aparato Hoffman medir la dureza al rayado	aparelho m Hoffman para medida da dureza por arranhadura	Hoffman ritshårdhetsmätare	Hoffmann`s kras-hardheidsmeter
1420	aparato m para probar la resistencia inicial en húmedo	aparelho m para ensaios de resistência inicial com amostra húmida	provapparat för initial våtstyrka	meter van de oorspronkelijke natte sterkte
1421	máquina f Schopper (dinamómetro)	máquina f Schopper (para ensaios de resistência à tracção)	Schopper dragprovmaskin	trekbank (Schopper)
1422	máquina f de Scott para ensayos de tracción	dinamómetro m Scott para ensaios de tracção	Scott dragprovmaskin	Scott trekbank
1423	aparato m para ensayos de rigidez	aparelho m para ensaios de rigidez	styvhetsprovare	stijfheidsmeter
1424	máquina f para ensayos de desgarramiento	aparelho m para ensaios de rasgamento	rivprovare	scheurvastheidsmeter
1425	equipo m de ensayo	aparelhagem f de ensaio	provningsutrustning	beproevingsapparatuur
1426	instrumento eléctrico de ensayo	instrumento m eléctrico para testes	elektriskt provinstrument	electrische beproevingsapparatuur
1427	tetrafluoroetileno - TFE	tetrafluoroetileno - TFE	tetrafluoreten - TFE	tetrafluorethylene - TFE
1428	textil m, teijido m	têxtil m, tecido m	textil	textiel
1429	estructura f	textura f, estrutura f	textur	structuur
1430	descomposición f térmica	decomposição f térmica	termisk nedbrytning	thermische ontleding
1431	deformación f térmica	deformação f térmica	termisk deformation	vervorming door warmte
1432	expansión f térmica, dilatación f térmica	dilatação f térmica, expansão f térmica	termisk utvidgning	thermische uitzetting
1433	inestablidad termica	instabilidade térmica	termisk instabilitet	thermische instabiliteit
1434	termoplástico	termoplástico	termoplastisk	thermoplastisch
1435	termoplasticidad f	termoplasticidade f	termoplasticitet	thermoplasticiteit
1436	termoendurecible, termoestable	termo-endurecido m	värmehärdande	warmhardend
1437	artículos mpl de paredes gruesas	artigos mpl de paredes espessas	tjockväggigt gods	dikwandige artikelen
1438	tixotrópico	tixotrópico	tixotropisk	thixotroop
1439	baldosa f	ladrilho m	golvplatta	vloertegel
1440	baldosa f de caucho	ladrilho m de borracha	gummiplatta	rubber tegel
1441	tiempo m de vulcanización	tempo m de vulcanização	vulktid	vulcanisatietijd, vulcanisatieduur
1442	tiempo de permanencia	tempo de retenção	stopp tid	verblijftijd
1443	tiempo m de vulcanización	tempo m de vulcanização	vulktid	vulcanisatietijd
1444	tiempo de vulcanización optimo	tempo de vulcanização óptimo	optimal vulktid	optimale vulcanisatie tijd (TC90)
1445	tiempo de solidificación	tempo de soldificação	omställningstid	instel tijd

T

timing belt

Nr	ENGLISH	FRANÇAIS	DEUTSCH	ITALIANO
1446	timing belt	courroie f crantée	Zahnriemen m, Steuerriemen m	cinghia di trasmissione
1447	tire (solid), tyre	bandage m plein	Vollgummireifen m	pneumatico m pieno
1448	tire (tyre) cord dip	mélange m de trempage pour câblés de pneus	Reifenkordimprägniermischung f	mescola f per immersione di corda per pneumatici
1449	tire (tyre) tread	bande f de roulement de pneu	Reifenlauffläche f	battistrada m del pneumatico
1450	titanium dioxide	dioxyde m de titane	Titandioxid n	biossido m di titanio
1451	tolerance	tolérance f	Toleranz f	tolleranza f
1452	toluene	toluène m	Toluol n	toluolo m
1453	ton	tonne f	Tonne f	tonnellata f
1454	torque	couple m de torsion	Drehmoment n	coppia f di torsione
1455	torque (90%)	couple à 90% (MC90)	Drehmoment n bei Vernetzungsgrad 90% im Rheometer (M90)	coppia al 90%
1456	torque (maximum) (MH)	couple maximum (MH)	Drehmomentmaximum n im Rheometer (MH)	coppia massima
1457	torque (minimum) (ML)	couple minimum (ML)	Drehmomentminimum n im Rheometer (ML)	coppia minimo
1458	torsion	torsion f	Torsion f	torsione f
1459	torsional stress	contrainte f de torsion	Torsionsspannung f	sforzo a torsione
1460	toughness	ténacité f	Festigkeit f, Zähigkeit f	tenacità f
1461	toxicity	toxicité f	Toxizität f, Giftigkeit f	tossicità f
1462	transition	transition f	Übergang m (z.B. Zustandsänderung)	transizione f
1463	transmittance (light transmittance)	transmittance f de la lumière	Lichtdurchlässigkeit f	trasmittanza f della luce
1464	transmitted energy	énergie f transmise	übertragene Energie f	energia f trasmessa
1465	transparent (adj)	transparent	transparent	trasparente
1466	trap (radical trap)	piège m à radicaux	Radikalfänger m	accettore di radicali
1467	tread	bande f de roulement, chape f	Reifenlauffläche f	battistrada m
1468	treadwear	usure f de la bande de roulement	Laufflächenabnutzung f	usura f del battistrada
1469	treatment	préparation f, traitement m	Behandlung f	trattamento m
1470	truck cover	bâche f pour camion	Lastwagenplane f	telone m per autocarro
1471	truck tire (tyre)	pneu m pour camion	Lastwagenreifen m	pneumatico m per autocarro
1472	twin-screw	double vis	Doppelschnecke f	doppia vite
1473	twisting moment	moment m de torsion	Drehmoment n	momento m torcente
1474	typical measured values	valeurs fpl de mesures typiques	Meßwerte fpl (typisch)	valori m tipici misurati
1475	tyre (tire passenger car)	pneu m tourisme	Personenwagenreifen m	pneumatico m per automobili
1476	U-ring	anneau m en U	Nutringdichtung f	anello a U
1477	ultimate tensile strength	résistance f à la rupture	Zerreißfestigkeit f	carico m di rottura
1478	ultra violet lamp	lampe f à rayonnement ultraviolet	Ultraviolettlampe f	lampada f ultra-violetta
1479	ultra violet light	lumière f ultraviolette	Ultraviolettlicht n	luce f ultra-violetta
1480	under-cure	sous - vulcanisation f	Untervulkanisation f	sotto-vulcanizzato
1481	under-cured	sous - vulcanisé	untervulkanisiert	sotto-vulcanizzata
1482	unit	unité f	Einheit f	unità f
1483	urethane	uréthanne m	Urethan n	uretano m
1484	V-belt	courroie f trapézoïdale, courroie f en V	Keilriemen m	cinghia f trapezoidale
1485	V-belt pulley	poulie f pour courroies trapézoïdales	Keilriemenscheibe f	puleggia f per cinghia trapezoidale
1486	vacuum brake parts	pièces fpl pour freins pneumatiques	Vakuumbremsenteile npl	parti f per freni a depressione
1487	vacuum extruder	extrudeuse f (sous vide)	Vakuumextruder m	estrusore m (sotto vuoto)
1488	value	valeur f	Wert m	valore m
1489	value (limiting value)	valeur f limite	Grenzwert m	valore m limite
1490	value-in-use	valeur f à l'usage, rapport m qualité/prix	Gebrauchswert m	valore m in uso

Nr	ESPAÑOL	PORTUGUÊS	SVENSKA	NEDERLANDS
1446	correa dentada	correia dentada	kuggrem	getande riem
1447	bandaje m	roda/pneumaciço, bandage f	massivdäck	massieve band
1448	mezcla f para inmersión de manguera de neumáticos	solução f para imersão de filamentos para pneus	kordimpregneringsanläggning	het dompelen van bandenkoord
1449	banda f de rodamiento de neumático	banda f de rodagem de pneumático	däckslitbana	loopvlak
1450	dióxido m de titanio	dióxido m de titânio	titandioxid	titaandioxide
1451	tolerancia f	tolerância f	tolerans	tolerantie
1452	tolueno m, toluol m	tolueno m, toluol m	tolven	tolueen
1453	tonelada f	tonelada f	ton	ton
1454	momento m de torsión	torque m, momento de torção	vridmoment	draaimoment
1455	par al 90%	torque a 90%	90% av moment maximum (MC90)	90% draaimoment
1456	par máximo	torque máximo	moment maximum (MH)	maximum koppel (MH)
1457	par mínimo	torque mínimo	moment minimum (ML)	minimum koppel (ML)
1458	torsión f	torção f	torsion	torsie
1459	esfuerzo m de torsión	esfôrço m de torção	torsionspåkänning	torsiespanning
1460	tenacidad f	tenacidade f	seghet	taaiheid
1461	toxicidad f	toxicidade f, toxidez f	giftighet	giftigheid
1462	transición f	transição	övergång	overgang
1463	transmitancia (de la luz)	transmitância luminosa	ljusgenomsläpplighet	lichtdoorlaatbaarheid
1464	energía f transmitida	energia f transmitida	överförd energi	overgedragen energie
1465	transparente	transparente	transparent	transparant
1466	radical estabilizado	radical estabilizado	radikal absorberare	radikaalvanger
1467	banda f de rodamiento	banda f de rodagem	slitbana	loopvlak
1468	desgaste m de la banda de rodamiento	desgaste m de banda de rodagem	slitbanenedslitning	slijtage van loopvlak
1469	tratamiento m	tratamento m	behandling	behandeling
1470	toldo m para camión	encerado m para camião	lastbilsöverdrag	dekkleed
1471	neumático m para camión	pneu m, pneumático m para camião	lastbilsdäck	vrachtautoband
1472	doble husillo	duplo fuso	dubbelskruv	dubbel-schroef
1473	momento m de torsión	momento m de torção, torque m	torsionsmoment	draaimoment
1474	valores medidos típicos	valores típicos medidos	typiska test värden	specifiek gemeten waarden
1475	neumático m de turismo	pneu m para automóvel de turismo	personbilsdäck	personenauto band
1476	junta en U	ó-ring, anel de vedação, junta tórica	U-ring	U-ring
1477	carga f de rotura	resistência f máxima à ruptura	brotthållfasthet	maximale treksterkte
1478	lámpara f ultravioleta	lâmpada f ultravioleta	ultraviolett lampa	ultraviolette lamp
1479	luz f ultravioleta	luz f ultravioleta	ultraviolett ljus	ultraviolet licht
1480	subcurado	insuficientemente vulcanizado	undervulkad	niet uitgevulcaniseerd
1481	subvulcanizado	sub-vulcanizado	undervulkad	niet uitgevulcaniseerd
1482	unidad	unidade	enhet	eenheid
1483	uretano m	uretano m	uretan	urethaan
1484	correa f trapezoidal, correa f en V	correia f em V, correia f trapezoidal	kilrem	V-snaar
1485	polea f para correa en V o trapezoidales	polia f para correias trapezoidais	kilremsskiva	V-snaarschijf
1486	piezas fpl para frenos al vacío	peças fpl para travões a vácuo	vakuumbromsdelar	delen van vacuumremmen
1487	extrusion de vacio	extrusora (vacuo)	sträng spruta (vakum)	spuitmachine (vacuum)
1488	valor	valor	värde	waarde
1489	valor m límite	valor m limite	gränsvärde	grenswaarde, limiet
1490	valor al uso	valor em uso	använder värde	gebruikswaarde

T

valve (diaphragm control valve)

V

Nr	ENGLISH	FRANÇAIS	DEUTSCH	ITALIANO
1491	valve (diaphragm control valve)	vanne f de contrôle à diaphragme	Membransteuerventil n	valvola f di controllo del diaframma
1492	VAMAC ethylene/acrylic elastomer	élastomère éthylène acrylique VAMAC	VAMAC Ethylen/Acrylatkautschuk m	VAMAC elastomero etilen/acrilico
1493	vaporisation	vaporisation f	Verdampfung f	vaporizzazione f
1494	vehicle (in paint)	véhicule m	Träger m (in Farben)	veicolo m (in verniciatura)
1495	velocity	vitesse f, vélocité f, rapidité f	Geschwindigkeit f	velocità f
1496	vent	évent m	Entlüftung f	sfogo dell'aria
1497	vibration damper	amortisseur m de vibrations	Vibrationsdämpfer m	ammortizzatore m di vibrazioni
1498	vibration damping	amortissement m des vibrations	Vibrationsdämpfung f	smorzamento m delle vibrazioni
1499	vinyl chloride	chlorure m de vinyle	Vinylchlorid n	cloruro di vinile
1500	vinylidene fluoride	fluorure m de vinylidène	Vinylidenfluorid n	fluoruro di vinilidene
1501	viscometer (ball viscometer)	viscosimètre m à bille	Kugelfallviskosimeter n	viscosimetro m a sfera
1502	viscosity	viscosité f	Viskosität f	viscosità f
1503	viscosity (Brookfield viscosity)	viscosité f Brookfield	Viskosität f (Brookfield)	viscosità f Brookfield
1504	viscosity (melt viscosity)	viscosité du polymère à l'état fondu	Schmelzviskosität f	viscocità f alla stato fuso
1505	viscosity (Mooney)	viscosité f Mooney	Viskosität f (Mooney)	viscosità f Mooney
1506	viscosity (relative)	viscosité f relative	Viskosität f (relativ)	viscosità f relativa
1507	viscous-elastic material	matériau m visco-élastique	visko-elastisches Material n	materiale m visco-elastico
1508	VITON fluoroelastomer	élastomère fluoré VITON	VITON Fluorelastomer m	VITON fluoroelastomero
1509	volatile (adj)	volatil	flüchtig	volatile (agg.)
1510	volatility	volatilité f	Flüchtigkeit f	volatilità f
1511	volume	volume m	Volumen n	volume m
1512	volume change	changement m de volume	Volumenveränderung f	cambiamento m di volume
1513	volume resistance	résistance f volumique	Volumenbeständigkeit f	resistenza f di volume
1514	volume resistivity	résistivité f volumique	spezifischer Durchgangswiderstand m	resistività f di volume
1515	volume swell	gonflement m en volume	Volumenzunahme f durch Quellung	aumento m di volume, rigonfiamento
1516	vulcanisate	vulcanisat m	Vulkanisat n	vulcanizzato
1517	vulcanisation	vulcanisation f, cuisson f	Vulkanisation f	vulcanizzazione f
1518	vulcanisation coefficient	coefficient m de vulcanisation	Vulkanisationskoeffizient m	coefficiente m di vulcanizzazione
1519	vulcanisation rate	vitesse f de vulcanisation	Vulkanisationsgeschwindigkeit f	velocità f di vulcanizzazione
1520	vulcanising mould	moule m de vulcanisation	Vulkanisationsform f	stampo m per vulcanizzazione
1521	warpage	gauchissement m	konvexe Verformung f	deformazione f di planarità, imbarcamento m
1522	washer	rondelle f	Dichtungsring m, Unterlegscheibe f	rondella f
1523	water extract	extrait m aqueux	wässriger Auszug m	estratto m acquoso
1524	water permeability tester	appareil m pour essais de perméabilité à l'eau	Wasserdurchlässigkeitsprüfer m	apparecchio f per prove di permeabilità all'acqua
1525	water permeability vapour	perméabilité f à la vapeur d'eau	Wasserdampfdurchlässigkeit f	permeabilità f al vapor d'acqua
1526	water suction hose	tuyau m d'aspiration d'eau	Wassersaugschlauch m	tubo m aspiratore per acqua
1527	water swell	gonflement m dans l'eau	Quellung in Wasser f	rigonfiamento m in acqua
1528	water vapour	vapeur f d'eau	Wasserdampf m	vapore m d'acqua
1529	waterproof fabric	tissu m imperméable	Gewebe n (wasserdicht)	tessuto m impermeabile
1530	waterproofing	imperméabilisation f, étanchéité	wasserdicht Imprägnieren n	impermeabilizzazione f
1531	wax (microcrystalline wax)	cire f microcristalline	mikrokristallines Wachs n	cera f microcristallina
1532	wear	usure f	Abrieb m, Abnutzung f	usura f, consumo m
1533	wear (abrasive wear)	usure f (par abrasion)	Abrieb m, Abnutzung f	usura f per abrasione
1534	wear at break	usure à la rupture	Abrieb m bei Reißkraft	usura f a rottura
1535	wear at maximum load	usure f à la charge maximale	Abrieb m bei Höchstkraft	usura f a carico massimo

Nr	ESPAÑOL	PORTUGUÊS	SVENSKA	NEDERLANDS
1491	válvula f de control de diafragma	válvula f de contrôle de diafragma	membranreglerventil	membraanregelklep
1492	VAMAC elastómero de etileno acrílico	VAMAC elastómero etileno acrílico	VAMAC eten/akryl elast	VAMAC, ethylene/acrylaat elastomeer
1493	vaporización f	vaporização f	förångning	verdamping
1494	vehículo m	veículo m, excipiente m	bindemedel	drager, bindmiddel
1495	velocidad f	velocidade f, rapidez f	hastighet	snelheid
1496	salida de gases venteo	respiro	ventil	ontluchting
1497	amortiguador m de vibraciones	amortecedor m de vibrações	vibrationsdämpare	trillingsdemper
1498	amortiguación f de vibraciones	amortecimento m de vibrações	vibrationsdämpning	het dempen van trillingen
1499	cloruro de vinilo	cloreto de vinilo	vinyl klorid	vinyl chloride
1500	fluoruro de vinilideno	fluoreto de vinilideno	vinyl fluorid	vinylideen fluoride
1501	viscosímetro m de bola	viscosímetro m de esfera	fallviskosimeter	kogelvalviscosimeter
1502	viscosidad	viscosidade	viskositet	viscositeit
1503	viscosidad f Brookfield	viscosidade f Brookfield	Brookfield viskositet	Brookfield viscositeit
1504	viscosidad de la masa fundida	viscosidade de fusão	smältviskositet	smelt viscositeit
1505	viscosidad f Mooney	viscosidade f Mooney	Mooney viskositet	Mooney viscositeit
1506	viscosidad f relativa	viscosidade f relativa	relativ viskositet	relatieve viscositeit
1507	material m viscoso y elástico	material m viscoso-elástico	visko-elastiskt material	viskeus elastisch materiaal
1508	VITON m fluoroelastomero	VITON m fluoroelastómero - elastómero fluorado	VITON fluorelast	VITON fluorelastomeer
1509	volátil	volátil	flyktig	vluchtig
1510	volatilidad f	volatilidade f	flyktighet	vluchtigheid
1511	volúmen m	volume m	volym	volume
1512	cambio m volumétrico	alteração f volumétrica	volymändring	volumeverandering
1513	resistencia f en volumen	resistência f específica, resistividade f	volymbeständighet	soortelijke weerstand
1514	resistividad volumétrica	resistividade em volume	volym resistivitet	volume weerstand
1515	hinchamiento m en volumen	dilatação f volumétrica/inchamento	volymsvällning	volume toename door zwelling
1516	vulcanizado m	vulcanizado m	vulkanisat	vulkanisaat
1517	vulcanización f	vulcanização f	vulkning	vulcanisatie
1518	coeficiente m de vulcanización	coeficiente m de vulcanização	vulkaniseringskoefficient	vulcanisatie coefficient
1519	velocidad f de vulcanización	velocidade f de vulcanização	vulkhastighet	vulcanisatie snelheid
1520	molde m de vulcanización	molde m de vulcanização	vulkform	vulcaniseervorm, vulcaniseermatrijs
1521	alabeo	tramagem, trama	buktighet	verforming
1522	arandela f plana	arruela f, anilha f	bricka	afdichting
1523	extracto m acuoso	extrato m aquoso	vattenextrakt	waterextract
1524	aparato m para pruebas de permeabilidad al agua	aparelho m para ensaio de permeabilidade à água	provapparat för vattengenomsläpplighet	apparatuur om waterdoorlaatbaarheid te testen
1525	permeabilidad al vapor de agua	permeabilidade ao vapor de água	permeabilitet av vattenånga	permeabiliteit van water damp
1526	manguera f para aspiración de agua	mangueira f para aspiração de água	vattensugslang	water (aan) zuigslang
1527	hinchamiento m por agua	inchamento m pela água	vattensvällning	zwelling door water
1528	vapor m de agua	vapor m de água	vattenånga	waterdamp
1529	tela f impermeable	tecido m impermeável	vattentät väv	waterdicht weefsel
1530	impermeabilización f	impermeabilização f	impregnering	waterdichtmaken
1531	cera microcristalina	cera micro cristalina	mikrokristallint vax	microkristallijne was
1532	desgaste	uso, desgaste	slitage	slijtage
1533	desgaste m por abrasión	degaste m por abrasão; (Port.) degaste m abrasivo	avnötning	slijtage
1534	desgaste a la rotura	desgaste à roptura	slitage vid brott	slijtage bij breuk
1535	desgaste a carga maxima	desgaste a carga maxima	slitage vid maximal belastning	slijtage bij maximale belasting

W

Nr	ENGLISH	FRANÇAIS	DEUTSCH	ITALIANO
1536	wear at yield	usure à la limite élastique	Abrieb m bei Streckkraft (Kraft bei Streckspannung)	usura f allo snervamento
1537	weather (exposure to weather)	exposition f aux intempéries	Bewitterung f	esposizione f agli agenti atmosferici
1538	weather strip	joint d'étanchéité, joint m	Fensterdichtung f, Türdichtung f	trafilato m di tenuta, profilo
1539	weathering	effet m des intempéries, vieillissement climatique	Bewitterung f	invecchiamento m, deterioramento m agli agenti atmosferici
1540	weathering exposure	exposition f aux intempéries	Bewitterung f	esposizione f agli agenti atmosferici
1541	web (continuous web of material in the machine)	trame f	endlose Bahn f eines Materials in der Maschine	foglia f continua
1542	web (woven fabric)	tissu m avec trame	gewobenes Tuch n	tessuto m
1543	weight gain (specific statement)	gain m en poids (état spécifique)	Gewichtzunahme f	aumento m di peso (termine specifico)
1544	welding	soudage m	Schweißen n	saldatura f
1545	wettability	mouillabilité f	Benetzbarkeit f	bagnabilità f
1546	wheel	roue f	Rad n	ruota f
1547	wheel (abrasive wheel)	disque m abrasif	Schleifscheibe f	mola f abrasiva
1548	wheel (buffing wheel)	disque m à polir	Polierscheibe f, Schwabbelscheibe f	mola f pulitrice
1549	wheel (grinding wheel)	meule f abrasive	Schleifrad n	mola f abrasiva
1550	whiting	craie f	Kreide f	carbonato di calcio
1551	windscreen rubber	caoutchouc m pour joints de pare-brises	Frontscheibendichtung f	gomma f per spazzole per parabrezza
1552	wiping	balayage m	Wegschieben n, Wegwischen n	strofinamento
1553	wire (insulated wire)	fil m isolé	Draht m (isoliert)	conduttore m isolato
1554	wood rosin	colophane f	Kolophonium n	colofonia f
1555	work clothing	vêtements mpl de travail	Arbeitskleidung f	abito m da lavoro
1556	worker's footwear	chaussures fpl de travail	Arbeitsschuhe mpl	scarpe f da lavoro
1557	worker's gloves	gants mpl de travail	Arbeitshandschuhe mpl	guanti m da lavoro
1558	worm (endless)	vis f sans fin	Schnecke f (endlos)	vite f senza fine
1559	yarn	fil m textile, fil m	Garn n	filo m
1560	yield (elongation at yield)	limite f d'allongement	Dehnung f bei Streckspannung	allungamento m a snervamento
1561	yield point	limite f élastique	Streckgrenze f	punto di snervamento
1562	Young's modulus	module m de Young	Youngscher Modul m	modulo m di Young
1563	zinc oxide	oxyde m de zinc	Zinkoxid	ossido m di zinco

W

NR	ESPAÑOL	PORTUGUÊS	SVENSKA	NEDERLANDS
1536	desgaste al límite elástico	desgaste no limite elástico	slitage vid brott	slijtage bij vloei
1537	exposición f a la intemperie	exposição f à intempérie	exponering för väder	blootstelling aan weersinvloeden
1538	burlete m para puertas y ventanas	tira f de vedação	tätningslist	tochtstrip
1539	envejecimiento a la intemperie	envelhecimento à intempérie	väder åldring	verweren
1540	exposición f a la intemperie	exposição f à intempérie	väderexponering	blootstellen aan weersinvloeden
1541	banda f continua	teia f	väv	baan (in een machine)
1542	banda continua con trama	tecido m	väv	weefsel
1543	ganancia de peso	ganho de peso	vikt ökning	gewichtsvermeerdering
1544	soldadura f	solda f, soldadura f	svetsning	lassen
1545	humectabilidad	molhabilidade	vätbarhet	vochtopname
1546	rueda f	roda f	hjul	wiel
1547	muela f abrasiva	roda f abrasiva, rebôlo m	slipskiva	slijpschijf
1548	rectificadora, esmeriladora	roda f de polir, disco m polidor	uppruggare	polijstmachine
1549	muela f de esmerilar o pulir	rebôlo m, roda f abrasiva	slipskiva	slijpwiel
1550	creta	cré	krita	krijt
1551	perfiles mpl para parabrisas	borracha f para párabrisas	vindrutegummi	voorruit profielrubber
1552	limpieza	limpeza	avtorkning	wegschuiven
1553	alambre m aislado	fio m isolado	isolerad tråd	geisoleerd ader
1554	resina f de pino, colofonia f	resina f de pinho	träharts	houthars
1555	ropa f de trabajo	roupa f de trabalho	arbetskläder	werkkleding
1556	calzado m para obreros	calçado m para operários	arbetsskodon	werkschoenen
1557	guantes mpl para obreros	luvas fpl para operários	arbetshandskar	industriële handschoenen
1558	tornillo m sin fin	parafuso m sem fim	ändlös skruv	worm (eindloze)
1559	hilo m, hilado m	fio m têtil, fio m	garn	garen
1560	alargamiento al límite elástico	alongamento	töjning vid sträckspänningen	maximum rek
1561	limite elástico	limite elástico	brottpunkt	rekgrens
1562	módulo m de Young	módulo m de Young	Young's modul	Young's modulus
1563	óxido de zinc	oxido de zinco	zink oxid	zink oxide

INDEX DU VOCABULAIRE A-Z

À double - filet 423
à sec 384
à un filet 1154
abaissement m, réduction f 671
abrasif m 12
abrasimètre, m Du Pont 3
abrasimètre m 2
abrasimètre m, machine f pour essais d'abrasion 11
abrasion (indice d') 5
abrasion f 4
abrasion f Taber 8
absorbeur m de choc en caoutchouc 1073
absorption (capacité f d') 14
absorption (machine f d'essai d') 16
absorption f 13
absorption f d'eau 17
absorption f d'humidité 15
accepteur m 20
accepteur m d'acide 21
accélérateur m 19
accélération f 18
accumulateur m 23
accumulateur m hydraulique 22
acides (résistant aux) 30
acides (résistant aux) 32
acide m inorganique 28
acide m organique 29
acide m phtalique 821
acide m stéarique 33
activateur m 36
activation f 35
activité f de surface 1306
additif m 38
adhérence f au sol, tenue f de route 543
adhérence f caoutchouc-tissu 1085
adhérent 40
adhérent 41
adhérer à 39
adhésif 49
adhésif m 48
adhésif m à base de latex 645
adhésif m sensible à la pression 51
adhésion (appareil m pour essais d') 47
adhésion f, collage m 42
adhésion f, résistance f au pelage 45
adhésivité f 1330
adhésivité f, collant m 1325
adhésivité f, poisseux m 1244
adsorption f 53
agent protecteur m contre le vieillissement 62
agent m anticollant 101
agent m antigrillant 63
agent m antioxygène 104
agent m antiozone 105
agent m antisolaire 64
agent anti-vieillissement m 100
agent m collant, de collage 74
agent m d'adhérisation 179
agent m de liaison 65
agent m de rétification 67
agent m de vulcanisation 68
agent m déodorant 69
agent m diluant, d'extension 468
agent m gonflant 66
agent m gonflant 73
agent m stabilisateur 71
agent m tensio-actif 72
agglomérant m, liant m 152
agglomérat m 75
agiter, remuer 1248
agrégat m 76
air m emprisonné 81
ajustage m 52
ALCRYN caoutchouc transformable par fusion 90
allongement m 449
allongement m 1287
allongement m à la rupture 187

allongement m à la rupture 453
allongement m par traction 1252
allongement m préalable 1288
âme f du câble 341
amorçage m du déchirement 1346
amortissement m 386
amortissement m acoustique 1188
amortissement m des vibrations 1498
amortissement m du bruit 1189
amortisseur m 385
amortisseur m, tampon m de choc, pare-chocs m 1146
amortisseur m de vibrations 1497
amplitude f 95
ampèremètre m 94
angle m de cisaillement 96
angle m de contact 98
angle m de perte 97
anneau m 1039
anneau m d'étanchéité 1042
anneau m de cylindre 1040
anneau m en U 1476
antigrillant m 106
antioxygène m 102
antioxygène m tachant 103
apex m 107
appareil m de mesure 693
appareil m (de mesure m de résistance à la flexion) 1416
appareil m de vieillissement 54
appareil m Hoffman pour la mesure de la dureté par usure 1419
appareil m pour essais de perméabilité à l'eau 1524
appareil m pour la mesure de la contrainte 1256
appareil m (pour la mesure de la fragilité à basse température) 1414
appareil m pour la mesure de la résistance initiale à l'état humide 1420
appareil m (pour la mesure de la rigidité) 1423
appareil m pour mesurer l'adhésivité 1327
appareil m pour mesures de fragilité 1412
appareillage m 108
appareillage m d'essai 1425
appliquer au pinceau 200
appui m amortisseur 133
appui m 132
aptitude à la mise en œuvre 897
aptitude f à l'extrusion 471
aptitude f à la dispersion 418
aptitude f au déchirement 1353
arbre m 1132
arrachage m, pincement m 844
article m moulé 113
article m moulé 738
article m trempé sans soudure 1116
articles mpl à parois épaisses 1437
articles mpl caoutchoutés 1090
articles mpl chirurgicaux 1310
articles mpl de ménage en caoutchouc 422
articles mpl découpés 405
articles mpl en caoutchouc de papeteries 1238
articles mpl en caoutchouc pour le ménage 585
articles mpl en caoutchouc-amiante 115
articles mpl en latex au trempé 651
articles mpl extrudés 473
articles mpl extrudés 1190
articles mpl industriels 695
articles mpl moulés 739

articles mpl sanitaires 114
assemblage m 117
assortiment m, jeu m 1128
attache f de courroie, agrafe f de courroie 139
auto-cicatrisant 1120
autovulcanisant 1119
auxiliaire m de mise en œuvre 896

Bâche f 1344
bâche f pour camion 1470
balata f 120
balayage m 1552
ballon m 122
ballon m météorologique 703
ballon m météorologique 704
bandage m plein 1447
bandage m pour chariot 550
bande f 123
bande f 1289
bande f de roulement de pneu 1449
bande f de roulement, chape f 1467
batteuse f (dans l'industrie de la mousse de latex) 135
biseauté 150
bitume m 153
blanchiment m 158
boîte f de batterie 128
boîte f de dérivation 329
boîte f de dérivation pour câbles 213
bombe f (appareil de vieillissement) 173
bombe f à air 77
bombe f à oxygène 174
bottes fpl en caoutchouc 183
boudinage m, extrusion f 478
boudiner, extruder 472
boudineuse f à plomb pour câbles sous plomb 653
bourre f de coton 353
boursouflures fpl 163
boyau m à alésage lisse 1165
bulle f d'air 162
buse f 764
butée f en caoutchouc 1081

Câble (enveloppe f de) 212
câble m 211
câble m aérien 216
câble m conducteur pour haute fréquence 1379
câble m de transmission de puissance 217
câble m pour pneumatiques, corde f 340
câble m sous plomb 214
calandrage m 225
calandre à désaxage 221
calandre f 220
calandrer 223
calfatage m 243
calfater 242
calorimétrie f 226
canal m d'écoulement 1324
canal m (largeur) 255
canal m (profondeur) 254
canaux mpl régulés 1092
canevas m, toile f de chanvre 227
canot m pneumatique 172
caoutchouc hydrocarboné NORDEL 762
caoutchouc synthétique HYPALON 593
caoutchouc synthétique Néoprène m 753
caoutchouc-butyle m 1056
caoutchouc m 1053
caoutchouc m à usage chirurgical 1311

caoutchouc m brut   939
caoutchouc m brut   1062
caoutchouc m cellulaire   246
caoutchouc m (cellulaire pour usage chirurgical)   1082
caoutchouc m chloré   1059
caoutchouc m en granulés   805
caoutchouc m en poudre   868
caoutchouc m micro-cellulaire   706
caoutchouc m micro-poreux   707
caoutchouc m mousse   1065
caoutchouc m naturel   750
caoutchouc m nerveux   1076
caoutchouc m (nitrile acrylique)   1054
caoutchouc m oxydé   1071
caoutchouc m plastifié à l'huile   1070
caoutchouc m pour joints de pare-brises   1551
caoutchouc m pour pistons   826
caoutchouc m régénéré   949
caoutchouc m régénéré   1072
caoutchouc m résistant à l'huile   1069
caoutchouc m (résistant aux produits chimiques)   1058
caoutchouc m étendu à l'huile   771
caoutchouc m étendu à l'huile   772
caoutchouc m sec   1063
caoutchouc m spongieux   1079
caoutchouc m spongieux   1080
caoutchouc m synthétique   1322
caoutchoutage m, gommage m   1091
capacité f   228
capacité f calorifique   230
capacité f de charge   229
capacité f de ramollissement   1
capot m (automobile)   182
capsulage m   201
caractéristique f, propriété f   906
caractéristiques fpl (principales)   904
carotte f   1216
catalyseur m   241
cavité f (de moule)   244
cellulaire, alvéolaire   245
cendre f   116
centipoise f   252
chaleur f de combustion   564
chaleur f de fusion   565
chaleur f spécifique   1192
chambre f à air   622
changement m de volume   1512
charbon m actif   235
charbon m activé   234
charge f   486
charge f   487
charge f à la rupture   188
charge f claire   758
charge f dynamique   433
charge f renforçante   957
charge f statique   661
charger en pigment   823
chaussure f de protection   791
chaussures fpl   513
chaussures fpl de pluie   916
chaussures fpl de travail   1556
cheville f de plateau de distribution   845
chloroprène m   260
chlorure m de vinyle   1499
chute f de température   1365
ciment latex m   647
ciment m (dans la construction)   249
cire f microcristalline   1531
cisaillement (vitesse f de)   1139
clapet m à bille   121
claquage m, décharge f disruptive   190
co-agent m   271
coagulation f   273

coefficient de dilatation thermique linéaire x-y °C   291
coefficient m d'élasticité   286
coefficient m de conductibilité thermique   293
coefficient m de dilatation   287
coefficient m de friction   285
coefficient m de friction   288
coefficient m de friction (dynamique) 3 m/min, 0,7 MPa   289
coefficient m de friction (statique)   290
coefficient m de perméabilité   292
coefficient m de perméabilité   811
coefficient m de viscosité   294
coefficient m de vulcanisation   1518
coefficient m de vulcanisation   295
cohésion f   296
collage m   43
collage m du caoutchouc   180
collant m de confection   203
colle f à base de caoutchouc   1057
colle f à base de celluloïd   247
colle f à prise rapide   250
collier m (chauffant)   301
colloïde m   302
colonne f de guidage   425
colophane f   1554
commande f (à vitesse variable)   1197
compatibilité f   307
composition f (chimique)   309
compression f   313
conducteur m de câble   324
conduction f   321
conductivité f   322
conductivité thermique   323
conduit m garni en caoutchouc   326
conduite f   325
confection f   202
confection f   204
configuration f   327
conique   1340
conseiller m (technique)   331
constante f diélectrique   409
constante f diélectrique, permittivité f   815
container m, réservoir m   332
contraction f   333
contrainte f   1272
contrainte f dans la couche de collage   178
contrainte f de cisaillement   1255
contrainte f de cisaillement   1282
contrainte f de compression   1276
contrainte f de compression   1286
contrainte f de flexion   1275
contrainte f de flexion   1278
contrainte f de torsion   1279
contrainte f de torsion   1459
contrainte f initiale   1254
contrainte f sous compression   1283
contrepression f d'extrusion   479
contrôlé par ordinateur   319
copolymère m   338
copolymérisation f   339
coprécipitation f   272
corona (décharge f)   346
corona (effet m)   345
corona (effet m)   347
corona (résistivité f)   349
corona interne (effet m)   348
corrosion f   350
corrosion f par les acides   26
couche primaire f   895
couleur   305
coupe f en biais   151

couple à 90% (MC90)   1455
couple maximum (MH)   1456
couple minimum (ML)   1457
couple m de torsion   1454
courant m alternatif   93
courant m continu   414
courbe f contrainte-déformation   1285
courbe f d'allongement   451
courbe f de compression   314
courbe f de déformation par compression   1284
courbe f de déformation sous charge   660
courbe f tension-déformation à vitesse réduite   664
courbure f   505
courroie f crantée   1446
courroie f transporteuse   336
courroie f trapézoïdale à bords tranchés (sans enveloppage)   141
courroie f trapézoïdale, courroie f en V   1484
courroie striée   142
courroie synchrone de distribution   144
courroies fpl   146
courroies fpl transporteuses pour grains   147
coussin m amortisseur en caoutchouc spongieux   799
coussin m amortisseur en caoutchouc spongieux   1204
coussin m en caoutchouc spongieux   800
coussin m gonflable, coussin m pneumatique   616
coussin m pneumatique   80
coussin m pneumatique, coussin m gonflable   849
CPE - polyéthylène chloré   360
crêpe m   368
craie f   1550
craquelage m   361
craquelure f due à la lumière solaire   1302
craquelure f par la lumière solaire   1300
craquelures fpl par flexion   362
crémage m du latex   365
cristallisation f   373
cuisson f   119
curative m, agent de cuisson, agent de vulcanisation   377
cycle m   382
cycle m d'hystérésis   596
cycle m d'hystérésis   663
cylindre m d'impression   606
cylindre m d'imprimerie, rouleau m d'imprimerie   1046
cylindre m d'imprimerie, rouleau m d'imprimerie   1047
cylindre m de filature   1044
cylindre m de malaxage   1045
cylindre m recouvert de caoutchouc   1048
cylindres (revêtement m de)   1049

D

dalle f de sol   1439
dalle f en caoutchouc   1440
déchets mpl   1106
déchirement m   1354
déchirement m Groves   1345
déchirer   1352
déchirure f, déchirement m   1347
décollage m   1123
décollage m, décollement m   176
décoloration f   417
décoloration f par la lumière solaire   1303
décomposition f thermique   1430
déformation f   389
déformation f, contrainte f   1251

déformation f initiale , fluage   620
déformation f maximale   691
déformation f permanente   1129
déformation f résiduelle, déformation f rémanente   1130
déformation f sous charge   659
déformation f sous charge ... °C, ... MPa, ... h   390
déformation f thermique   1431
dégagement m (de gaz)   464
dégagement m de chaleur   562
dégazeur m   391
dégorgement m (de couleurs)   160
dégorger   159
dégradation f (physique-chimique)   192
dégradation f, décomposition f   392
dégraissant m   393
degré m   394
degré m de cristallinité   372
démoulage m   395
densité f apparente, masse volumique apparente   109
dentelé   1125
dépolymérisation f   397
dérangement m, panne f   191
déraper, patiner   1161
déraper, patiner, glisser   1164
désodorisation f   396
dessin m gaufré   802
détermination de la dureté f par pénétration   399
détermination f de la résistance f au déchirement amorcé   1350
détérioration f   398
dévidoir m   954
diamètre m de la filière   401
diamètre m de la vis   1108
diaphragme m   400
diaphragme m à disque   415
diélectrique m   407
dimensions fpl normalisées   1159
dioxyde m de titane   1450
direction f de la machine   674
direction f perpendiculaire à l'appareil   369
dispersion f   419
dispositif m d'aspiration   1296
disque m à polir   1548
disque m abrasif   1547
dissolution f, colle f, adhésif m   248
dissolution f de caoutchouc   1077
dissolvant   1182
distorsion f   421
donneur m de soufre   1299
double vis   1472
drap m (d'hôpital, alèse f)   1143
draps mpl de lit   137
durabilité f   430
durcissant à chaud, thermodurcissable   1436
durée f de repos   808
durée f de repos   1030
durée f de service   1126
durée f de stockage   1145
dureté IRHD f   553
dureté f   551
dureté f par pénétration (Rockwell)   609
dureté f Rockwell   554
dureté f Shore   555
dureté f Shore A   552
duromètre m   431
duromètre m   556
dynamomètre m   434

É

ébarbage m   388
ébonite (bac m d'accumulateur en)   1067

écart-type m, écart m standard 1232
échantillon m 1097
échauffement m interne 561
éclatomètre m 1413
écumage m 519
effet m des intempéries, vieillissement climatique 1539
effet m Joule 637
effet m thermique 563
efficacité f 435
efflorescence f 165
efflorescence f 171
efflorescence f de soufre 166
efflorescence f de soufre 1297
effort d'allongement m 452
effort m de traction, contrainte de traction 1378
éjecteur m 436
éjecteur m 437
élasticité f 442
élastique 438
élastomère éthylène acrylique VAMAC 1492
élastomère fluoré VITON 1508
élastomère m 443
élastomère m synthétique 1320
électriquement 446
électrolyte m 447
ELVAX (résines d'EVA) 454
émail m 457
emporte-pièce m 267
emporte-pièce m 403
émulsifiant m 70
émulsifiant m 455
émulsion f 456
émulsion f latex-bitume 646
encapsulation f 458
encrassement m du moule 736
enduction f 1213
enduction f au couteau 640
enduction f par trempé 280
enduire à la brosse 1210
enduire à la râcle 276
enduire au métier à gommer 1211
enduire par trempé 413
énergie f transmise 1464
engrenages fpl pour étirer les courroies 143
enjoliveur m 587
équipement de laboratoire 641
erreur f standard (racine carrée de la variance) 1233
essai m 467
essai m, test m 1384
essai m d'abrasion 9
essai m (d'endurance) 1401
essai m (de cisaillement) 1403
essai m (de compression permanente) 1402
essai m (de corrosion) 1390
essai m de courbure 506
essai m (de déchirement) 1407
essai m (de déchirement) 1408
essai m de déchirement (méthode du haricot) 1391
essai m (de déformation par compression) 1389
essai m de flexibilité 1393
essai m de flexibilité 1394
essai m (de flexion sur éprouvette entaillée) 1400
essai m (de gonflement) 1405
essai m (de laboratoire) 1397
essai m (de maintien) 1396
essai m (de perméabilité à l'eau) 1409
essai m (de propagation du déchirement) 1406
essai m (de résistance à la coupure par compression) 1388
essai m de résistance aux flexions répétées 1392
essai m (de service simulé) 1404

essai m de vieillissement 61
essai m (de vieillissement accéléré) 1385
essai m (de vieillissement accéléré à la lumière) 1386
essai m (de vieillissement accéléré aux intempéries) 1387
essai m en service 1127
essai m (mécanique) 1398
essai m pratique pour la mesure de l'abrasion 7
estampage m à chaud 1230
étanche aux gaz 534
éthylène tétrafluoré ou tétrafluoroéthylène 1427
étirage m, dessin m 426
étuve f 790
étuve f à air 84
étuve f Geer-Evans 789
étuve f de vieillissement 58
étuve f de vieillissement Geer 788
évent m 1496
exigence f 961
expansion f, dilatation f 465
expansion f thermique, dilatation f thermique 1432
exposition f à l'extérieur 786
exposition f aux intempéries 1537
exposition f aux intempéries 1540
exposition f aux radiations 912
exsudation f 480
exsudation f 1313
extensibilité f 469
extrait m aqueux 110
extrait m aqueux 1523
extrait m sec 1173
extrapolation f 470
extrudeuse 475
extrudeuse f 477
extrudeuse f 675
extrudeuse f (sous vide) 1487
extrudeuse f alimentée à chaud 583
extrudeuse f alimentée à froid 299
extrusion/soufflage f 168

Facteur de pertes diélectriques (tan δ) 420
facteur de puissance (sin δ) 870
facteur m de sécurité 1094
faible gauchissement m 670
fatigue f 484
fatigue f à la flexion 498
fatigue f diélectrique 410
feuilles fpl (en caoutchouc-liège) 1144
feuilles fpl en caoutchouc spongieux 1205
feuille f (calandrée) 1141
feuille f de caoutchouc 1075
feuille f de liège caoutchoutée 344
feuille f (extrudée) 1142
feuille f métallique 1229
fil m aggloméré 1217
fil m élastique extrudé 1084
fil m isolé 1553
fil m métallique (pneus) 1242
fil m textile, fil m 1559
filière d'extrudeuse 476
film m 488
finissage m 490
fissure sous contrainte 1277
fixation f 840
flexion f 495
flexion f 501
flexions fpl répétées 499
flexomètre m 502
floculation f 508
fluage m 363

fluage m, écoulement m 366
fluage m à froid 298
fluage m à froid 300
fluage m à long terme 662
fluage m initial 619
fluage m relatif 958
fluide m hydraulique 589
fluorure m de vinylidène 1500
fluor m 511
force de jointoiement f 1114
force f d'adhérence 177
force f de cisaillement 1266
force f de collage 44
force f de dénudage 1290
force f de fermeture 263
formation f du film 489
forme f, profil m 1134
formule f 945
fourreau m 125
fragile, cassant 194
fragilité f à basse température 196
fragilité f 195
frette f 765
friable 575
friction à la calandre 224
friction f, frottement m 516
frictionnage m, friction f 518
frittage m 1155
frottement m interne 630
frottement m statique 1237
fréquence f 514

Gaînage m de câble 1140
gain m en poids (état spécifique) 1543
gaine f 634
gallon m 528
galoche f 793
gamme f de température 1369
gamme f (de températures) 920
gants mpl de ménage 584
gants mpl de protection 907
gants mpl de travail 1557
gants mpl pour l'industrie 613
garnissage m (de roue) 359
garniture f 798
garniture f d'étanchéité 532
gauchissement m 1521
géométrie f de la vis 1109
gilet m de sauvetage 207
gommage m à la calandre, enduction f par calandrage 283
gomme ester f 463
gommer, caoutchouter 1088
gommer (à la calandre) 275
gonflé 1319
gonflement m 1316
gonflement m à la sortie de la filière 404
gonflement m à la sortie de la filière 934
gonflement m après boudinage 1317
gonflement m dans l'eau 1527
gonflement m en volume 1515
gonfler 1314
gonfler (se) 1315
goudron m 1342
grain m, granulés 538
grain m de calandrage 222
gramme m 539
granulé m 804
granulométrie 1157
granulométrie f 1158
graphique m 256
graphique m 540
gravité f 541
grillage m 1101
grillage m Mooney 1102
guillotine f 547
guillotine f, découpeur m 261
guillotine f, machine f à trancher à guillotine 548

guillotine f, machine f à trancher à guillotine 677

Halogène m 549
hausse f de température 1371
haute fréquence f 575
hexafluoropropylène 574
homogénéisation f 579
huile f (combustible) lourde 773
huile f aromatique f 112
huile f aromatique f 769
huile f minérale 774
huile f naphténique 775
huile f paraffinique 776
huile f siccative 427
huile f siccative 770
humidité f 588
humidité f 732
humidité f atmosphérique 82
humidité f de l'air 83
humidité f relative 959
hydrocarbure m aliphatique 91
hydrocarbure m aromatique 111
hydrolyse f 591
hygroscopique 592
hypothèse f 594
hystérésis f 595
HYTREL élastomère thermoplastique technique 598

Imperméabilisation f, étanchéité 1530
imperméabilité f 603
imprégnation f 605
imprégnation f au caoutchouc 604
impédance f 601
impulseur m 602
impulseur m de pompe 909
indice d'inflammabilité f 492
indice d'oxygène m 795
indice de fluidité m 510
indice de fluidité à l'état fondu 698
indice de réfraction 955
indice de stabilité thermique 1433
indice m de température 1367
indice limite d'oxygène 796
indice limite d'oxygène m 610
indice m 611
indice m d'acétyle 24
indice m d'iode 633
inductance f 612
inflammabilité f 493
inflammable 615
influence f du temps sur la plastification 833
infrarouge 617
ingrédients actifs 37
inhibiteur de moisissures m 710
inhibiteur m 618
inhibiteur m d'adhésivité 1328
inhibiteur m de craquelure par la lumière solaire 1301
inhibiteur m de vieillissement thermique 559
ininflammable 491
injection d'air (moulage) 88
injection-soufflage f 169
injection-soufflage f 741
inorganique 623
insert m 624
insonorisant 1187
installation f de dégraissage 828
instrument d'essai m électrique 1426
instrument m de mesure électrique 445
instrument m de mesure électrique 694
interpolation f 631
intervalle m de temps 632

isolant m 628
isolation f 627
isoler 625

J
auge f d'épaisseur Elcometer 444
jauge f de broyage Hegman 573
jauge f de contrainte 1253
joint d'étanchéité, joint m 1538
jointure f, soudure f 1200
joint m 636
joint m à lèvres 656
joint m à recouvrement 643
joint m bout à bout 210
joint m bout à bout 1115
joint m (d'arbre) 1112
joint m d'étanchéité en liège 343
joint m de dilatation 466
joint m (de queue de soupape) 1113
joint m (de vitrage) 1110
joint m (étanche à l'huile) 1111
joint m homocinétique 330
joint m torique 1041

K
ALREZ (pièces fpl en élastomère perfluoré) 638
kaolin m 264
kaolin m 265
kaolin m dur 266
kilogramme m 639

L
aiton m 186
lame f ajustable, réglable 156
lamination f 642
lampe f à rayonnement ultraviolet 1478
canal m (largeur) 255
latex m 644
latex m concentré par crémage 648
latex m concentré par des méthodes électriques 649
latex m concentré par évaporation 650
latex m de Néoprène 751
latex m synthétique 1321
latex pré-vulcanisé 894
liaison f 175
liège m 342
limite f d'allongement 450
limite f d'allongement 1560
limite f d'élasticité 440
limite f élastique 1561
liquide m d'essai 1395
liège m 342
litharge f 657
litre m 658
livre f 865
livre f par pied carré 866
livre f par pouce carré 867
longueur f de la filière 402
lubrifiant m 672
lumière f ultraviolette 1479

M
âchoire f 262
mâchoire f d'accouplement, accouplement m à serrage 635
machine f de flexion 500
machine f de flexion 676
machine f et outillage m pour insérer les attaches de courroie 140
machine f (pour essais de déchirement) 1424
machine f (pour essais de flexion) 1415
machine f pour essais de flexions 1411

machine f pour essais de résistance 681
machine f pour essais de résistance 1270
machine f pour la fabrication de câbles 215
machine f pour la fabrication de câbles 683
machine f pour tester le caoutchouc 678
machine f pour tester le caoutchouc 1083
machine f à enduire 282
machine à revêtir les câbles 356
machine à rubanner 682
machine à rubanner 1341
machine Schopper (pour essais de résistance à la traction) 1421
machine f (Scott pour essais de traction) 1422
macrostructure f 684
magnésie f 685
magnésie f calcinée 219
malaxage m 715
malaxeur, mélangeur m ouvert, mélangeur m à cylindres, m 780
manchon m 208
manchon m chauffant 124
manomètre m 686
manomètre m 892
marge f de sécurité 1095
masse volumique apparente 205
matières fpl solides totales 1175
matière f plastique, plastique m 829
matière f première 689
matière f première 938
matière f première 1250
matière f solide 690
matière f solide 1172
matériau m isolant 626
matériau m isolant 688
matériau m visco-élastique 1507
mélange m (pour mastic) 308
mélange m 310
mélange m caoutchouc-liège 1055
mélange m cru 312
mélange m de trempage pour câbles de pneus 1448
mélange m maître 687
mélange m pour gainage de câble 218
mélange-maître m curatif 378
mélange caoutchouc 311
mélangeage m 717
mélanger 161
mélangeur, plastificateur m Gordon 830
mélangeur Banbury m 716
mélangeur m à dissolution 1180
mélangeur m ouvert, malaxeur m 712
mélangeur m ouvert, malaxeur m 718
mélangeur m ouvert, mélangeur m à cylindres, malaxeur m 713
mémoire f élastique 441
mémoire f élastique 700
méthode d'essai 1399
métier à gommer m 673
métier à gommer 679
métier à gommer 680
métier à gommer 1160
métier à gommer 1212
métier à gommer 1214
mesure f 692
mesure f de retrait 334
mètre m 705
mettre à la terre 546
meule f abrasive 1549
microstructure f 708
migration f 709
milligramme m 714
minium m 951

mise f à la terre 328
modifié chimiquement 258
module d'élasticité en flexion 503
module de flexion 497
module m 719
module m d'allongement 729
module m d'élasticité 721
module m d'élasticité 724
module m d'élasticité en cisaillement 727
module m d'élasticité en compression 316
module m d'élasticité en compression 725
module m d'élasticité en flexion 726
module m d'élasticité en tension 728
module m d'élasticité en traction 1376
module m d'élasticité transversale 731
module m de cisaillement 730
module m de cisaillement, module d'élasticité en cisaillement 1135
module m de compression 315
module m de compression 723
module m de Young 1562
module m dynamique effectif 720
module m en haute fréquence 722
moment m d'inertie 614
moment m d'inertie 733
moment m d'inertie 734
moment m de torsion 1473
monofilament m 735
montée f, hausse f, élévation f 1043
mouillabilité f 1545
moulage par transfert 744
moulage m 740
moulage m par coulée 240
moulage m par injection 621
moulage m par injection 742
moulage m sous presse 743
mouler par coulée 238
moule m à plusieurs empreintes 737
moule m à plusieurs empreintes 749
moule m de vulcanisation 1520
mousse f de latex 652

N
éoprène n spongieux 752
nerf m 754
neutralisation f 755
neutraliser 756
noir EPC m 461
noir FEF (extrusion fine) 521
Noir FF m 522
Noir FT 523
noir GPF (usage général) 537
noir HAF (très bonne abrasion) 524
Noir HEF m (allongement élevé) 525
Noir HMF m (haut module) 526
noir HPC m 586
noir ISAF (intermédiaire) 629
noir m animal 154
noir m de carbone HAF 236
noir m de carbone activé 237
noir MPC m 747
noir MT 696
noir MT m 748
noir SAF (super) 1305
Noir SRF m (semi-renforçant) 1122
non tissé 757
non-décolorant 759
non-gauchi 761
non-tachant 760
normal, standard 1231

normalisé 1235
norme f 1234

O
bturateur m 1166
once f 785
oscillogramme m 782
oscillographe m 783
osmotique 784
outil m à estamper, à découper 910
oxyde m de fer rouge 794
oxyde m de zinc 1563
oxyde m métallique 702

P
are-choc m 206
particule f 801
pas de vis m 507
pas m 827
pas m de vis peu profond 1133
pâte f 424
PBTP, téréphtalate de polybutylène 803
pénétration f 806
performance f, comportement m 807
permittivité relative ... Hz 816
perméabilité f 810
perméabilité f à la vapeur d'eau 812
perméabilité f à la vapeur d'eau 1525
perméabilité f aux gaz 530
perméabilité f aux gaz 813
perméable 814
perméamètre m à air 1410
perpendiculaire au granulé 34
perte f 665
perte f d'énergie 459
perte f d'énergie 666
perte f de chaleur 667
perte f des substances solubles 668
perte f diélectrique 411
perte f par hystérésis 597
PETP, téréphtalate de polyéthylène 818
petrolatum (vaseline brute) 820
pièce f pour automobile 118
pièces fpl pour freins pneumatiques 1486
pied m 512
pied m cubique 374
piège m à radicaux 1466
pigment m 822
pile à papier f 136
pinte f 825
pistolet m à enduire 281
pistolet m à enduire 284
pistolet m à pulvériser 1206
pistolet m à pulvériser 1209
piston m (d'injection) 919
piston m 846
piston m 917
piston m (d'éjection) 918
plaque m 1162
plaque f chauffante 838
plaque f d'appui 134
plastifiant m 831
plastifiant m 1168
plastifiant m ester 832
plastification f, ramollissement 189
plateau m 842
plateau m chauffant 843
plateau m d'éjection 837
plateau m de fermeture 836
plateau m de fixation 835
plateau m de polissage 839
plateau m intermédiaire de démoulage 841
pliage m 148
plier 149
pliographe m 1417
pneu m pour camion 1471
pneu m tourisme 233

pneu m tourisme   1475
poche f d'air   87
poids m spécifique   1191
point de retenue   576
point m d'aniline   99
point m d'injection   824
point m d'injection (en moulage par injection)   535
point m de déchirement   1355
point m de fusion   527
point m de fusion   699
point m de fusion   850
point m éclair   494
poisseux   1245
poisseux   1246
poisseux   1331
polymère m   851
polymère m à structure en échelle   854
polymère m bloc   164
polymère m brut   855
polymère m cis-tactique   852
polymère m greffé   853
polymérisation en solution f   1181
polymérisation f   856
polymérisation f en masse   857
polystyrène m   858
polysulfure m   859
polyuréthanne m   860
porosimètre m pour gaz   1418
porosité f   861
porte-mandrin m   1198
post-réticulation f   862
post-vulcanisation f   863
pouce f cubique   375
pouce m   608
pouce m carré   1218
poulie f (pour courroies trapézoïdales)   908
poulie f pour courroies trapézoïdales   1485
pouvoir m adhésif   1259
pouvoir m collant initial   1326
pouvoir m d'adhésion, force f de collage   46
pouvoir m gonflant   1318
pré-cuit   871
préchauffé   873
précoce, grillant   1105
première couche f   126
préparation f, traitement m   1469
presse f (à plateaux multiples)   884
presse à vulcaniser f   876
presse f   874
presse f à double piston   877
presse f à estamper   406
presse f (à estamper)   885
presse f (à estamper), presse f (à découper)   888
presse f à estamper, presse f de découpage   875
presse f (à piston ascendant)   889
presse f (à plomb)   881
presse f (à plomb)   882
presse f à plomb pour tuyaux   880
presse f (à vis piston)   886
presse f à vulcaniser les courroies   145
presse f (col de cygne)   883
presse f crocodile   879
presse f descendante   878
presse f (mono-poste)   887
presser, essorer   1219
pression f atmosphérique   891
pression f d'air   890
pression f de moulage, de préformage   872
pression f manométrique   536
prise f rapide   1243
prix m, tarif m, taux m   928
produit m de condensation (aldéhyde-aniline)   320
produit m pour le traitement des surfaces   1309

produits mpl chimiques pour l'industrie du caoutchouc   259
profilé m, article m profilé   899
propagation f du déchirement   1348
proportion f, degré m, taux m   929
propriétés fpl (cryogéniques)   901
propriétés fpl (des vulcanisats)   905
propriétés fpl (dynamiques)   902
propriétés fpl (mécaniques)   903
pulvérisation f   1208
pulvériser,   869
pulvériser, arroser   1207

R acle f   157
  radical (libre) m   914
ramollir   1167
rapport L/D m   936
rapport m   932
rapport m (de tirage)   935
rayonne f   940
réactif m   941
rebond m   942
rebondir   944
réduction, décroissance, diminution   387
réduction f   953
réduire   952
refroidissement par air   337
régénération f du caoutchouc   956
régénéré (caoutchouc m)   948
régénéré m (acide)   946
régénérer   947
régulateur m de tension de tirage   1380
relaxation f   960
relaxation f de contrainte   1281
relaxation f de tension, de contrainte   1280
repousser   170
repousser, faire reculer   911
reprise f   950
reprise f élastique   834
réservoir m, citerne f   1332
résilience f   962
résilience f   963
résilient   964
résine f   965
résine f (acrylique)   966
résine f (alkyde)   967
résine f (d'urée)   969
résine f de mélamine   697
résine f époxy   462
résine f (phénolique)   968
résistance à la flexion   504
résistance f (à la flexion)   1262
résistance f (à la rupture par traction)   1268
résistance au cisaillement   1137
résistance f (aux rayonnements)   988
résistance f   970
résistance f   1258
résistance f (électricité)   1029
résistance f à basse température   1370
résistance f (à cru)   985
résistance f (à cru)   1263
résistance f à l'abrasion   6
résistance f (à l'affaissement)   974
résistance f (à l'arc)   972
résistance f à l'eau   1015
résistance f (à l'effet corona)   975
résistance f (à l'essence)   982
résistance f à l'exposition   999
résistance f à l'huile   1004
résistance f à l'oxydation   1005
résistance f (à l'ozone)   987

résistance f à l'ozone   1006
résistance f (à l'état humide)   1271
résistance f à l'usure et au déchirement   1016
résistance f à la chaleur   566
résistance f à la compression   317
résistance f (à la compression)   1261
résistance f (à la corrosion)   976
résistance f à la coupure, à l'entaille   998
résistance f (à la cristallisation)   978
résistance f à la flamme   1000
résistance f (à la flexion)   980
résistance f à la lumière   1003
résistance f à la lumière solaire   1304
résistance f (à la propagation de l'entaille)   979
résistance f à la rupture   193
résistance f à la rupture   1260
résistance f à la rupture   1477
résistance f à la séparation des plis   1007
résistance f (à la traction)   1265
résistance f (à la traction à la limite élastique)   1269
résistance f à la traction à la rupture   1377
résistance f à la vapeur   1011
résistance f au choc   599
résistance f (au choc)   986
résistance f au choc   1147
résistance f au choc avec entaille   763
résistance f au choc avec entaille, Izod   600
résistance f au cisaillement   1010
résistance f au cisaillement   1136
résistance f au déchirement   1014
résistance f au déchirement   1349
résistance f au déchirement   1351
résistance f au fluage   367
résistance f (au froid, au gel)   981
résistance f au gonflement   1013
résistance f au grattage, aux éraflures   1009
résistance f (au pelage)   1264
résistance f au ramollissement par la chaleur   1002
résistance f (au roulement)   990
résistance f (au roulement)   991
résistance f (au roulement)   1008
résistance f (au roulement)   1050
résistance f au soleil   1012
résistance f au vieillissement   60
résistance f aux acides   31
résistance f aux acides   993
résistance f aux bases   994
résistance f (aux craquelures)   977
résistance f aux craquelures   995
résistance f aux craquelures   997
résistance f aux flexions   1001
résistance f aux flexions répétées   496
résistance f (aux matières grasses)   984
résistance f (aux moisissures)   989
résistance f aux produits chimiques   996
résistance f (chimique)   973
résistance f de la couleur à la lumière   304
résistance f spécifique, résistivité f   992
résistance f volumique   1513
résistant   1017
résistante f (à l'essence)   983
résistant à l'huile   777

résistant à l'huile   1020
résistant (à l'ozone)   1021
résistant à la chaleur   567
résistant à la chaleur et aux huiles   560
résistant à la corrosion   351
résistant à la vapeur   1241
résistant aux bases   92
résistant aux combustibles liquides   1022
résistant (aux matières grasses)   1018
résistant aux moisissures   711
résistant aux solutions salines   1023
résistant aux solvants   1024
résistivité f   1025
résistivité f électrique   1027
résistivité f en courant continu   1026
résistivité f superficielle   1028
résistivité f volumique   1514
ressort m   1215
ressort m en caoutchouc   1074
retardateur m   1031
réticulation f   370
rétifications fpl   371
rétraction f   1033
retrait m   1149
retrait m (au moulage)   1151
retrait m au niveau du diamètre   1150
réversion f   1034
revêtement m   655
revêtement m de courroie   354
revêtement m de cuve   1333
revêtement m (de cylindre), garniture f (de cylindre)   358
revêtement m de paroi en caoutchouc   1086
revêtement m (de sol)   357
revêtement m de sol   509
revêtement m de sol en caoutchouc   1064
revêtement m, enduit m   279
revêtir, gommer, enduire   274
revêtir au pinceau   199
revêtu, gommé, enduit   277
revêtu au pinceau   198
rhéomètre à disque oscillant - ODR   1037
rhéomètre m   1036
rhéomètre m à piston   1038
rigidification f simple   1247
rigidité f diélectrique (... mm, ISO = rigidité électrique)   412
robe f extérieure de tuyau   355
robe f intérieure de tuyau   581
rondelle f   1522
rondelle f, grommet m   544
rotor m   1052
roue f   1546
roue f en caoutchouc   1087
rouleau m d'impression   607
roulette f   239
roulette f lisse, machine f à brocher   1249
ruban m, bande f   1334
ruban m adhésif   1335
ruban m adhésif sensible à la pression   893
ruban m adhésif sensible à la pression   1339
ruban m de frein   1257
ruban m isolant   1338
ruban m pour câbles   1336
ruban m pour friction   1337
rupture f   1093
rupture f, échec   482
rupture f d'une émulsion   1124
rupture f par fatigue   485

S ablage m   1099
  sabler   1100
séchoir m   127
section f d'alimentation   1117

French index 223

section f de dosage 1118
sécurité f (de mise en œuvre) 1096
semelle f 1170
semelle f (en caoutchouc spongieux) 1171
semelles fpl 1176
séparateur m d'accumulateur 129
séparation f entre plis 847
séparation f entre plis 848
Shore (duromètre) m 1148
silice f précipitée 1153
silice f sublimée 1152
site m (de rétification) 1156
solides mpl non-volatiles 1174
solidification f, durcissement m 1131
solidité à la lumière, résistance f à la lumière 483
solubiliser 1177
solubilité f 1178
soluble 1179
solvant m, dissolvant m 1184
solvant m aromatique 1183
solvant m pour caoutchouc 1078
soudage m 1544
soudure f à la chaleur 571
soudure f bout à bout 209
soudure f en phase solvant 1186
soufflage m 167
soufflet m de direction 138
soufflet m de joint homocinétique, de transmission 181
soumettre à l'essai de cisaillement 1138
soupape f de contrôle 335
souplesse f 1169
sous - vulcanisation f 1480
sous - vulcanisé 1481
spécification f 1193
spécimen m, échantillon m 1194
spongieux m 1201
spongieux m 1202
spongieux m à cellules fermées 1203
stabilisation f 1220
stabilité f 1222
stabilité à la chaleur 569
stabilité f (au stockage) 1227
stabilité f (chimique) 1223
stabilité f de couleur 303
stabilité f de couleur 1226
stabilité f (dimensionnelle) 1224
stabilité f (mécanique) 1225
stabilité f (thermique) 1228
stabilizzante m 1234
stator m de pompe 1239
structure f 1291
structure f 1429
structure f de chaîne 253
structure f (du noir) 1292
support m 746
support m antivibratoire 745
support m moteur 460
survulcanisé 792
suspension colloïdale 1312
système m de vulcanisation 1323
système m de vulcanisation à base d'oxyde métallique 701
système m de vulcanisation organique 781

Tackifiant m 1329
talon m en caoutchouc 1068
tampon m en caoutchouc, butoir m en caoutchouc 201

tapis m pour auto 232
taux m (de compression) 933
taux de réduction 937
taux m de déchets 930
technicien m 1357
technique f 1356
tellure m 1358
température d'auto-allumage (ISO = température d'allumage spontané) 1121
température f (d'auto-inflammation) 1366
température de fragilisation 197
température de fragilisation 1362
température f de ramollissement Vicat, point Vicat 1374
température de transition 1373
température f 1359
température f ambiante 1360
température f ambiante 1372
température f de fléchissement sous charge 1364
température f (de fusion) 1368
température f de service en continu 1363
température f du fourreau 1361
temps de grillage (TS2) 1103
temps optimum de vulcanisation (TC90) 1444
temps f de vulcanisation 1443
temps m de grillage 809
temps m de maintien 577
temps m de maintien 1442
temps m de solidification 1445
temps m de vulcanisation 1441
ténacité f 1375
ténacité f 1460
tendance f au grillage 1104
teneur f en caoutchouc sec 1061
tension, contrainte f de rupture 1273
tension - déformation f 1382
tension disruptive 408
tension f de contrainte f à la rupture 1274
tension f de surface critique 1381
tension f superficielle 1308
tenue f (aux acides) 971
tenue f de la couleur 306
tenue f de la couleur 1032
ternir 1343
terpolymère m 1383
terre f 545
test m de rebond 943
thermoplasticité f 1435
thermoplastique 1434
thermosensible 568
thixotrope 1438
tissu m 481
tissu m, textile m, toile f 1428
tissu m adhésif 50
tissu m avec trame 1542
tissu m doublé 268
tissu m élastique 439
tissu m enduit 278
tissu m gommé, tissu caoutchouté 1089
tissu m imperméabilisé 900
tissu m imperméable 1529
tissu m pour friction 517
toile f de chanvre, coutil m, canevas m 428
toile f de Hollande 432
toile f de Hollande 578
tolérance f 1451
toluène m 1452
tonne f 1453
torsion f 1458
tourniquet m dégage-talon 130
tours mpl par minute, r.p.m., t/min. 1035

toxicité f 1461
trame f 1541
transfert m de chaleur 570
transition f 1462
transmission f des gaz 531
transmittance f de la lumière 1463
transparent 1465
travail m, usinage m, mise f en œuvre 898
trémie f 580
tresseuse f 185
tringle f de talon 131
trouble m 557
tuyau m à acétylène 25
tuyau m à essence 520
tuyau m à essence 533
tuyau m à gaz 529
tuyau m à serpentin 297
tuyau m circonvoluté, cannelé 352
tuyau m d'aspiration d'eau 1526
tuyau m d'aspiration et de dépottage 1294
tuyau m d'aspiration et de dépottage d'huile 779
tuyau m d'aspiration pour ciment 251
tuyau m d'aviation 89
tuyau m de chauffage 572
tuyau m de frein pneumatique 78
tuyau m de gaz basse pression 669
tuyau m de vidange 416
tuyau m extrudé 474
tuyau m hydraulique 590
tuyau m pour abattoirs 1163
tuyau m pour acides 27
tuyau m pour compresseur 318
tuyau m pour crèmeries 364
tuyau m pour laiterie 383
tuyau m pour produits chimiques 257
tuyau m pour radiateur 913
tuyau m pour sable et gravier 1098
tuyau m pour wagons de chemin de fer 915
tuyau m résistant (à l'essence) 1019
tuyau m résistant à l'essence et à l'huile 819
tuyau m résistant à l'huile 778
tuyau m résistant à l'huile et à l'essence 767
tuyau m résistant à l'huile et à l'essence 768
tuyau m résistant aux solvants 1185
tuyau m spiralé 1199
tuyau m tressé 184
tuyauterie f d'air 429

Unité f 1482
uréthanne m 1483
usure (essais m d') sur route 10
usure à la limite élastique 1500
usure à la rupture 1534
usure f 1532
usure f à la charge maximale 1535
usure f de la bande de roulement 1468
usure f (par abrasion) 1533

Valeur f 1488
valeur à l'usage, rapport m qualité/prix 1490

valeur f limite 1489
valeurs fpl de mesures typiques 1474
vanne f de contrôle à diaphragme 1491
vapeur f d'eau 1528
vapeur f vive 1240
vaporisation f 1493
véhicule m 1494
ventouse f 1293
ventouse f 1295
vessie f 155
vêtement m de protection 269
vêtement m en caoutchouc 1066
vêtements mpl de protection contre le vent et l'eau 270
vêtements mpl de travail 1555
vêtements mpl en caoutchouc 1060
vie f en pot 864
vieillissement en étuve 85
vieillissement m à l'ozone 797
vieillissement m dans l'huile 57
vieillissement m dans l'huile 766
vieillissement m du caoutchouc 56
vieillissement m dynamique 55
vieillissement m en étuve 59
vieillissement m en étuve 787
vieillissement m statique 1236
vieillissement m thermique 558
vis f 1107
vis f sans fin 1558
viscosimètre m à bille 1501
viscosité du polymère à l'état fondu 1504
viscosité f 1502
viscosité f Brookfield 1503
viscosité f Mooney 1505
viscosité f relative 1506
vitesse (d'extrusion) 1196
vitesse f, rapidité f, vélocité f 931
vitesse f, vélocité f, rapidité f 1495
vitesse f d'écoulement 926
vitesse f de cisaillement 927
vitesse f (de cristallisation) 921
vitesse f de cristallisation 924
vitesse f de propagation des craquelures 923
vitesse f (de vulcanisation) 922
vitesse f de vulcanisation 925
vitesse f de vulcanisation 1519
vitesse f égale (des cylindres) 1195
volatil 1509
volatilité f 1510
volume m 1511
vulcanisable 376
vulcanisat m 1516
vulcanisation f, cuisson f 379
vulcanisation f, cuisson f 381
vulcanisation f, cuisson f 1517
vulcanisation f à l'air chaud 582
vulcanisation f à température ambiante 1051
vulcanisation f au peroxyde 817
vulcanisation f au soufre 1298
vulcanisation f en étuve 79
vulcanisation f en étuve 86
vulcanisation f sous électrons 448
vulcanisation f sous plomb 654
vulcanisé (non-) 380

Zone f superficielle 1307

WÖRTER-VERZEICHNIS
A-Z

Abbau m 189
Abbau m 192
Abbau m 392
Abdeckung f 231
abdichten 242
Abdichten n 243
Abdichtmasse f 308
Abdichtungsring m 1042
Abfall m, Ausschuß m 1106
abgekantet 150
Abietinsäureester m 463
Abnahme f 387
Abnutzungsbeständigkeit f 1016
Abnutzungsprüfung f 10
Abrieb m bei Höchstkraft 1535
Abrieb m bei Reißkraft 1534
Abrieb m bei Streckkraft (Kraft bei Streckspannung) 1536
Abrieb m nach Taber 8
Abrieb m 4
Abrieb m, Abnutzung f 1532
Abrieb m, Abnutzung f 1533
Abriebbeständigkeit f 6
Abriebindex m 5
Abriebprüfer (Du Pont-) m 3
Abriebprüfgerät n 11
Abriebprüfung f in der Praxis 7
Abriebprüfung f 9
Absorption f 13
Absorptionsprüfer m 16
Absorptionsvermögen n 14
Abstreifkraft f 1290
Abstreifplatte f 841
Abzweigdose f, Verbindungsstück n 329
Acetylenschlauch m 25
Acetylzahl f 24
Achsmanschette f 181
Acrylharz n 966
Adhäsion f, Bindungskraft f 42
Adhäsionskraft f, Klebkraft f 44
Adhäsionsprüfer m 47
Adsorption f 53
Agglomerat n 75
Aggregat n 76
Akkumulator m (hydraulischer) 22
Akkumulator m 23
Akkumulatorgehäuse n, Batteriegehäuse n 128
Akkumulatorscheidewand f 129
aktive Bestandteile 37
Aktivierung f 35
Aktivkohle f 234
Aktivkohle f 235
Akzeptor m 20
ALCRYN thermoplastisches Elastomer 90
Aldehyd-Anilin-Kondensationsprodukt n 320
alkalibeständig 92
Alkalienbeständigkeit f 994
Alkydharz n 967
Alterung f (dynamische) 55
Alterung f (statisch) 1236
Alterungsbeständigkeit f 60
Alterungsgerät n 54
Alterungsprüfung f (beschleunigt) durch Lichteinwirkung 1386
Alterungsprüfung f (beschleunigt) 1385
Alterungsprüfung f 61
Alterungsschrank m 58
Alterungsschutzmittel n 62
Alterungsschutzmittel n 100
Amperemeter n 94
Amplitude f 95
anfängliches Kriechen n 619
Anfangsdeformation f (Kriechen) 620
Anfangsdehnung f 1254
Anfangsnaßfestigkeitsprüfer m 1420
Anforderung f 961
Anguß m 1216
angußlos 1092
Anilinzahl f 99
anklebend 1245
anlaufen 170
anlaufen 1343
Anlaufen n 171
anorganisch 623
Anschnittkanal m 535
Anstieg m 1043
anstreichen, beschichten (mit Pinsel) 200
anstreichen, mit dem Pinsel auftragen 199
Anvulkanisation f 1101
Anvulkanisationsneigung f 1104
Anvulkanisationszeit f im Rheometer (TS2) 1103
Anvulkanisationszeit f 809
Apparat m, Gerät n 108
Appretur (Ausrüstungs-)maschine (für Textilausrüstung) f 679
Appretur (Ausrüstungs-)maschine (für Textilausrüstung) f 1160
Arbeitshandschuhe mpl 1557
Arbeitskleidung f 1555
Arbeitsschuhe mpl 1556
aromatischer Kohlenwasserstoff m 111
aromatisches Lösungsmittel n 1183
Artikel mpl (chirurgische), Chirurgiebedarf m 1310
Artikel mpl (gummierte) 1090
Artikel mpl (starkwandig), dickwandige Artikel mpl 1437
Artikel mpl (technische) 695
Asche f 116
Aufhängung f 746
aufquellen 1315
Aufrahmen n von Latex 365
Aufspannkörper m 835
Aufspannplatte f 836
Aufspannplatte f 842
Ausblühen n 165
ausbluten 159
Ausbluten n (von Farben) 160
Ausdehnung f 465
Ausdehnungskoeffizient m 287
Ausdrückplatte f 837
Ausdrückstift m 437
ausgekleidet, angestrichen, beschichtet 277
Auskleiden n, Anstreichen n, Beschichten n 279
auskleiden, streichen, beschichten 274
Auskleidung f 655
Ausreißen n 844
Ausschußquote f 930
Ausschwitzen n 1313
Ausschwitzung f 480
Ausstoßkolben m, Auswerfer m 436
Auswerfer m (hydraulisch) 911
Auswerferkolben m 918
Automatte f 232
Autoreifen m 233

Balata f 120
Balg m, Stulpe f 182
Ballon m 122
Banburymischer m, Innenmischer m 716
Band n 123
Band n 1334
Bandwickelmaschine f 682
Bandwickelmaschine f 1341
Baumwoll-Linters pl 353
Behälter m 332
Behandlung f 1469
Behandlung f, Verarbeitung f 898
Belastung f (statisch) 661
Benetzbarkeit f 1545
benzinbeständig 983
benzinbeständiger Schlauch m 1019
Benzinbeständigkeit f, Kraftstoffbeständigkeit f 982
Benzinschlauch m 533
Berstfestigkeitsprüfer m, Falzfestigkeitsprüfer m (Papier) 1413
Berührungswinkel m 98
beschichten im Streichverfahren n (mit Rakel) 1211
beschichten mittels Kalander 275
Beschichtung n auf dem Kalander 283
Beschichten n mit Messer oder Rakel, Streichen n 640
beschichtet (mit Pinsel aufgetragen) 198
Beschichtungsmaschine f 282
Beschleuniger m 19
Beschleuniger m, Aktivator m 36
Beschleunigung f 18
beständig gegen flüssigen Treibstoff 1022
beständig gegen Lösungsmittel 1024
beständig gegen Salzlösungen 1023
beständig, widerstandsfähig 1017
Beständigkeit gegen Oberflächenrißbildung 995
Beständigkeit f gegen dynamische Beanspruchung 496
Beständigkeit f gegen Lagertrennung f 1007
Beständigkeit f gegen Sonnenlicht 1304
Beständigkeit f 970
Betteinlage f 137
Betteinlage f 1143
Bewitterung f im Freien 786
Bewitterung f 1537
Bewitterung f 1539
Bewitterung f 1540
Bewitterungsprüfung f (beschleunigt) 1387
Biege-E-Modul m 503
Biegebeanspruchung f 1275
Biegebeständigkeit f 980
Biegebeständigkeit f 1001
Biegeermüdung f 498
Biegefestigkeit f 504
Biegefestigkeit f 1262
Biegefestigkeitsprüfer m 1416
Biegemodul m 497
biegen 149
Biegeprüfer m 502
Biegeprüfer m 1411
Biegeprüfer m, Biegeprüfgerät n 500
Biegeprüfer m, Biegeprüfgerät n 676
Biegeprüfgerät n 1415
Biegeprüfung f 506
Biegeprüfung f 1393
Biegeprüfung f 1394
Biegerisse mpl 362
Biegespannung f 1278
Biegung f 148
Biegung f 495
Biegung f 501
Biegung f 505
Biegung f, Biegen n 499
Bindemittel n 65
Binder m, Bindemittel n 152
Bindung f, Haftung f 175
Bitumen n 153
Blase f (Luftblase) 162
Blase f, Heizschlauch m 155
Blasen n, Blähen n 167
blasenbildend 163
Blasformverfahren n 169
Blasformverfahren n 741
Bleichung f 158
Bleiextruder m für Bleikabel 653
Bleiglätte f 657
Bleikabel n 214

Bleimantelpresse f 882
Bleipresse f 881
Bleipressenvulkanisation f 654
Bleiumschlagmaschine f, -presse f
  für Schläuche 880
Block-Polymer n 164
Blockpolymerisation f 857
Blutkohle f 154
Bodenbelag m 357
Bodenbelag f 509
Bodenplatte f, Fußbodenbelag m
  1439
Bombe f (Alterungsapparat) 173
Brechungsindex m 955
Bremsband f 1257
Brennbarkeitsklasse 492
Bruch m 1093
Bruchdehnung f 187
Bruchdehnung f 453
Bruchfestigkeit f 193
Bruchfestigkeit f 1260
Bruchlast f 188
Bruchspannung f 1273
Bruchspannung f 1274
Buchse f, Lagerbuchse f 208
Butylkautschuk m 1056

C
ellulosekleb stoff m 247
Centipoise n 252
Chemikalien für die
  Kautschukindustrie 259
Chemikalienbeständigkeit f 996
Chemikalienbeständigkeit f,
  chemische Beständigkeit f
  973
Chemikalienschlauch m 257
chemisch modifiziert 258
chirurgische Gummiwaren fpl
  1311
chlorinierter Kautschuk m 1059
Chloropren n, Chlorbutadien n
  260
Coagens n 271
computergesteuert 319
Copolimerisation f 339
Copolymer(isat) n,
  Mischpolymer(isat) n 338
CPE - chloriertes Polyethylen n
  360
Crêpe m 368

D
ämpfung f 386
  Dämpfungslager n 133
Dämpfungslager n 745
DAM trocken 384
dampfbeständig 1241
Dampfbeständigkeit f 1011
Darstellung f (graphisch) 540
Dauerbiegeprüfung f 1392
Deformation f unter Last ... °C, ...
  MPa, ... h 390
Deformation n, Verformung f 389
Dehnbarkeit f, Streckbarkeit f
  469
Dehnung f bei Streckspannung
  450
Dehnung f bei Streckspannung
  1560
Dehnung f 449
Dehnung f 1287
Dehnung f, Beanspruchung f,
  Längenänderung f 1251
Dehnungsfuge f, Kompensator m
  466
Dehnungskurve f 451
Dehnungsmodul f 729
Depolymerisierung f 397
Desodorisierung f 396
Desodorisierungsmittel n 69
Diagonalschnitt m 151
Diagramm n, Aufstellung f 256
Dichtkraft f 1114
Dichtung f 532

Dichtungsring m,
  Unterlagsscheibe f 1522
Dichtungsring m 544
Dielektrikum n 407
dielektrischer Verlustfaktor m
  (tan δ) 420
Dielektrizitätskonstante f 815
Dielektrizitätskonstante f,
  Dielektrizitätszahl f ... Hz 409
Dielektrizitätszahl ... Hz 816
Dispergierbarkeit f 418
Dispersion f 419
Doppelkolbenpresse f 877
Doppelschnecke f 1472
Doppeltuch n 268
Dornhalter m mit Stegen 1198
Draht m (isoliert) 1553
Drehmoment n bei
  Vernetzungsgrad 90% im
  Rheometer (M90) 1455
Drehmoment n 1454
Drehmoment n 1473
Drehmomentmaximum n im
  Rheometer (MH) 1456
Drehmomentminimum n im
  Rheometer (ML) 1457
Druck m (atmosphärisch) 891
Druck-E-Modul m 316
Druckbelastung f 1276
Druckbelastung f,
  Druckspannung f 1286
Druckfestigkeit f 317
Druckfestigkeit f 1261
Druckluftbremsschlauch m 78
Druckluftflasche f 77
Druckmodul m 315
Druckmodul m 723
Druckverformung f 659
Druckverformungskurve f 660
Druckverformungsprü-
  fung f 1389
Druckwalze f 607
Druckwalze f 1046
Druckwalze f 1047
Druckzylinder m, Druckwalze f
  606
Düse f 764
Düse f, Spritzmundstück n 476
Düsendurchmesser m 401
Düsenkopf m 765
Düsenlänge f 402
Durchdringung f 806
durchlässig 814
Durchlässigkeit f 810
Durchlässigkeitskoeffizient f 811
Durchlässigkeitskoeffi-
  zient m 292
Durchmesserschwindung f 1150
Durchschlag m 190
Durchschlagfestigkeit f, ... mm
  412
Durchschlagspannung f 408
dynamische Belastung f 433
Dynamometer m 434

E
ffektiver dynamischer
  Modul m 720
Eigenschaft f 906
Eigenschaften fpl (mechanisch)
  903
Eigenschaften fpl (dynamisch)
  902
Eigenschaften f bei
  Tiefsttemperaturen f 901
Einbettung f, Kapselung f 458
einbrennen 871
Eindruckhärteprüfung f 399
einem Scherversuch m
  unterziehen 1138
einfache Versteifung f 1247
Einfülltrichter m 580
eingängig 1154
Einheit f 1482
Einreißpunkt m 1355

Einrichtungszeit f 1445
Einsatz m 624
Einsatztemperatur f
  (kontinuierlich) 1363
Einsatzverhalten n,
  Gebrauchstüchtigkeit f 807
Einspannklemme f, Klaue f 635
einstellbares Messer n, Klinge f
  156
Einstellung f, Justierung f 52
Einzelpresse f 887
Eisenbahnwagenschlauch m 915
Eisenoxid rot 794
elastisch 438
elastisch 964
elastische Erholung f 834
elastische Nachwirkung f 700
Elastizität f gespeichert 441
Elastizität f 442
Elastizität f 963
Elastizitätsgrenze f 440
Elastizitätskoeffizient m 286
Elastizitätsmodul bei Zug 728
Elastizitätsmodul m bei Biegung
  726
Elastizitätsmodul m bei Druck
  725
Elastizitätsmodul m 721
Elastizitätsmodul m 724
Elastomer f 443
Elcometer-Dickemesser m 444
elektrisch 446
elektrisches Meßinstrument n
  445
elektrisches Prüfinstrument n
  1426
Elektrolyt m 447
Elektronenstrahlvernet-
  zung f 448
ELVAX Ethylen-Vinyl-Acetat
  Copolymer n 454
Emaille f 457
Emulgator m 455
Emulgiermittel m, Emulgator m
  70
Emulsion f 456
endlose Bahn f eines Materials in
  der Maschine 1541
Energieverlust m 459
Energieverlust m 666
Entfettungsanlage f 828
Entfettungsmittel n 393
entflammbar 615
Entflammbarkeit f 493
entformen 395
Entgaser m 391
Entgraten n 388
Entladungsschlauch m 416
Entlüftung f 1496
Entspannung f 960
Entwicklung f (von Gasen) 464
EPC-Ruß m, Channell-Ruß für
  leichte Verarbeitbarkeit 461
Epoxidharz n 462
Erde f, Masse f 545
erden 546
Erdverbindung f 328
Erhitzen n, Einbrennen n, Härten n
  119
Erholung f, elastische 950
Ermüdung f (dielektrisch) 410
Ermüdung f 484
Erniedrigung f 671
Erstarrung f, Erhärtung f 1131
erwärmte Platte f 843
erweichen 1167
Experiment n, Versuch n 467
Extrapolation f 470
Extruder m, Spritzmaschine f
  475
Extrudierbarkeit f, Spritzbarkeit f
  471
extrudieren, spritzen 472
extrudierter Artikel mpl 1190
extrudierter Gummifaden m
  1084

Extrusion f 478
Extrusionsblasverfahren n 168
Extrusionsgeschwindigkeit f,
  Spritzgeschwindigkeit f 1196
Extrusionsstaudruck m 479

F
altenschlauch m 352
  Falzzahlprüfgerät n 1417
Farbbeständigkeit f 1226
Farbe f 305
Farbechtheit f 303
Farbechtheit f 1032
Feder f 1215
FEF-Ruß, feiner
  Extrusions-Furnace-Ruß 521
Feinheitsprüfer m (Hegman) 573
feinporiges Zellgummi f 706
feinporiges Zellgummi n 707
Fensterdichtung f, Türdichtung f
  1538
Festigkeit f unvulkanisierter
  Mischung f 985
Festigkeit f, Zähigkeit f 1460
Festigkeitsprüfer m 681
Festigkeitsprüfer m 1270
Festkautschuk m 1063
Festkörper m 1172
Festkörpergehalt m 1173
Festkörpergehalt m 1174
Fett n 542
fettbeständig 1018
Fettbeständigkeit f 984
Feuchtigkeit f (relativ) 959
Feuchtigkeit f 588
Feuchtigkeit f 732
Feuchtigkeitsaufnahme f 15
Film m, Folie f 488
flachgängig 1133
flammbeständig 491
Flammbeständigkeit f 1000
Flammpunkt m 494
Flammpunkt m,
  Entzündungstemperatur f
  1366
flankenoffener Keilriemen m 141
Fließgeschwindigkeit m 926
Fließindex m 510
Fließnase f 1324
Flockung f, Ausflockung f 508
flüchtig 1509
Flüchtigkeit f 1510
Flugzeugschlauch m 89
Fluor n 511
Förderband n, Transportband n
  336
Folie f (extrudiert) 1142
Form f, Gestalt f 1134
Formartikel mpl 739
Formartikel m 113
Formartikel m 738
Formartikelherstellung f, Formen f
  740
Formbeständigkeit f in der Wärme
  1364
Formbeständigkeit f 1224
Formenschwund m 1151
Formhöhlung f, Formnest n 244
Formverschmutzung f 736
Freidampf m 1240
Frequenz f, Häufigkeit f 514
Friktionierung f mittels Kalander
  224
Friktionierung f 518
Friktionsband n 1337
Friktionsgewebe n 517
Frontscheibendichtung f 1551
Frostbeständigkeit f 981
Füllstoff m (verstärkend) 957
Füllstoff m 486
Füllstoff m, Streckmittel n 468
Füllstoffdosierung f 487
füllstofffreie Kautschukmischung f
  311
Fütterzone f 1117
Furnaceruß m (fein), FF-Ruß 522

Fuß m

Fuß m 512
Fußbekleidung f 513

Gallone f 528
Galoschen fpl 916
Gangbreite f 255
Gangtiefe f 254
Garn n 1559
gasdicht 534
Gasdurchlässigkeit f 530
Gasdurchlässigkeit f 531
Gasdurchlässigkeit f 813
Gasdurchlässigkeitsprüfer m 1418
Gasschlauch m 529
Gebläserad n, Pumpenflügelrad n 602
Gebrauchsdauer f 1126
Gebrauchsfähigkeitsprüfung f, Einsatzprüfung f 1401
Gebrauchsprüfung f 1127
Gebrauchswert m 1490
Geer-Alterungsofen m 788
Geer-Evans-Ofen m 789
Gefüge n, Struktur f 1429
Gelenk n 636
gemeinsame Ausfällung f 272
genormt 1231
genormt 1235
gequollen 1319
geräuschabsorbierend 1187
geräuschdämpfend 1188
Geräuschdämpfung f 1189
geringer Verzug m 670
Gerinnung f, Koagulation f 273
Geschwindigkeit f 931
Geschwindigkeit f 1495
Getreideförderband n 147
Gewebe n (beschichtet), gummiertes Gewebe n 278
Gewebe n (elastisch) 439
Gewebe n (gummiert) 1089
Gewebe n (gummiert), imprägniertes Gewebe n 900
Gewebe n (selbstklebend) 50
Gewebe n (wasserdicht) 1529
Gewebe n 481
Gewicht n (spezifisch), Dichte f 1191
Gewichtzunahme f 1543
gewobenes Tuch n 1542
gezahnt, geriffelt 1125
gießen 238
Gießen n 240
Gleichlaufgelenk n, homokinetisches Gelenk n 330
gleichmäßige Geschwindigkeit f (der Walzen) 1195
Gleichstrom m 414
Gleichstromwiderstand m 1026
gleiten, laufen 1164
Gordon-Kneter m 830
Grad m 394
Gramm m 539
Granulat n 804
Granulatkautschuk m 805
Grenzwert m 1489
Griff m, Griffigkeit f 543
Größe f (genormt) 1159
Groves-Weiterreißprüfung f 1345
Grundierung f 895
Grundstrich m, Grundierung f 126
Gummi m, Kautschuk m 1053
Gummi n mit hoher Elastizität 1076
Gummi-Asbest-Artikel fpl 115
Gummiabsatz m 1068
Gummialterung f 56
Gummibekleidung f 1060
Gummibekleidung f 1066
Gummibodenbelag m 1064
Gummibodenplatte f 1440
gummieren 1088
Gummieren n 1091

gummierte Korkplatte f 344
Gummifeder f 1074
Gummigewebebindung f 1085
Gummilösung f 1077
Gummilösungsmittel n 1078
Gummiprüfmaschine f 678
Gummiprüfmaschine f 1083
Gummipuffer m 201
Gummipuffer m 1081
Gummirad n 1087
Gummischwamm m 1079
Gummistiefel mpl 183
Gummistoßfänger m 1073
Gummiwalze f 1048
Gummiwandverkleidung f 1086
Gummiwaren fpl des Schreibwarenhandels 1238
Gurtverbindereindrückmaschine und -werkzeuge f 140

Hackmaschine f 548
Hackmaschine f 677
Härte f (Rockwell) 554
Härte f (Shore) 555
Härte f Durometer A 552
Härte f IRHD 553
Härte f 551
Härteprüfer m 556
Härteprüfer m, Durometer n 431
HAF-Ruß m, Furnace-Ruß für hohe Abriebfestigkeit 524
haften, kleben, verkleben 39
haftend, klebend 40
haftend, klebend 41
Haftfähigkeit f, Klebrigkeit f 1330
Haftfestigkeit f 177
Haftfestigkeit f 1265
Haftkleber m 51
Haftspannung f 178
Haftung f, Bindung f, Adhäsion f 43
Haftvermittler m 179
Haftvermögen f 46
Haftvermögen f 1259
Halogen n 549
Haltbarkeit f 430
Halteplatte f 840
Haltestift m 824
Handkarrenbereifung f 550
Harnstoffharz n 969
Hartgummi-Batteriekasten m 1067
Hartkaolin m 266
Harz n 965
Haspel f, Spule f 954
Haupteigenschaften fpl 904
Haushaltsgummiwaren fpl 422
Haushaltsgummiwaren fpl 585
Haushaltshandschuhe mpl 584
HEF-Ruß m, Furnace-Ruß für hohe Dehnung 525
Heftmaschine f 1249
Heißfütterextruder m 583
Heißluftofen m 790
Heißluftvulkanisation f 86
Heißluftvulkanisation f 582
Heißprägen n 1230
Heizband n 124
Heizkragen m 301
Heizöl n (schwer) 773
Heizplatte f 838
Heizschlauch m 572
Hexafluorpropylen n 574
hitze- und ölbeständig 560
Hitzealterung f 558
Hitzealterungsschutzmittel n 559
hitzebeständig 567
Hitzebeständigkeit f, Wärmebeständigkeit f 566
Hitzebeständigkeit f, Wärmebeständigkeit f 569
hitzeempfindlich 568
Hitzeerweichungsbeständigkeit f 1002
hitzehärtend 1436
Hitzeverschweißen n 571

HMF-Ruß m, Furnace-Ruß für hohen Modul 526
Hochfrequenz f 575
Hochfrequenzmodul m 722
Hochfrequenzspannungskabel n 1379
Holländer m 136
Homogenisierung f 579
HPC-Ruß m, Channell-Ruß hochverstärkend 586
Hydraulikflüssigkeit f 589
Hydraulikschlauch m 590
Hydrolyse f 591
hygroskopisch 592
HYPALON Synthesekautschuk, chlorsulfoniertes Polyethylen 593
Hypothese f 594
Hysterese f 595
Hysterese-Schleife f, für langsame Verformung 664
Hystereseschleife f 596
Hystereseschleife f 663
Hystereseverlust m 597
HYTREL elastischer Konstruktionswerkstoff 598

Impedanz f 601
Imprägnieren n mit Kautschuk 604
Imprägnierung f 605
in Fliessrichtung f 674
Indexzahl f 611
Induktanz f 612
Industriehandschuhe mpl 613
infrarot 617
Inhibitor m, Verzögerer m 618
innere Koronawirkung f 348
Instabilwerden n einer Emulsion 1124
Interpolation f 631
ISAF-Ruß m, Furnace-Ruß für mittlere Abriebfestigkeit 629
Isolator m 628
Isolierband n 1338
isolieren 625
Isoliermaterial n 626
Isoliermaterial n 688
Isolierung f 627

Jodzahl f 633
Joule-Effekt m 637

Kabel n 211
Kabelband n 1336
Kabelherstellungsmaschine f 215
Kabelherstellungsmaschine f 683
Kabelleiter f, Kabelader f 324
Kabelmantel m 212
Kabelmantelmischung f 218
Kabelseele f 341
Kabelüberführungskasten m, Kabelverteiler m 213
Kabelummantelung f 1140
Kabelummantelungsmaschine f 356
Kälteprödigkeitsprüfgerät n 1414
Kalander m mit Schränkung einer Walze f 221
Kalander m 220
Kalandereffekt m 222
Kalandern m 225
kalandrieren, kalandern 223
kalandrierte Platte f 1141
Kalorimetrie f 226
KALREZ Teil n aus Perfluorkohlenstoffelastomer 638
Kaltfluß m 298
Kaltfluß m 300
Kaltfütterextruder m 299

Kaolin n 264
Kaolin n 265
Kapazität f, Fassungsvermögen n 228
Katalysator m 241
Kautschuk m (chemisch beständiger Kautschuk) 1058
Kautschuk m (ölbeständig) 1069
Kautschuk m (ölgestreckt) 771
Kautschuk m (ölgestreckt) 772
Kautschuk m (oxidiert) 1071
Kautschuk m (regeneriert), Kautschukregenerat n 949
Kautschuk m (regeneriert), Regenerat n 1072
Kautschuk-Korkmischung f 1055
Kautschukbindung f, Kautschukhaftung f 180
Kautschukkleber m 1057
Kautschukmischung f (ölplastizierte) 1070
Kautschukregenerat m 948
Kautschuktrockengehalt m 1061
Keilriemen m 1484
Keilriemenscheibe f 908
Keilriemenscheibe f 1485
Keilrippenriemen m 142
Kerbbiegeprüfung f 1400
Kerbdruckversuch m 1388
Kerbschlagzähigkeit f (eingekerbt) 763
Kerbschlagzähigkeit f Izod (eingekerbt) 600
Kerbzähigkeitstest m mittels nierenförmigen Prüfkörpers 1391
Kettenstruktur f 253
Kieselsäure f, hochaktiv/pyrogen 1152
Kieselsäure, gefällt 1153
Kilogramm n 639
Klebeband f 1335
klebend 49
Kleber m, Klebstoff m 248
klebrig 1246
klebrig 1331
Klebrigkeit f 1244
Klebrigkeit f 1325
Klebrigkeitsmesser m 1327
Klebrigkeitsverminderer m, Antihaftmittel n 101
Klebrigmacher m 74
Klebrigmacher m 1329
Klebstoff m, Kleber m 48
Kleidung f (wind- und wasserfest) 270
Klemme f 262
Klöppelmaschine f, Umflechtmaschine f 185
Kohäsion f 296
Kohlenwasserstoff m (aliphatisch) 91
Kolben m 846
Kolbengummi m 826
Kolbenrheometer n 1038
Kolloid n 302
Kolophonium n 1554
Kompression f 313
Kompressionskurve f 314
Kompressionszone f 1118
Kompressorschlauch m 318
Konfektionierung f, Aufbau m, Montage f 204
Konfektionierung f, Zusammenbau m 202
Konfektionsklebrigkeit f 203
konisch 1340
Kontraktion f, Zusammenziehung f 333
konvexe Verformung f 1521
Kord (Festigkeitsträger) m 340
Kork m 342
Korkdichtung f 343
Korn n 538
Korona f 345
Koronabeständigkeit f 975

Koronaeffekt m 347
Koronaentladung f 346
Koronafestigkeit f 349
Korrosion f 350
korrosionsbeständig 351
Korrosionsbeständigkeit f 976
Korrosionsprüfung f 1390
Kraft f, Stärke f, Festigkeit f 1258
Kraftfahrzeugteil n 118
Kraftstromkabel n 217
Kratzbeständigkeit f 1009
Kratzhärteprüfer m (Hoffman) 1419
Kreide f 1550
Kriechbeständigkeit f, Dauerstandfestigkeit f, Druckfestigkeit f 367
Kriechen n in langen Zeiträumen 662
Kriechen n, Fließen n 366
Kristallinität f 372
Kristallisation f 373
Kristallisationsbeständigkeit f 978
Kristallisationsgeschwindigkeit f 921
Kristallisationsgeschwindigkeit f 924
kritische Oberflächenspannung f 1381
Kubikfuß m 374
Kubikzoll m 375
Kühlerschlauch m 913
Kugeldruckhärte f (Rockwell) 609
Kugelfallviskosimeter n 1501
Kugelventil n 121
Kunststoff m 829

L /D-Verhältnis n 936
Laboratoriumsausrüstung f, -ausstattung f 641
Laboratoriumprüfung f 1397
Lagentrennung f 176
Lagentrennung f 848
Lager n 132
Lagerbeständigkeit f 1145
Lagerbeständigkeit f 1227
Lagerplatte f 134
Laminierung f 642
Lastwagenplane f 1470
Lastwagenreifen m 1471
Latex m (vorvulkanisiert) 894
Latex m 644
Latex m, durch Aufrahmen konzentriert 648
Latex m, durch Eindampfen konzentriert 650
Latex m, durch elektrische Methoden konzentriert 649
Latex-Bitumen-Emulsion f 646
Latexkleber m 645
Latexkleber m 647
Latexschaum m 652
Laufflächenabnutzung f 1468
Laufrolle, Streubüchse f 239
leicht anvulkanisierend 1105
leicht zerbröckelnd, leicht zu zerreiben 515
Leinen n (ungebleicht) 578
Leistungsfaktor (sin δ) 870
Leiterpolymer n 854
Leitfähigkeit f 322
Leitung f (mit Gummi ausgekleidet) 326
Leitung f 325
Lenkungsmanschette f 138
Lichtbeständigkeit f 483
Lichtbeständigkeit f 1003
Lichtbogenbeständigkeit f 972
Lichtdurchlässigkeit f 1463
Lichtechtheit f der Farbe, Lichtbeständigkeit f 304
Lichtechtheit f, Farbenbeständigkeit f 306
Lichtriß m 1300

Lichtrißbildung f 1302
Lichtrißschutzmittel n 1301
Lichtschutzmittel n 64
Lippendichtung f 656
Liter m 658
Lösemittel n, Lösungsmittel n 1184
lösen, in Lösung bringen 1177
lösend 1182
löslich 1179
Löslichkeit f 1178
Lösungsmischer m 1180
Lösungsmittelverschweißen n 1186
Lösungspolymerisation f 1181
Luftblase f 88
Luftdruck m 890
Luftdurchlässigkeitsprüfer m 1410
Lufteinschluß m 81
Lufteinschluß m 87
Luftfeuchtigkeit f 82
Luftfeuchtigkeit f 83
Luftkabel n, Freiluftkabel n 216
Luftkissen n 80
Luftkissen n 616
Luftkissen n 849
Luftkühlung f 337
Luftleitung f 429
Luftofen m 84
Luftofenalterung f 85

M agnesiumoxid n kalziniert 219
Magnesiumoxid m 685
Makrostruktur f 684
Manometer n 686
Manometer n 892
manometrischer Druck m 536
Masterbatch m, Vormischung f 687
Materialprobe f, Muster f, Prüfkörper m 1134
Materialprobe f, Muster m 1097
Materialversagen n durch Ermüdung 485
maximale Verformung f 691
Mehrfach-Etagenpresse f 884
Mehrfachform f 737
Mehrfachform f 749
Melaminharz f 697
Membran f 400
Membransteuerventil n 1491
Mennige f 951
Messing m 186
Meßinstrument n (elektrisch) 694
Meßinstrument n 693
Messung f, Maß n 692
Meßwerte fpl (typisch) 1474
Metalloxid n 702
Metalloxid-Vulkanisationssystem n 701
Meter n 705
Microstruktur f 708
mikrokristallines Wachs n 1531
Milligramm n 714
Mineralöl n 774
Mischen n 717
mischen, verschneiden 161
Mischung f 310
Mischwalze f 712
Mischwalze f 713
Mischwalze f 718
Mischwalze f 780
mit einer Rakel auftragen 276
Mittel n (klebrigkeitsverhindernd) 1328
Modul m 719
Molkereischlauch m 364
Molkereischlauch m 383
Monofilament n 735
Montieren n 117
Mooney-Anvulkanisation f 1102
Moosgummi/Schwammgummi n 1202

Moosgummidämpfungspolster n 799
Moosgummidämpfungspolster n 1204
Moosgummisohle f 1171
Motorlagerung f 460
MPC-Ruß m 747
MT-Ruß m, Thermalruß mittlerer Teilchengröße 748
Mundstückquellung f, Spritzquellung f 404
Mundstückquellung f, Spritzquellung f 934

N achheizen, tempern 863
Nachvernetzung f 862
nahtloser Tauchartikel m 1116
Naßfestigkeit f 1271
Naturkautschuk m 750
Neoprene Synthesekautschuk m, Polychloroprenkautschuk m 753
Neoprene-Latex m 751
Neoprene-Moosgummi m 752
Nerv m 754
Neutralisation f 755
neutralisieren 756
nichtfleckend, nicht kontaktverfärbend 760
nichtverfärbend 759
Niederdruckgasschlauch m 669
Nitrilkautschuk m 1054
NORDEL Kohlenwasserstoff-Ethylen-Propylen-Dien-Kautschuk m 762
Norm f 1234
normaler Fehler m (Quadratwurzel der Abweichungen) 1233
Nutringdichtung f 1476

O -Ring m, O-Ringdichtung f 1041
Oberfläche f (spezifisch) 1307
oberflächenaktive Substanz f 72
Oberflächenaktivität f 1306
Oberflächenbehandlung f, Oberflächenbeschaffenheit f 490
Oberflächenbehandlungsprodukt n 1309
Oberflächenspannung f 1308
Öl n (trocknend) 427
Öl n (trocknend) 770
Ölabdichtung f 1111
Ölalterung f 57
Ölalterung f 766
ölbeständig 777
ölbeständig 1020
Ölbeständigkeit f 1004
Ofenalterung f 59
Ofenalterung f 787
ohne Verzug m 761
organische Säure f 29
organisches Vulkanisationssystem n 781
osmotisch 784
Oszillogramm n 782
Oszillograph m 783
Oxidationsbeständigkeit f 1005
Oxidationsschutz m, Alterungsschutzmittel n 102
Oxidationsschutzmittel n 104
Ozonalterung f 797
ozonbeständig 1021
Ozonbeständigkeit f 987
Ozonbeständigkeit f 1006
Ozonschutzmittel n 105

P ackung f 798
Partikel fpl 801
Paßstift m 425
Paste f 424
Peroxidvernetzung f 817

Personenwagenreifen m 1475
Pfropfpolymer n 853
Pfund n pro Quadratfuß 866
Pfund n pro Quadratzoll 867
Pfund n 865
Phenolharz f 968
Phthalsäure f 821
Pigment n 822
Pigmentfüllung f 823
Pinte f 825
Plane f, Persenning f 1344
Platte f, Prüfklappe f 1162
Platten fpl aus einer Gummi-Korkmischung 1144
Polierblech n 839
Polierscheibe f, Schwabbelscheibe f 1548
Polybutylenterephthalat n (PBTP) 803
Polyethylenterephthalat n 818
Polymer n (cis-taktisch) 852
Polymer n 851
Polymerisation f 856
Polystyrol n 858
Polysulfid n 859
Polyurethan n 860
Porosität f 861
Prägefolie f 1229
Prägemuster n 802
Preis m, Tarif m 928
Presse f mit grosser Hublänge f 883
Presse f mit obenliegendem Druckzylinder m, Oberkolbenpresse f 878
Presse f mit untenliegendem Druckzylinder m 889
Presse f 874
Preßformen n 743
Profil n 899
Prüfgeräte npl 1425
Prüfmethode f 1399
Prüfung der Rückprallelastizität f 943
Prüfung f (mechanisch) 1398
Prüfung f der bleibenden Druckverformung f 1402
Prüfverfahren n unter nachgeahmten Praxisbedingungen 1404
Puffer m, Stoßstange f 206
pulverisieren 869
Pulverkautschuk m 868
Pumpenflügelrad n, Pumpenrotor m 909
Pumpenstator m 1239

Q uadratzoll m 1218
Quellbeständigkeit f 1013
quellen 1314
Quellkraft f, Quellstärke f 1318
Quellmedium n 73
Quellprüfung f 1405
Quellung in Wasser f 1527
Quellung f 1316
quer zur Faser 34
quer zur Verarbeitungsrichtung f 369
Querschneider m 547
quetschen 1219

R ad n 1546
Radikal n (frei) 914
Radikalfänger m 1466
Radkappe f 587
Radüberzug m, Radverkleidung f 359
Rakel f 157
Raumtemperatur f 1372
Raumtemperaturvulkanisation f 79
Rayon m, Kunstseide f 940
Reagenz n 941
Reckverhältnis n 935

Reduktionsverhältnis n

Reduktionsverhältnis n   937
reduzieren, vermindern   952
Reduzierung f, Verminderung f   953
Regelantrieb m   1197
regenerieren   947
Regenerierung von Kautschuk m   956
Regulierventil n, Steuerventil n   335
Reibung f (innere)   630
Reibung f (statisch)   1237
Reibung f   516
Reibungskoeffizient m   285
Reibungskoeffizient m   288
Reibungszahl f (dynamisch) oder Gleitreibungszahl f 3 m/min, 0,7 MPa   289
Reibungszahl f (statisch) oder Haftreibungszahl f   290
Reifenhebel m, Wulstlösegerät n   130
Reifenkordimprägniermischung f   1448
Reifenlauffläche f   1449
Reifenlauffläche f   1467
Reifenschlauch m   622
reißen   1352
Reißen n   1354
Reißfestigkeitsprüfer m   1424
Reißversuch m   1408
relativer Fluß n   958
Retraktion f   1033
Reversion f   1034
Rezept n   945
Rheometer n   1036
Riemenmaterial n   146
Riemenstreckgetriebe n   143
Riemenverbinder m   139
Riemenvulkanisierpresse f, Förderbandpresse f   145
Ring m   1039
Riß m   1347
Rißbeständigkeit f   977
Rißbeständigkeit f   997
Rißbildung f   361
Rißeinleitung f   1346
Rißwachstum f   1348
Rißwachstumsgeschwindigkeit f   923
Rohfestigkeit f   1263
Rohkautschuk m   939
Rohkautschuk m   1062
Rohmaterial n   689
Rohmaterial n   938
Rohmischung f   312
Rohpolymer n   855
Rollwiderstand m   990
Rollwiderstand f   991
Rollwiderstand f   1008
Rollwiderstand m   1050
Rotor, Läufer m   1052
Rückprall m   942
Rückprallelastizität f   962
rückprallen   944
rühren   1248
Ruheperiode f   808
Ruheperiode f   1030
Runzelbildung f (Farbe), Kriechen n (Kunststoff)   363
Ruß m (Furnace), MT-Ruß m (mittlerer Teilchengröße)   696
Ruß m (GPF), Allzweck-Furnace-Ruß   537
Ruß m (HAF)   236
Ruß m (SAF), Furnace-Ruß mit sehr hoher Abriebbeständigkeit   1305
Ruß m aktiviert   237
Rußstruktur f   1292
rutschen   1161

S

Säure f (anorganische)   28
Säureakzeptor m   21
säurebeständig   32
Säurebeständigkeit f   31
Säurebeständigkeit f   993
säurefest   30
Säurebeständigkeit f   971
Säurekorrosion f   26
Säureregenerat n   946
Säureschlauch m   27
Sand- und Kiesschlauch m   1098
sandstrahlen   1100
Sandstrahlen n   1099
sanitäre Artikel mpl   114
Satz m zusammengehöriger Stücke   1128
Sauerstoffbombe f   174
Sauerstoffindex m (LOI)   610
Sauerstoffindex m   795
Sauerstoffindex m, LOI   796
Saug- und Zuleitungsschlauch m   1294
Saugeinrichtung f   1296
Saugnapf m   1293
Saugnapf m   1295
Schälfestigkeit f   45
Schälfestigkeit f   1264
Schäumen n   519
Schaumgummi m   1065
Scheibendichtung f   1110
Scheibenmembran f   415
scheinbare Dichte f, Schüttdichte f, Schüttgewicht n   109
Scheitel m   107
Scherbeanspruchung f   1255
Scherbeständigkeit f   1010
Scherfestigkeit f   1136
Scherfestigkeit f   1137
Scherfestigkeit f   1266
Schergeschwindigkeit f   927
Schergeschwindigkeit f   1139
Schermodul m   730
Schermodul m, Schubmodul m   1135
Scherspannung f   1282
Scherversuch m   1403
Scherwinkel m   96
Schichtaufbau m   489
Schichtenablösung f   847
schimmelbeständig   711
Schimmelbeständigkeit f   989
schimmelverhütendes Mittel n   710
Schlachthausschlauch m   1163
Schlagfestigkeit f   986
Schlagmaschine f   135
Schlagzähigkeit f   599
Schlauch m (benzin- und ölbeständig)   819
Schlauch m (extrudiert)   474
Schlauch m (lösungsmittelbeständig)   1185
Schlauch m (öl- und benzinbeständig)   767
Schlauch m (ölbeständig)   778
Schlauch m (umflochten)   184
Schlauch m mit glatter Innenwand   1165
Schlauch m, öl- und kraftstoffbeständig   768
Schlauch m, Ölansaug- und -zuleitungsschlauch   779
Schlauchboot n   172
Schlauchdecke f   355
Schlauchseele f   581
Schleifapparat m, Abriebsgerät n, Abriebsprüfgerät m   2
Schleifmittel n   12
Schleifrad n   1549
Schleifscheibe f   1547
Schließkraft f   263
Schmelzindex m   698
Schmelzpunkt m   527
Schmelzpunkt m   699
Schmelzpunkt m   850
Schmelztemperatur f   1368
Schmelzviskosität f   1504
Schmelzwärme f   565
Schmiermittel n, Gleitmittel n   672
Schnappverschluß m, Klemmverschluß m   1166
Schnecke f (endlos)   1558
Schnecke f   1107
Schneckendurchmesser m   1108
Schneckengeometrie f   1109
Schneidbeständigkeit f   998
Schneidwiderstand m   979
schnellabbindender Kleber m   250
schnelle Haftung f   1243
Schrumpfung f, Schwindung f, Schwund m   1149
Schubmodul m   727
Schubschneckenpresse f   886
Schüttdichte f, Schüttgewicht n   205
Schutzhandschuhe mpl   907
Schutzkleidung f   269
Schutzstiefel m   791
Schwamm m   1201
Schwammgummi m, Moosgummi n   1080
Schwammgummiplatte f, Moosgummiplatte f   1205
Schwammgummipolster n, Moosgummipolster n   800
Schwefelausblühung f   166
Schwefelausblühung f   1297
Schwefelspender m   1299
Schwefelvulkanisation f   1298
Schweißen n   1544
Schwere f   541
Schwimmweste f   207
Schwingelastometer n, Schwingscheibenrheometer n   1037
Schwundmaß n   334
Segeltuch n   227
Segeltuch n, Zeltleinwand f   428
Selbstentzündungstemperatur f   1271
selbstheilend   1120
Selbstklebeband n   893
Selbstklebeband n   1339
selbstvulkanisierend   1119
Shore-Härteprüfer m   1148
Sicherheitsfaktor m   1094
Sicherheitsspielraum m   1095
Sintern n   1155
Sohle f   1107
Sohlen fpl   1176
Sonnenlichtbeständigkeit f   1012
Spannprüfung f   1396
Spannung f unter Druck   1283
Spannung f, Beanspruchung f   1272
Spannung f, Spannungsrelaxation f   1281
Spannungsanzeiger m für Aufwickelmaschine f   1380
Spannungsbiegung f   1382
Spannungsdeformationskurve f   1284
Spannungsrelaxation f   1280
Spannungsriß m   1277
Spezifikation f   1193
spezifischer Durchgangswiderstand m   1514
spezifischer Oberflächenwiderstand m   1028
spezifischer Widerstand m   992
spezifischer Widerstand m   1025
spezifischer Widerstand m   1027
Spinnstuhlwalze f   1044
Spinnvlies n   1217
Spiralschlauch m   297
Spiralschlauch m   1199
Spleißstelle f   1200
Spritzartikel m, Extrusionsartikel m   473
Spritzgießen n   621
Spritzgießen n   742
Spritzkolben m   919
Spritzmaschine f, Extruder m   477
Spritzmaschine f, Extruder m   675
Spritzpistole f   281
Spritzpistole f   284
Spritzpistole f   1206
Spritzpistole f   1209
Spritzpressen n, Transferpressen n   744
Spritzquellung f   1317
spröde, brüchig   194
Sprödigkeit f in der Kälte   196
Sprödigkeit f, Brüchigkeit f   195
Sprödigkeitsprüfer m   1412
Sprühen n, Spritzen n   1208
sprühen, spritzen   1207
SRF-Ruß m (SRF) (halbverstärkend)   1122
Stabilisator m   1221
Stabilisiermittel n   71
Stabilisierung f   1220
Stabilität f (chemisch)   1223
Stabilität f (mechanisch)   1225
Stabilität f (thermisch)   1228
Stabilität f   1222
Stahlkord m   1242
Standardabweichung f   1232
Standfestigkeit f   974
Stanzartikel mpl   405
Stanzmesser m   267
Stanzmesser n   403
Stanzpresse f   406
Stanzpresse f   875
Stanzpresse f   879
Stanzpresse f   885
Stanzwerkzeug f   910
Stauung f   576
Stearinsäure f   33
Stegbreite f   507
Steifheitsprüfer f   1423
Steigung f   827
Stempel m   917
Stempelpresse f, Gesenkpresse f   888
Störung f   191
Stoßbeständigkeit f   1147
Stoßdämpfer m   385
Stoßdämpfer m   1146
Strahlenbeständigkeit f   988
Strahlung f (einer Strahlung aussetzen)   912
Streckgrenze f   1561
Streckspannung f   1269
streichen mit Pinsel   1210
Streichen, Beschichten, Auftragen f   1213
Streichmaschine f   673
Streichmaschine f   680
Streichmaschine f   1212
Streichmaschine f   1214
Streifen m, Leiste f   1289
Streuplatte f   845
Struktur f   327
Struktur f   1291
Stumpfnaht f   210
Stumpfnaht f   1115
Stumpfstoß m   209
Suspension f (kolloidale)   1312
Synthesekautschuk m   1322
synthetischer Latex m   1321
synthetisches Elastomer n   1320

T

Tankauskleidung f, Behälterauskleidung f   1333
Tauchartikel mpl (Latex)   651
tauchlackieren, tauchbeschichten   413
Tauchlackierung f, Tauchbeschichtung f   280
Technik f   1356
Techniker m, Technologe m   1357
technischer Berater m   331

Teer m 1342
Teilchengröße f, durchschnittlicher Durchmesser m 1157
Teilchengröße f, durchschnittlicher Durchmesser m 1158
Tellur n 1358
Temperatur f 1359
Temperaturabfall m 1365
Temperaturanstieg m 1371
Temperaturbereich m 920
Temperaturbereich m 1369
Temperaturindex 1367
Terpolymer n 1383
Testflüssigkeit f, Prüfflüssigkeit f 1395
Tetrafluorethylen n -TFE 1427
Textil f 1428
Thermalruß m (fein), FT-Ruß 523
thermische Instabilität f 1433
thermische Verformung f 1431
thermische Zersetzung f 1430
thermoplastisch 1434
Thermoplastizität f 1435
thixotrop 1438
Tieftemperaturbeständigkeit f 1370
Titandioxid n 1450
Toleranz f 1451
Toluol n 1452
Tonne f 1453
Topfzeit f 864
Torsion f 1458
Torsionsspannung f 1279
Torsionsspannung f 1459
Toxizität f, Giftigkeit f 1461
Träger m (in Farben) 1494
Trägheitsmoment n 614
Trägheitsmoment n 733
Trägheitsmoment n 734
Tragfähigkeit f 229
transparent 1465
Transportbanddecke f 354
Transversalelastizitätsmodul m 731
Treibmittel n, Blähmittel n 66
Treibstoffschlauch m, Kraftstoffschlauch m 520
Trennung f einer Klebverbindung 1123
Trockenklebrigkeit f 1326
Trockensubstanz f 690
Trockensubstanz f 1175
Trockner m 127
Trübung f 557

Ü bergang m (z.B. Zustandsänderung) 1462
Überlappungsverbindung f 643
Überschuh m 793
übertragene Energie f 1464
Übertragung f, Fortleitung f 321
übervulkanisiert 792
Ultraviolettlampe f 1478
Ultraviolettlicht n 1479
Umdrehungen pro Minute, 1/min 1035
Umgebungstemperatur f 1360
Ummantelung f 634
Umwälztank m 1332
Umwandlungstemperatur f 1373
Undurchlässigkeit f 603
ungebleichtes Leinen n 432
ungewebt 757

Untervulkanisation f 1480
untervulkanisiert 1481
unvulkanisiert 380
Unze f 785
Urethan n 1483

V akuumbremsenteile npl 1486
Vakuumextruder m 1487
VAMAC Ethylen/Acrylatkautschuk m 1492
Vaseline n 820
Ventilschaftdichtung f 1113
Verarbeitbarkeit f 897
Verarbeitungshilfsmittel m 896
Verarbeitungssicherheit f 1096
Verbindung f 309
Verbrennungswärme f 564
Verdampfung f 1493
Verdichtungsverhältnis n 933
Verdrehung f, Verformung f 421
verfärbendes Alterungsschutzmittel n 103
Verfärbung f im Sonnenlicht 1303
Verfärbung f 417
Verformung f (bleibend) 1129
Verformung f (bleibend) 1130
Verhältnis n 929
Verhältnis n 932
Verlust m (dielektrisch) 411
Verlust m an löslichen Bestandteilen 668
Verlust m 665
Verlustwinkel m 97
Vernetzung f bei Raumtemperatur 1051
Vernetzung f 370
Vernetzungsmittel n 67
Vernetzungsmittel n 377
Vernetzungsmittel Masterbatch m 378
Vernetzungsstelle f 1156
Vernetzungsstellen f 371
Vernetzungszeit f, Zeit für 90% Vernetzung im Rheometer (T90) 1444
Versagen n, Fehler m 482
Versprödungstemperatur f 197
Versprödungstemperatur f 1362
Verstärkerfüllstoff m (hell) 758
Versuch m, Prüfung f 1384
Verträglichkeit f 307
Verweilzeit f 577
Verweilzeit f 1442
Verzögerer m 1031
Vibrationsdämpfer m 1497
Vibrationsdämpfung f 1498
Vicat Erweichungspunkt m 1374
Vinylchlorid n 1499
Vinylidenfluorid n 1500
visko-elastisches Material n 1507
Viskosität f (Brookfield) 1503
Viskosität f (Mooney) 1505
Viskosität f (relativ) 1506
Viskosität f 1502
Viskositätskoeffizient m 294
VITON Fluorelastomer m 1508
Vollgummiplatte f 1075
Vollgummireifen m 1447
Volumen n 1511
Volumenbeständigkeit f 1513
Volumenveränderung f 1512

Volumenzunahme f durch Quellung 1515
Vorrat m, Lagerbestand m 1250
Vorspannung f (z.B. von Förderbändern während der Vulkanisation) 1288
Vorverformungsdruck m 872
vorwärmen 873
Vulkanisat n 1516
Vulkanisateigenschaften fpl 905
Vulkanisation f 1517
Vulkanisation f, Vernetzung f 379
Vulkanisationsdauer f 1441
Vulkanisationsform f 1520
Vulkanisationsgeschwindigkeit f 925
Vulkanisationsgeschwindigkeit f 1519
Vulkanisationsgeschwindigkeit f, Vernetzungsgeschwindigkeit f 922
Vulkanisationskoeffizient m 295
Vulkanisationskoeffizient m 1518
Vulkanisationsmittel n 68
Vulkanisationspresse f 876
Vulkanisationssystem n 1323
Vulkanisationsverzögerer m 63
Vulkanisationsverzögerer m 106
Vulkanisationszeit f 1443
vulkanisierbar 376
Vulkanisieren n, Vernetzen n 381

W ärme f (spezifisch) 1192
Wärmeausdehnung f 1432
Wärmeausdehnungs-, Längenausdehnungskoeffizient m (linear) x-y °C 291
Wärmeentwicklung f (innere) 561
Wärmeentwicklung f 562
Wärmekapazität f 230
Wärmeleitfähigkeit f 323
Wärmeleitzahl f, Wärmeleitungskoeffizient m 293
Wärmeübertragung f 570
Wärmeverlust m 667
Wärmewirkung f 563
wäßriger Auszug m 110
wässriger Auszug m 1523
Walze f 1045
walzen 715
Walzenbezug m 1049
Walzenüberzug m 358
Wanderung f, Migration f 709
Wasseraufnahme f 17
Wasserbeständigkeit f 1015
Wasserdampf m 1528
Wasserdampfdurchlässigkeit f 812
Wasserdampfdurchlässigkeit f 1525
wasserdicht Imprägnieren n 1530
Wasserdurchlässigkeitsprüfer m 1524
Wasserdurchlässigkeitsprüfung f 1409
Wassersaugschlauch m 1526
Wechselstrom m 93
Wegschieben n, Wegwischen n 1552
Weichheit f 1169
Weichmacher m (aromatisch) 112

Weichmacher m (aromatisch) 769
Weichmacher m (naphthenisch) 775
Weichmacher m (paraffinisch) 776
Weichmacher m 831
Weichmacher m 1168
Weichmacher-Zeiteffekt m 833
Weichmachungsfähigkeit f 1
Weiterreißfestigkeit f, Kerbzähigkeit f 1014
Weiterreißprüfung f 1406
Weiterreißwiderstand m 1349
Weiterreißwiderstand m 1351
Weiterreißwiderstandsprüfung f an einem eingeschnittenen Prüfkörper 1350
Weiterreißwiderstandsprüfung f 1407
Welle f, Achse f 1132
Wellendichtung f 1112
Wert m 1488
Wetterballon m 703
Wetterballon m 704
Widerstand m 1029
Wirkungsgrad m 435
Witterungsbeständigkeit f 999
Wulstdraht m 131

Y oungscher Modul m 1562

Z ähigkeit f 1375
Zahnriemen m 144
Zahnriemen m, Steuerriemen m 1446
Zeichnung f 426
Zeitraum m 632
Zell-, Moos-, Schwammgummi n 246
Zellgummi m (mit geschlossener Zellstruktur) 1203
Zellgummi n für chirurgische Zwecke 1082
zellig 245
Zement m 249
Zementsaugschlauch m 251
Zerkleinerungsmaschine f 261
Zerreißbarkeit f 1353
Zerreißfestigkeit f 1268
Zerreißfestigkeit f 1477
Zerreißfestigkeitsprüfer m (Schopper) 1421
Zerreißfestigkeitsprüfer m (Scott) 1422
Zersetzung f, Zerstörung f 398
Zinkoxid 1563
Zoll m 608
Zug-Dehnungskurve f 1285
Zugdehnung f 452
Zugdehnung f 1252
Zugfestigkeit f 1377
Zugschlagfestigkeit f 1267
Zugspannung f 1378
Zugspannungsmesser m 1253
Zugspannungsprüfer m 1256
Zugspannungswert m, Modul m 1376
Zusatz m, Zusatzmittel n 38
zweigängig 423
Zyklus m 382
Zylinder m 125
Zylinderring m 1040
Zylindertemperatur f 1361

INDICE DEL VOCABOLARIO
A-Z

A luce poco profonda 1133
a passo doppio (vite) 423
a secco 384
abito m da lavoro 1555
abrasimetro m 12
abrasione (indice d') 5
abrasione (prova f di) 9
abrasione f 4
abrasione f Taber 8
abrasivo (utensile m) 2
accelerante m 19
accelerazione f 18
accettore di radicali 1466
accettore m 20
accettore m di acidi 21
accumulatore m, batteria 23
accumulatore m idraulico 22
acidi (resistente agli) 30
acido m ftalico 821
acido m inorganico 28
acido m organico 29
acido m stearico 33
ad alta planarità f, planare (agg.) 670
additivo m 38
aderente 40
aderente 41
aderenza f al suolo 543
aderire 39
adesione (apparecchio m per prove di) 47
adesione f (attacco m) 43
adesione f (attrazione molecolare) 42
adesione f tra gomma e tessuto 1085
adesività f 1325
adesività f 1330
adesività f, appiccicosità f 1244
adesivo 49
adesivo (autoadesivo) 51
adesivo m 48
adesivo m, colla f 248
adesivo m a base di celluloide 247
adesivo m a base di gomma 1057
adesivo m a base di lattice 645
adesivo m a presa rapida 250
adesivo superficiale (potere m) 45
affioramento m 171
affiorare 170
agente di appiccicosità 1329
agente m agglomerante (legante) 152
agente m anti adesivo 101
agente m anti-adesivo 1328
agente m antiossidante 104
agente m d'incollaggio 179
agente m deodorante 69
agente m emulsionante 70
agente m gonfiante 66
agente m legante 65
agente m promotore di appiccicosità 74
agente m protettivo anti luce solare 64
agente m protettivo contro l'invecchiamento 62
agente m reticolante 67
agente m rigonfiante 73
agente m ritardante della prevulcanizzazione 63
agente m ritardante della prevulcanizzazione 106
agente m stabilizzante 71
agente m tensio-attivo 72
agente m vulcanizzante 68
agglomerato 75
aggregato, conglomerato 76
agitare 1248
albero m 1132
ALCRYN gomma lavorabile allo stato fuso 90
alimentatore, tramoggia 580
allungamento f iniziale 1254
allungamento m 449

allungamento m 1287
allungamento m a rottura 187
allungamento m a rottura 453
allungamento m a snervamento 450
allungamento m a snervamento 1560
allungamento m a trazione 1252
allungamento m preliminare 1288
alogeno 549
alta frequenza f 575
ammortizzamento m 386
ammortizzatore 385
ammortizzatore m 1146
ammortizzatore m di vibrazioni 1497
amperometro m 94
ampiezza f 95
anello a U 1476
anello m di tenuta 1042
anello m 1039
anello paraolio 1112
angolo di perdita 97
angolo m di contatto 98
angolo m di taglio 96
anima f del cavo 341
anti-crittogamico m 710
anti-invecchiante m 100
antiossidante 102
antiossidante m macchiante 103
antiozonante 105
apice m 107
apparecchio f per prove di permeabilità all'acqua 1524
apparecchio per la misura della resistenza iniziale ad umido 1420
apparecchio m di laboratorio 641
apparecchio m di prova 1425
apparecchio m Hoffman per misurare la durezza mediante raschiatura 1419
apparecchio m per misurare la deformazione 1256
apparecchio m per misurare la fragilità 1412
apparecchio m per misurare la permeabilità all'aria 1410
apparecchio m per prove d'invecchiamento 54
apparecchio m per prove di fragilità a bassa temperatura 1414
apparecchio m per prove di lacerazione 1424
apparecchio m per prove di resistenza 1270
apparecchio m per prove di resistenza 681
apparecchio m per prove di rigidità 1423
apparecchio m per staccare i talloni dei pneumatici 130
apparecchio m 108
appiccicosità f a secco 1326
appiccicoso 1245
appiccicoso 1246
appiccicoso 1331
appoggio m (cuscinetto) 132
appoggio m ammortizzante in spugna 799
appoggio m ammortizzante in spugna 1204
appoggio m di spugna 800
aria f intrappolata 81
armatura f dei talloni dei pneumatici 131
articoli m a pareti spesse 1437
articoli m casalinghi in gomma 585
articoli m chirurgici 1310
articoli m da immersione in lattice 651
articoli m di gomma-amianto 115
articoli m di gomma per cancelleria 1238

articoli m estrusi 473
articoli m estrusi 1190
articoli m gommati 1090
articoli m igienici 114
articoli m in gomma per la casa 422
articoli m stampati 739
articoli m tecnici 695
articoli m tranciati 405
articolo m per immersione, senza giuntura 1116
articolo m stampato 113
articolo stampato 738
assorbimento (apparecchio m per prove di) 16
assorbimento (capacità f di) 14
assorbimento d'umidità 15
assorbimento m 13
assorbimento m 53
assortimento m 1128
assorbimento m d'acqua 17
attivatore m 36
attivazione f 35
attività f superficiale 1306
attrito m interno 630
attrito m statico 1237
aumento m di peso (termine specifico) 1543
aumento m di temperatura 1371
aumento m di volume, rigonfiamento 1515
auto-vulcanizzante 1119
autocicatrizzante (agg.) 1120

B agnabilità f 1545
balata f 120
banda riscaldante 301
battistrada m 1467
battistrada m del pneumatico 1449
biossido m di titanio 1450
bitume m 153
bobina f, rotolo m di alimentazione 954
boccola f 208
bolla f, soffiatura 167
bolla f d'aria 87
bolla f d'aria 162
bomba (apparecchio per l'invecchiamento) f 173
bomba f ad aria 77
bomba f ad ossigeno 174

C alandra f 220
calandra f con cilindri ad assi incrociati 221
calandrare 223
calandratura f 225
calo m di temperatura 1365
calore m di combustione 564
calore m di fusione 565
calore m specifico 1192
calorimetria f 226
calzatura f, scarpa f 513
cambiamento m di volume 1512
camera f d'aria 155
camera f d'aria 622
camicia f, guaina f 634
campione m 1097
campione m, provino m 1194
canaletto 1324
candeggio m 158
canotto m pneumatico 172
caolino m 264
caolino m 265
caolino m duro 266
capacità f 228
capacità f di carico 229
capacità f di rammollimento 1
capacità f termica 230

Italian index

capping m, interruzione f crescita della catena molecolare 231
carbonato di calcio 1550
carbone attivo m 235
carbone m attivo 234
carica f 486
carica f con pigmento 823
carica f rinforzante 957
carica rinforzante non nera (bianca) 758
carico m a rottura 188
carico m di rottura 1477
carico m di rottura a trazione 1268
carico m dinamico 433
carico m statico 661
cassetta f di derivazione 329
cassetta f di derivazione per cavi 213
catalizzatore m 241
catrame m 1342
cavità dello stampo, impronta 244
cavo m 211
cavo m di potenza 217
cavo m per alta frequenza 1379
cavo m ricoperto in piombo 214
cavo m sospeso 216
cellulare 245
cemento m 249
ceneri f 116
centipoise m 252
cera f microcristallina 1531
chilogrammo m 639
chiusura a scatto 1166
ciclo m 382
ciclo m di isteresi 596
ciclo m di isteresi 663
ciclo m sforzo-deformazione a bassa velocità 664
cilindro 125
cilindro m di impressione 606
cilindro m di impressione 607
cilindro m per filatura 1044
cilindro m rivestito di gomma 1048
cinghia dentata (per trasmissioni sincrone) 144
cinghia di trasmissione 1446
cinghia f trapezoidale 1484
cinghia scanalata 142
cinghia trapezoidale a spigoli grezzi 141
cinghia f 146
cloroprene m 260
cloruro di vinile 1499
co-precipitazione 272
coadiuvante di lavorazione 896
coagente 271
coagulazione f 273
coefficiente m di attrito 288
coefficiente m di attrito (dinamico) 3 m/min, 0,7 MPa 289
coefficiente m di attrito (statico) 290
coefficiente m di conduttività termica 293
coefficiente m di dilatazione 287
coefficiente m di dilatazione termica lineare x-y °C 291
coefficiente m di elasticità 286
coefficiente m di frizione 285
coefficiente m di permeabilità 292
coefficiente m di permeabilità 811
coefficiente m di viscosità 294
coefficiente m di vulcanizzazione 295
coefficiente m di vulcanizzazione 1518
coesione f 296
colare, fondere 238
colata f per scorrimento 240
colloide m 302
colofonia f 1554
colore m 305

colpo m a vuoto di prova, (nello stampaggio) 88
comando m a velocità variabile 1197
compatibilità f 307
composto m 309
compressione f 313
condotto m 325
condotto m per aria 429
condotto m rivestito in gomma 326
conducibilità f termica 323
conduttività 322
conduttore m (cavo conduttore) 324
conduttore m isolato 1553
conduzione f 321
confezione f, costruzione 202
confezione f montaggio m 204
configurazione f 327
conico m 1340
consulente m 331
contenitore m 332
contenuto in solidi 1173
contenuto m in gomma solida 1061
contrazione f 333
contrazione f 1033
controllato da computer (calcolatore) 319
contropressione f di estrusione 479
copolimerizzazione 339
copolimero m 338
coppia al 90% 1455
coppia f di torsione 1454
coppia massima 1456
coppia minimo 1457
copriruota m, coppa f 587
corda f, filo m 340
corrente f alternata 93
corrente f continua 414
corrosione f 350
corrosione f da acidi 26
costante f dielettrica 409
cottura f 119
CPE - polietilene clorurato 360
creep m a freddo 298
crematura f (del lattice) 365
cristallinità f 372
cristallizzazione f 373
cucitrice f 1249
cuffia per giunti a velocità constante 181
curva f carico-flessione 660
curva f di allungamento 451
curva f di compressione 314
curva f sforzo deformazione 1285
curva f sforzo deformazione per compressione 1284
cuscinetto m assorbente 133
cuscino m d'aria 80
cuscino m gonfiabile 616
cuscino m pneumatico 849

Decomposizione f, deterioramento m 192
decomposizione f termica 1430
deformazione f 389
deformazione f (sotto carico) 1251
deformazione f di planarità, imbarcamento m 1521
deformazione f di taglio 1255
deformazione f iniziale (per scorrimento) 620
deformazione f massima 691
deformazione f per allungamento 452
deformazione f permanente 1129
deformazione f residua 1130
deformazione f sotto carico ... C, ... MPa, ... h 390

deformazione f termica 1431
degasatore m 391
degradazione f 392
densità f apparente 109
densità f apparente 205
dentellato (agg.) 1125
deodorizzazione f 396
depolimerizzazione f 397
deterioramento m 398
determinazione f della durezza per indentazione 399
determinazione f della resistenza alla lacerazione su provino intagliato 1350
deviazione f standard 1232
diaframma m 400
diaframma m a disco 415
diametro della vite 1108
diametro m dell'ugello 401
dielettrico 407
dilatazione f termica 1432
dimensione f agglomerati, diametro m medio 1157
dimensione f normalizzata 1159
diminuzione f 387
dinamometro 434
dinamometro m per prove di scoppio 1413
dinamometro m Scott 1422
direzione della macchina 674
direzione perpendicolare alla macchina 369
disegno m 426
disegno m goffrato 802
disperdibilità 418
dispersione f 419
dispositivo m d'aspirazione 1296
distacco dallo stampo 395
distacco m dell'incollaggio 176
distacco m parziale 844
distorsione f 421
donatore di zolfo 1299
doppia vite 1472
durata f 430
durata f di esercizio 1126
durata f di magazzinaggio 1145
durezza Durometro A 552
durezza IRHD 553
durezza f 551
durezza f di penetrazione (Rockwell) 609
durezza f Rockwell 554
durezza f Shore 555
durometro m 431
durometro m 556

Effetto Joule m 637
effetto m corona 347
effetto m corona, scarica f corona 345
effetto m corona interno 348
effetto m della calandra 222
effetto m elastico residuo (memoria elastica) 441
effetto m elastico residuo (memoria elastica) 700
effetto m plastificante-tempo 833
effetto m termico 563
efficienza f 435
efflorescenza f 165
efflorescenza f di zolfo 166
efflorescenza f di zolfo 1297
eiettore f 436
elasticità f 442
elastico (agg.) 438
elastomero m 443
elastomero m sintetico 1320
elettricamente 446
elettrolito m 447
ELVAX EVA 454
emulsionante 455
emulsione f 456
emulsione f di lattice-bitume 646

energia f trasmessa 1464
entità f della deformazione 1253
errore m standard (radice quadrata della varianza) 1233
esafluoro propilene 574
eseguire una prova di taglio 1138
espansione f 465
esperimento m 467
esposizione f agli agenti atmosferici 1537
esposizione f agli agenti atmosferici 1540
esposizione f all'aperto 786
esposizione f alle radiazioni 912
essiccatore m per mescole 127
essudamento m 480
essudazione f 1313
estensibilità f 469
estensore m 468
estrapolazione f 470
estratto m acquoso 110
estratto m acquoso 1523
estrudere 472
estrudibilità f 471
estrusione f 478
estrusione per soffiaggio 168
estrusore m (alimentato a caldo) 583
estrusore m (alimentato a freddo) 299
estrusore m (sotto vuoto) 1487
estrusore m, trafila f 475
estrusore m, trafila f 477
estrusore m, trafila f 675

Fascia f 123
fascia f riscaldante 124
fatica f 484
fatica f a flessione 498
fatica f dielettrica 410
fattore m di perdita (tan δ) (ISO = fattore m di perdita dielettrica) 420
fattore m di potenza (sen δ) 870
fattore m di sicurezza 1094
filato m agglomerato 1217
film m, pellicola f 488
filo m 1559
filo m di gomma estruso 1084
filo m metallico per pneumatici 1242
finitura f 490
flessione f 495
flessione f 501
flessione f 505
flessione f sotto carico 659
flessioni f successive 499
flessometro f 502
flocculazione f 508
fluido m idraulico 589
fluoro 511
fluoruro di vinilidene 1500
foglia f continua 1541
foglio f di sughero gommata 344
foglio m calandrato di supporto 137
foglio m di gomma solida 1075
foglio m di gomma spugnosa 1205
foglio m metallizzato 1229
fonoassorbente (agg.) 1187
forma f 1134
formazione di pellicole 489
forza di taglio 1137
forza di tenuta 1114
forza f dell'incollaggio, resistenza f dell'adesione 177
forza f di adesione 44
forza f di spellamento, strippaggio 1290
fragile 194
fragilità f 195
fragilità f a basse temperature 196

frequenza f   514
friabile (agg.)   515
frizionatura f   518
frizione   516
frizione f della calandra   224
frullatore m, nell'industria della schiuma di lattice   135

G

allone m   528
gamma f di temperature   920
gamma f di temperature   1369
ganascia f   635
geometria della vite   1109
ghigliottina   261
ghigliottina f   547
girante f, rotore   602
girante f di una pompa   909
giri m al minuto   1035
giubbetto f di salvataggio   207
giunto m   636
giunto m a sovrapposizione   643
giunto m di dilatazione   466
giunto m omocinetico   330
giunto m testa a testa   209
giunzione f, giunta f   1200
gomma cellulare f   246
gomma clorurata   1059
gomma in polvere   868
gomma pellettizzata   805
gomma solida   1063
gomma-spugna (gommapiuma)   1065
gomma f   1053
gomma f a base di esteri   463
gomma f acrilo-nitrilica   1054
gomma f butilica   1056
gomma f cellulare per chirurgia   1082
gomma f chirurgica   1311
gomma f con molto nervo   1076
gomma f crepe   368
gomma f estesa con olio   771
gomma f estesa con olio   772
gomma f estesa con olio   1070
gomma f greggia   939
gomma f greggia   1062
gomma f micro-cellulare   706
gomma f micro-porosa   707
gomma f naturale   750
gomma f ossidata   1071
gomma f per pistoni   826
gomma f per spazzole per parabrezza   1551
gomma f recuperata   949
gomma f resistente ai prodotti chimici   1058
gomma f resistente all'olio   1069
gomma f rigenerata   948
gomma f rigenerata   1072
gomma f sintetica   1322
gomma f spugnosa   1080
gommare   1088
gommatura f   1091
grado m   394
graffa f per cinghie   139
grafico m   256
grafico m   540
grammo m   539
granulo m   538
granulo m, pellet m   804
granulometria (dimensione della particella)   1158
grasso m   542
gravità f   541
guanti m da lavoro   1557
guanti m industriali   613
guanti m per uso casalingo   584
guanti m protettivi   907
guarnizione di tenuta per olio   1111
guarnizione f   532
guarnizione f a labbro   656
guarnizione f di sughero   343
guarnizione f per vetrate   1110
guasto m, avaria f   191

guidavalvola m, guarnizione f guidavalvola f   1113

H

YPALON gomma sintetica   593
HYTREL elastomero termoplastico   598

I

drocarburo m alifatico   91
idrocarburo m aromatico   111
idrolisi f   591
igroscopico m   592
imballaggio   798
impasto m   424
impasto m, mescolazione   715
impedenza f   601
impermeabile ai gas   534
impermeabilità f   603
impermeabilizzazione f   1530
impianto m per sgrassare   828
impregnazione f   605
impregnazione f con gomma   604
impressione f a caldo   1230
incapsulamento   458
incollaggio m della gomma   180
incrinatura f per sforzo   1277
indice, velocità f di scorrimento   926
indice, velocità f di taglio   927
indice, velocità f di vulcanizzazione   925
indice d'ossigeno   795
indice di scorrimento   510
indice di scorrimento a fusione   698
indice m di infiammabilità   492
indice m di rifrazione   955
indice m di temperatura   1367
indice m limite di ossigeno   610
indice m limite di ossigeno   796
indumenti m impermeabili   270
indumento m di gomma   1060
indumento m di gomma   1066
indumento m protettivo   269
induttanza f   612
infiammabile   615
infiammabilità f   493
infrarosso m   617
ingobbatura f dell'isolante (nei cavi elettrici)   911
ingranaggio m per tirare le cinghie   143
ingredienti m attivi   37
inibitore di invecchiamento a caldo   559
inibitore m di screpolature causate dalla luce solare   1301
inibitore m   618
ininfiammabile (agg.)   491
inizio m della lacerazione   1346
inorganico   623
inserto m, prigioniero m   624
instabilità f termica   1433
interpolazione f   631
intervallo m di tempo   632
invecchiamento m, deterioramento m agli agenti atmosferici   1539
invecchiamento m a caldo   558
invecchiamento m all'ozono   797
invecchiamento m della gomma   56
invecchiamento m dinamico   55
invecchiamento m in olio   57
invecchiamento m in olio   766
invecchiamento m in stufa   59
invecchiamento m in stufa   787
invecchiamento m in stufa ad aria calda   85
invecchiamento m statico   1236
involucro m d'ebanite per batterie   1067
ipotesi f   594
irrigidimento m semplice   1247
isolamento m   627

isolare   625
isolatore m   628
isteresi f   595

K

ALREZ articoli finiti in perfluoroelastomero   638

L

acerabilità f   1353
lacerare   1352
lacerazione f   1344
lacerazione f Groves   1345
lama f regolabile   156
laminazione f   642
lampada f ultra-violeta   1478
larghezza del canale   255
lastra f   1162
lastra f calandrata   1141
lastra f di miscela di gomma e sughero   1144
lastra f estrusa   1142
lattice pre-vulcanizzato   894
lattice m   644
lattice m cemento   647
lattice m concentrato elettricamente   649
lattice m concentrato per crematura   648
lattice m concentrato per evaporazione   650
lattice m di Neoprene   751
lattice m sintetico   1321
lavorabilità   897
lavorazione f   898
legame m   175
libbra f   865
libbra f per piede quadrato   866
libbra f per pollice quadrato   867
limite m elastico   440
linter m di cotone   353
liquido m di prova   1395
litargirio m   657
litro m   658
lubrificante m   672
luce f ultra-violetta   1479
lunghezza f dell'ugello   402

M

acchina f Du Pont per prove d'abrasione (abrasimetro Du Pont)   3
macchina e utensili per applicare le graffe alle cinghie   140
macchina f incollatrice   679
macchina f incollatrice   1160
macchina f per far prove sulla gomma   678
macchina f per far prove sulla gomma   1083
macchina f per la fabbricazione dei cavi   215
macchina f per la fabbricazione dei cavi   683
macchina f per prove di abrasione   11
macchina f per prove di flessione   500
macchina f per prove di flessione   676
macchina f per prove di flessione   1411
macchina f per prove di flessione (flessometro)   1415
macchina f per prove di piegamento   1417
macchina f per prove di resistenza a flessione   1416
macchina f per rivestire i cavi   356
macchina f Schopper (dinamometro)   1421
macchina f spalmatrice   282
macchina f spalmatrice a coltello   673
macchina f trecciatrice   185
macrostruttura   684

magnesia calcinata f   219
mano f di fondo   126
mano f di fondo   895
manometro m   686
manometro m   892
margine m di sicurezza   1095
masterbatch m, mescola madre f   687
matarozza   1216
materia f prima   689
materia f prima   938
materia f prima, mescola f, materiale   1250
materia f solida, secca   690
materiale m isolante   626
materiale m isolante   688
materiale m visco-elastico   1507
mattonella f per pavimenti   1439
mescola di gomma   311
mescola madre a base di agenti vulcanizzanti   378
mescola f cruda   312
mescola f di gomma e sughero   1055
mescola f per immersione di corda per pneumatici   1448
mescola f per rivestimento di cavi   218
mescola f per stucco   308
mescola f   310
mescolare   161
mescolatore m aperto   712
mescolatore m aperto   713
mescolatore m aperto   718
mescolatore m aperto   780
mescolatore m Banbury   716
mescolatore m Gordon   830
mescolatore m olandese nell'industria della carta   136
mescolatore m per soluzioni   1180
mescolazione f, mescolamento m   717
messa f a terra   328
metodo m di prova   1399
metro m   705
mettere a terra   546
microstruttura   708
migrazione f   709
milligrammo m   714
minio m   951
misura f   692
misura f della contrazione   334
misuratore m di adesività   1327
misuratore m di granulometria Hegman   573
modificato chimicamente   258
modulo a flessione   497
modulo m   719
modulo m a compressione   316
modulo m a compressione   723
modulo m a flessione   503
modulo m a trazione   1376
modulo m ad alta frequenza   722
modulo m di allungamento   729
modulo m di compressione   315
modulo m di elasticità   721
modulo m di elasticità trasversale   731
modulo m di taglio   730
modulo m di taglio   1135
modulo m di Young   1562
modulo m dinamico effettivo   720
modulo m elastico   724
modulo m elastico a compressione   725
modulo m elastico a flessione   726
modulo m elastico a taglio   727
modulo m elastico a trazione   728
mola f abrasiva   1547
mola f abrasiva   1549
mola f pulitrice   1548
molla f   1215
molla f di gomma   1074
momento m di inerzia   614
momento m di inerzia   733

momento m di inerzia   734
momento m torcente   1473
mono-passo   1154
monofilamento m, monobava m   735
montaggio m   117
morbidezza f   1169

N astratrice f   682
nastratrice f   1341
nastri m trasportatori per granuli   147
nastro auto-adesivo   893
nastro m   1334
nastro m adesivo   1335
nastro m autoadesivo   1339
nastro m isolante   1338
nastro m per cavi   1336
nastro m per freni   1257
nastro m per frizionatura   1337
nastro trasportatore   336
Neoprene m gomma sintetica   753
nero fumo fine tipo thermal m, carbon black FT m   523
nero fumo fornace MT (medium thermal)   696
nero fumo m   236
nero fumo m fine tipo fornace, carbon black FF m   522
nero fumo m fornace FEF (per estrusione fine)   521
nero fumo m fornace GPF (per uso generale)   537
nero fumo m fornace HAF (ad alta abrasione)   524
nero fumo m fornace ISAF (intermedio)   629
nero fumo m fornace SAF (super)   1305
nero fumo m semi-rinforzente tipo fornace, carbon black SRF   1122
nero fumo m tipo fornace ad alto allungamento HEF m   525
nero fumo m tipo fornace ad alto modulo, carbon black HMF   526
nero m animale   154
nerofumo m attivo   237
nerofumo m tipo channel a facile dispersione   461
nerofumo m tipo channel difficile da disperdere   586
nerofumo m tipo thermal a lavorabilità media, carbon black MT   748
nervo m   754
neutralizzare   756
neutralizzazione f   755
non ordito (agg.), senza trama   761
non-discolorante   759
non-macchiante   760
non-vulcanizzato   380
NORDEL gomma idrocarbonica   762
norma f   1234
norma f, specifica f   1193
normalizzato   1235
numero m d'indice   611
numero m di iodio   633

O -ring m   1041
olio m aromatico   112
olio m combustibile pesante   773
olio m essicativo   427
olio m essicativo   770
olio m minerale   774
olio m naftenico   775
olio m paraffinico   776
omogeneizzazione f   579
oncia f   785
opacizzato   1343
oscillografo m   783

oscillogramma m   782
osmotico   784
ossido m di ferro rosso   794
ossido m di magnesio   685
ossido m di zinco   1563
ossido m metallico   702
ottone m   186

P allone m   122
pallone m metereologico   703
pallone m metereologico   704
paraurti m   206
paraurti m di gomma   1073
parti f per freni a depressione   1486
particella   801
passo   827
passo della vite   507
penetrazione f   806
pennellare   199
perdita f   665
perdita f di calore   667
perdita f di energia   459
perdita f di energia   666
perdita f di materiali solubili   668
perdita f dielettrica   411
perdita f per isteresi   597
periodo m di riposo   808
periodo m di riposo   1030
permeabile (agg.)   814
permeabilità f   810
permittività f   815
permeabilità f ai gas   530
permeabilità f ai gas   813
permeabilità f al vapor d'acqua   812
permeabilità f al vapor d'acqua   1525
permittività f relativa ... Hz   816
perno m d'inserzione   824
perno m di centratura, di guida   425
perno m espulsore   437
perpendicolarmente alla venatura   34
peso m specifico   1191
petrolato m   820
pezzo m per autovettura   118
piano m di pressa   842
piastra f di appoggio   134
piastra f di chiusura   836
piastra f di espulsione   837
piastra f di estrazione, di smontaggio   841
piastra f di fissaggio, di staffaggio   835
piastra f di fissaggio dello stampo   840
piastra f di lucidatura   839
piastra f distributrice omogeneizzante   845
piastra f riscaldante   838
piastrella f di gomma   1440
piatto m riscaldato   843
piede m   512
piede m cubo   374
piegamento m (flessione f)   148
piegare   149
pigmento m   822
pinta f   825
pistola f per spruzzatura   281
pistola f per spruzzatura   284
pistola f spruzzatrice   1206
pistola f spruzzatrice   1209
pistone d'eiezione   918
pistone   917
pistone m   846
pistone m di iniezione   919
(plastica) materia f plastica   829
plastificante m   831
plastificante   1168
plastificante m a base di esteri   832

pneumatici m per carrelli   550
pneumatico m per autocarro   1471
pneumatico m per automobili   1475
pneumatico m per autovetture   233
pneumatico m pieno   1447
polibutilene tereftalato   803
polietilenetereftalato   818
polimerizzazione   856
polimerizzazione f in massa   857
polimerizzazione f in soluzione   1181
polimero graffato   853
polimero m   851
polimero m a blocchi   164
polimero m a scala   854
polimero m cis-tattico   852
polimero m greggio   855
polisolfuro m   859
polistirolo m, polistirene   858
poliuretano m   860
pollice m   608
pollice m cubo   375
pollice m quadrato   1218
polverizzare   869
porosimetro m per gas   1418
porosità   861
porta mandrino m   1198
porta ugello m   765
post-reticolazione f   862
post-vulcanizzazione   863
potere m adesivo   203
potere m adesivo   1259
potere m adesivo, forza f di adesione   46
potere m di presa rapida   1243
potere m rigonfiante   1318
pre-riscaldamento   873
premasticazione f   189
preriscaldamento   871
pressa   874
pressa f a bilanciere   879
pressa f a corsa lunga   883
pressa f a doppio pistone   877
pressa f a piani multipli   884
pressa f a piombo   881
pressa f a piombo per tubi   880
pressa f a vite punzonante   886
pressa f ascendente   889
pressa f discendente   878
pressa f per applicare guaine di piombo   882
pressa f per punzonare   888
pressa f per tranciare   875
pressa f per vulcanizzare   876
pressa f per vulcanizzare le cinghie   145
pressa f punzonante   885
pressa f singola   887
pressione f atmosferica   891
pressione f d'aria   890
pressione f di chiusura, di stampo   263
pressione f di preformatura   872
pressione f manometrica   536
prestazione   807
prezzo m, tariffa f   928
prodotti m chimici per l'industria della gomma   259
prodotto m di condensazione a base di aldeide-anilina   320
prodotto m per il trattamento superficiale   1309
profilo m   899
profondità del canale   254
propagazione f della lacerazione   1348
proporzione f, rapporto m   929
proprietà f   906
proprietà f criogeniche   901
proprietà f dei vulcanizzati   905
proprietà f di base   904
proprietà f dinamiche   902
proprietà f meccaniche   903
prova abrasione su strada   10

prova d'invecchiamento accelerata agli agenti atmosferici   1387
prova d'invecchiamento accelerata dalla luce   1386
prova di permeabilità all'acqua   1409
prova di resilienza, resa f elastica   943
prova di rigonfiamento   1405
prova f   1384
prova f d'esercizio   1127
prova f d'invecchiamento   61
prova f d'invecchiamento accelerato   1385
prova f di corrosione   1390
prova f di deformazione per compressione   1389
prova f di deformazione permanente a compressione   1402
prova f di esercizio simulato   1404
prova f di flessibilità   506
prova f di flessione   1394
prova f di flessione   1393
prova f di flessione con provini incisi   1400
prova f di laboratorio   1397
prova f di lacerazione   1407
prova f di lacerazione   1408
prova f di prestazione   1401
prova f di propagazione della lacerazione   1391
prova f di propagazione della lacerazione   1406
prova f di resistenza al taglio per compressione   1388
prova f di resistenza alle flessioni ripetute   1392
prova f di taglio   1403
prova f di tenuta   1396
prova f meccanica   1398
prova f pratica di abrasione   7
puleggia f per cinghia trapezoidale   908
puleggia f per cinghia trapezoidale   1485
punto di infiammabilità   494
punto di reticolazione   1156
punto di snervamento   1561
punto m di anilina   99
punto m di fusione   527
punto m di fusione   699
punto m di fusione   850
punto m di iniezione (nello stampaggio a iniezione)   535
punto m di lacerazione   1355
punto m di ristagno   576

R adicale (libero)   914
raffreddamento m (aria di)   337
rammollire   1167
rapporto di scarto   930
rapporto m   932
rapporto m di compressione   933
rapporto m di riduzione   937
rapporto m di rigonfiamento nell'estrusione   404
rapporto m di rigonfiamento nell'estrusione   934
rapporto m di stiro   935
rapporto m lunghezza/diametro   936
rayon m   940
reagente   941
recipiente m per batterie, batteria f   128
recupero m   950
recupero m elastico (in prove di plasticità)   834
regolatore m di tensione per il riavvolgimento   1380

regolazione f

regolazione f 52
reometro a disco oscillante - ODR 1037
reometro a pistone 1038
reometro m 1036
resiliente 964
resilienza f 962
resilienza f 963
resina f 965
resina f acrilica 966
resina f alchidica 967
resina f epossidica 462
resina f fenolica 968
resina f melamminica 697
resina f ureica 969
resistente (agg.) 1017
resistente agli acidi 32
resistente agli alcali 92
resistente ai combustibili liquidi 1022
resistente ai grassi 1018
resistente ai solventi 1024
resistente al calore 567
resistente al calore ed agli olii 560
resistente al vapore 1241
resistente all'olio 777
resistente all'olio 1020
resistente all'ozono 1021
resistente alla benzina 983
resistente alla corrosione 351
resistente alle muffe 711
resistente alle soluzioni saline 1023
resistenza a trazione di mescola non vulcanizzata 1263
resistenza allo spellamento 1264
resistenza chimica f 973
resistenza f 970
resistenza f, forza 1258
resistenza f a flessione 504
resistenza f a flessione 1262
resistenza f a flessioni ripetute 496
resistenza f a scarica corona 349
resistenza f a scarica corona 975
resistenza f agli acidi 31
resistenza f agli acidi 971
resistenza f agli acidi 993
resistenza f agli agenti chimici 996
resistenza f agli alcali 994
resistenza f agli urti 1147
resistenza f ai grassi 984
resistenza f ai tagli 998
resistenza f al calore 566
resistenza f al distacco degli strati 1007
resistenza f al freddo 981
resistenza f al rammollimento a caldo 1002
resistenza f al rigonfiamento 1013
resistenza f al rotolamento 990
resistenza f al rotolamento 991
resistenza f al rotolamento 1008
resistenza f al rotolamento 1050
resistenza f al taglio 1010
resistenza f al taglio 1136
resistenza f al taglio 1266
resistenza f al vapore 1011
resistenza f all'abrasione 6
resistenza f all'acqua 1015
resistenza f all'arco 1012
resistenza f all'esposizione 999
resistenza f all'incisione (di isolante) 979
resistenza f all'invecchiamento 60
resistenza f all'olio 1004
resistenza f all'ossidazione 1005
resistenza f all'ozono 987
resistenza f all'ozono 1006
resistenza f all'urto 599
resistenza f all'urto 986
resistenza f all'urto (con intaglio) 763

resistenza f all'urto con intaglio Izod 600
resistenza f all'urto con trazione 1267
resistenza f all'usura e alla lacerazione 1016
resistenza f alla benzina 982
resistenza f alla compressione 317
resistenza f alla compressione 1261
resistenza f alla corrosione 976
resistenza f alla cristallizzazione 978
resistenza f alla fiamma 1000
resistenza f alla lacerazione 1014
resistenza f alla lacerazione 1349
resistenza f alla lacerazione 1351
resistenza f alla luce 1003
resistenza f alla luce solare 1012
resistenza f alla luce solare 1304
resistenza f alla raspatura 1009
resistenza f alla rottura 193
resistenza f alla rottura 977
resistenza f alla rottura 1260
resistenza f alla trazione 1377
resistenza f alle basse temperature 1370
resistenza f alle flessioni 980
resistenza f alle flessioni 1001
resistenza f alle muffe 989
resistenza f alle radiazioni 988
resistenza f alle rotture 997
resistenza f alle screpolature 995
resistenza f allo schiacciamento 974
resistenza f allo scorrimento 367
resistenza f allo snervamento 1269
resistenza f allo strappo 1265
resistenza f di volume 1513
resistenza f in condizioni di umidità 1271
resistenza f specifica, resistività f 992
resistività f 1025
resistività f alla corrente continua 1026
resistività f di volume 1514
resistività f elettrica 1027
resistività f superficiale 1028
resistore 1029
respingente m di gomma 201
reticolazione f, cross linking m 370
reticoli f, legami incrociati 371
reversione m (ad un altro stato) 1034
ricetta f, formula f 945
richiesta 961
ridurre 952
riduzione f 671
riduzione f 953
riempimento m con cariche 487
rigenerare 947
rigenerazione f con acidi 946
rigenerazione f della gomma 956
rigidità f dielettrica 412
rigonfiamento m 163
rigonfiamento m 1316
rigonfiamento m dopo l'estrusione 1317
rigonfiamento m in acqua 1527
rigonfiare 1314
rigonfiarsi 1315
rigonfiato 1319
rilassamento m 960
rilassamento m dello sforzo 1280
rilassamento m dello sforzo 1281
rimbalzare 944
rimbalzo f 942
ritardante f 1031
ritenzione f del colore 1032
ritiro m 1149
ritiro m del diametro 1150
ritiro m dopo lo stampaggio 1151
rivestimento in calandra 283

rivestimento in gomma di pavimenti 1064
rivestimento f per immersione 280
rivestimento m 279
rivestimento m 655
rivestimento m del cavo 212
rivestimento m del cavo 1140
rivestimento m di cilindri 1049
rivestimento m di cilindro 358
rivestimento m di pavimenti 357
rivestimento m di un tubo 355
rivestimento m di un tubo 581
rivestimento m esterno delle cinghie 354
rivestimento m in gomma per pareti 1086
rivestimento m per pavimenti 509
rivestimento m per ruote o copriruota 359
rivestimento m per serbatoi 1333
rivestire 274
rivestire a calandra 275
rivestire a pennello 200
rivestire a spatola 276
rivestire con spalmatura 1211
rivestire per immersione 413
rivestito 277
rivestito a pennello 198
rondella f 544
rondella f 1522
rotella f orientabile 239
rotore m 1052
rottura f 361
rottura f 1093
rottura f, guasto m 482
rottura f per fatica 485
rullo m di mescolazione 1045
rullo m per stampanti 1046
rullo m stampatore 1047
ruota f 1546
ruota f di gomma 1087

Sabbiare 1100
sabbiatura f 1099
saldatura a caldo 571
saldatura con solventi 1186
saldatura f 1544
salita f, aumento m 1043
sbavatura 388
scarica f corona 346
scarica f perforante 190
scarpe f da lavoro 1556
scarto 1106
schiuma di lattice 652
schiuma f 519
sciogliere 1177
scollaggio m 1123
scoloramento m 417
scolorimento m causato dalla luce solare 1303
scorrimento m 366
scorrimento m a freddo 300
scorrimento m a lungo termine 662
scorrimento m iniziale 619
scorrimento m relativo 958
scottabile 1105
scottabilità f 1101
scottabilità f Mooney 1102
screpolatura f causata dalla luce solare 1300
screpolatura f causata dalla luce solare 1302
screpolature f da flessione 362
segmento m del pistone 1040
senza canali 1092
separatore m per batterie 129
separazione di un'emulsione 1124
separazione f tra gli strati 847
separazione f tra gli strati 848
serbatoio m di circolazione 1332

sfogo dell'aria 1496
sforzo a torsione 1459
sforzo m a compressione 1276
sforzo m d'incollaggio 178
sforzo m a compressione 1286
sforzo m di flessione 1275
sforzo m di taglio 1282
sforzo m di trazione 1378
sforzo m sotto compressione 1283
sgrassante 393
Shore durometro m 1148
sicurezza f di lavorazione 1096
silice calcinata 1152
silice f precipitata 1153
sinterizzazione 1155
sistema di vulcanizzazione organico 781
sistema m di vulcanizzazione con ossidi metallici 701
sistema m di vulcanizzazione 1323
slittare 1164
slittare, frenare 1161
smalto m 457
smorzamento del suono 1188
smorzamento m delle vibrazioni 1498
smorzante del rumore 1189
smussato, bisellato 150
soffietto m, cuffia f 182
soffietto per scatola guida 138
solidi non volatili 1174
solidi totali 1175
solidificazione f, indurimento m 1131
solido m 1172
solubile 1179
solubilità f 1178
soluzione f di gomma 1077
solvente m 1182
solvente m 1184
solvente m aromatico 1183
solvente m per gomma 1078
soprascarpa f 791
soprascarpa f 793
soprascarpe f da pioggia 916
sospensione f (colloidale) 1312
sotto-vulcanizzata 1481
sotto-vulcanizzato 1480
sovravulcanizzato (agg.) 792
spalmare a pennello 1210
spalmatrice f 680
spalmatrice f 1212
spalmatrice f 1214
spalmatura f 1213
spalmatura f a spatola 640
spatola f per spalmatura 157
spessimetro m Elcometer 444
sporcamento stampi 736
spruzzare 1207
spruzzatura f 1208
spugna f 1201
spugna f a celle chiuse 1203
spugna f di gomma 1079
spugna f di Neoprene 752
spugna f 1201
stabilità chimica f 1223
stabilità f 1222
stabilità f al calore 569
stabilità f al magazzinaggio 1227
stabilità f alla luce 483
stabilità f del colore 303
stabilità f del colore 306
stabilità f del colore 1226
stabilità f del colore alla luce 304
stabilità f della soluzione 864
stabilità f dimensionale 1224
stabilità f meccanica 1225
stabilità f termica 1228
stabilizzatore m 1221
stabilizzazione f 1220
staffa f 262
stampaggio 740
stampaggio a compressione 743
stampaggio a iniezione 621
stampaggio a iniezione 742

Italian index

stampaggio in transfer (stampaggio per trasferimento) 744
stampaggio per soffiaggio 169
stampaggio per soffiaggio 741
stampo a impronte multiple 737
stampo a impronte multiple 749
stampo m per punzonare 403
stampo m per tranciare 267
stampo m per tranciare 910
stampo m per vulcanizzazione 1520
standard 1231
statore di pompa 1239
stivali m di gomma 183
strappo m 1354
striscia f 1289
strisciamento m 363
strizzare 1219
strofinamento 1552
strumento m di misura 693
strumento m per misure elettriche 445
strumento m per misure elettriche 694
strumento m per prove elettriche 1426
struttura del carbon black (del nero) 1292
struttura f 1291
struttura f 1429
struttura f della catena 253
stuccare 242
stuccatura f 243
stufa f ad aria calda 84
stufa f di invecchiamento Geer 788
stufa f Geer-Evans 789
stufa f per prove d'invecchiamento 58
stufa f riscaldata 790
sughero m 342
suola f 1170
suola f 1176
suola f di gomma spugnosa 1171
superficie f, area superficiale 1307
supporto m 746
supporto m anti vibrante 745
supporto m per motori 460
sviluppo (di gas) m 464
sviluppo m di calore 562
sviluppo m di calore interno 561

T acco m di gomma 1068
tagliato secondo un angolo, tagliato diagonalmente 151
tampone m di gomma 1081
tappetino m per autovetture 232
tecnica f 1356
tecnologo m 1357
tela olandese f 432
tela olandese f 578
tela f olona 227
tela olona f 428
tellurio m 1358
telone m 1344
telone m per autocarro 1470
teloni m per ospedali 1143
temperatura del cilindro 1361
temperatura di fusione 1368
temperatura f 1359

temperatura f ambiente 1360
temperatura f ambiente 1372
temperatura f di autoaccensione (ISO = temp. di accensione spontanea) 1121
temperatura f di deflessione 1364
temperatura f di esercizio continuo 1363
temperatura f di infiammabilità 1366
temperatura f di infragilimento 197
temperatura f di infragilimento 1362
temperatura f di rammollimento Vicat 1374
temperatura f di transizione 1373
tempo di permanenza 577
tempo di permanenza 1442
tempo di scottabilità (TS2) 1103
tempo di scottatura 809
tempo di vulcanizzazione ottimale (TC90) 1444
tempo m di solidificazione 1445
tempo m di vulcanizzazione 1441
tempo m di vulcanizzazione 1443
tenacità f 1375
tenacità f 1460
tenacità f a crudo 985
tendenza f alla scottabilità 1104
tensione f, sforzo m 1272
tensione f a rottura 1273
tensione f a rottura 1274
tensione f deflessione f 1382
tensione f di scarica dielettrica (ISO = tensione disruttiva) 408
tensione f per flessione 1278
tensione f per torsione 1279
tensione f superficiale 1308
tensione f superficiale critica 1381
termoindurente (agg.) 1436
termoplasticità f 1435
termoplastico (agg.) 1434
termosensibile 568
terpolimero 1383
terra f, massa f 545
tessuto 1428
tessuto m 481
tessuto m 1542
tessuto m adesivo 50
tessuto m di frizionatura 517
tessuto m doppio 268
tessuto m elastico 439
tessuto m gommato 1089
tessuto m impermeabile 1529
tessuto m impermeabilizzato 900
tessuto m non tessuto 757
tessuto m rivestito 278
tetrafluoroetilene 1427
tipo channel a lavorabilità media, carbon black MPC 747
tissotropico (agg.) 1438
tolleranza f 1451
toluolo m 1452
tonnellata f 1453
torsione f 1458
tossicità f 1461
trafila f a piombo per cavi 653
trafilato m di tenuta, profilo 1538

trancia f 406
trancia f a ghigliottina 548
trancia f a ghigliottina 677
transizione f 1462
trasmissione f dei gas 531
trasmissione f del calore 570
trasmittanza f della luce 1463
trasparente 1465
trasudamento m (dei colori) 160
trasudare 159
trattamento m 1469
tubo a spirale 297
tubo m aspiratore per acqua 1526
tubo m con interno liscio 1165
tubo m corrugato 352
tubo m d'aspirazione e di scarico 1294
tubo m d'aspirazione e di scarico dell'olio 779
tubo m d'aspirazione per cemento 251
tubo m di scarico 416
tubo m estruso 474
tubo m idraulico 590
tubo m per acetilene 25
tubo m per acidi 27
tubo m per aerei 89
tubo m per benzina 533
tubo m per carburante 520
tubo m per carri ferroviari 915
tubo m per compressore 318
tubo m per freni ad aria compressa 78
tubo m per gas 529
tubo m per latterie 383
tubo m per macelli 1163
tubo m per prodotti caseari 364
tubo m per prodotti chimici 257
tubo m per radiatore 913
tubo m per riscaldamento 572
tubo m per sabbia e ghiaia 1098
tubo m resistente ai solventi 1185
tubo m resistente all'olio 778
tubo m resistente all'olio ed al petrolio 819
tubo m resistente all'olio ed alla benzina 767
tubo m resistente all'olio ed alla benzina 768
tubo m resistente alla benzina 1019
tubo m rinforzato con treccia 184
tubo m spiralato 1199
tubo per gas a bassa pressione 669

U gello 764
ugello m dell'estrusore 476
umidità 588
umidità 732
umidità f atmosferica 82
umidità f dell'aria, (umidità dell'aria) 83
umidità f relativa 959
unione f testa a testa 210
unione f testa a testa 1115
unità f 1482
uretano m 1483

usura f, consumo m 1532
usura f a carico massimo 1535
usura f a rottura 1534
usura f allo snervamento 1536
usura f del battistrada 1468
usura f per abrasione 1533

V alore m 1488
valore m di acetile 24
valore m in uso 1490
valore m limite 1489
valori m tipici misurati 1474
valvola f a sfera 121
valvola f di controllo 335
valvola f di controllo del diaframma 1491
VAMAC elastomero etilen/acrilico 1492
vapore m d'acqua 1528
vapore m vivo 1240
vaporizzazione f 1493
veicolo m (in verniciatura) 1494
velo m 557
velocità di estrusione 1196
velocità f 931
velocità f 1495
velocità f di cristallizzazione 921
velocità f di cristallizzazione 924
velocità f di propagazione delle screpolature 923
velocità f di taglio 1139
velocità f di vulcanizzazione 922
velocità f di vulcanizzazione 1519
velocità f eguale (dei cilindri) 1195
ventosa f 1293
ventosa f 1295
viscosimetro m a sfera 1501
viscosità f 1502
viscosità f alla stato fuso 1504
viscosità f Brookfield 1503
viscosità f Mooney 1505
viscosità f relativa 1506
vite 1107
vite f senza fine 1558
VITON fluoroelastomero 1508
volatile (agg.) 1509
volatilità f 1510
volume m 1511
vulcanizzabile 376
vulcanizzante 377
vulcanizzato 1516
vulcanizzazione a perossidi 817
vulcanizzazione a temperatura ambiente 1051
vulcanizzazione a zolfo 1298
vulcanizzazione f 379
vulcanizzazione f 381
vulcanizzazione f 1517
vulcanizzazione f a piombo 654
vulcanizzazione f con fasci di elettroni 448
vulcanizzazione f in aria 79
vulcanizzazione f in aria calda 582
vulcanizzazione f in stufa ad aria calda 86

Z ona di alimentazione 1117
zona di misurazione 1118

Part 4

SI units and conversion factors

$$\frac{5000 A}{sp\,in}$$

$1\,sp\,in = 6.45\,cm^2$

...place
...ple multi-
p... ...tion in this Section
do... be complete but it does
sur... most of the presentation and
conversion data required by the rubber industry. Complete details of the SI system can be obtained by reference to such publications as ISO 31, ISO 1000, BS 5555, BS 5775, DIN 1301 and similar international and national standards.

plementary SI units

based on seven base units:

	Unit	Symbol
	mole	mol
	ampere	A
	metre	m
...ous intensity	candela	cd
Mass	kilogram	kg
Thermodynamic temperature	kelvin	K
Time	second	s

and two supplementary units:

Plane angle	radian	rad
Solid angle	steradian	sr

The two supplementary units may be considered as either base or derived units and ISO 31 treats them as dimensional derived units with the value of unity.

Derived SI units

All other units are derived from the base and supplementary units and are expressed by means of the mathematical symbols of multiplication and division. For example, velocity is expressed as metre per second (m/s), acceleration as metre per second squared (m/s²) and mass density as kilogram per cubic metre (kg/m³).

Certain of the more important of the derived units have been named after scientists who were prominent in the particular field. For example, the unit for pressure, the newton per square metre, is known as the pascal (Pa).

Some of the derived units, special names and symbols are listed below:

Quantity	Special name of derived SI unit	Symbol	Expressed in terms of base or supplementary SI units or in terms of other derived SI units
Celsius temperature	degree Celsius	°C	°C = 1 K
electric capacitance	farad	F	$F = C/V = A^2.s^4/kg.m^2$
electric charge, flux, quantity of electricity	coulomb	C	$C = A.s$
electric conductance	siemens	S	$S = \Omega^{-1} = A/V = A^2.s^3/kg.m^2$
electric field strength	volt per metre	–	$V/m = kg.m/A.s^3$
electric polarization, displacement, surface density of charge	ampere second per metre squared	–	$A.s/m^2$
electric potential, potential difference, tension, electromotive force	volt	V	$V = J/C = A.\Omega = kg.m^2/A.s^3$
electric resistance	ohm	Ω	$\Omega = V/A = kg.m^2/A^2.s^3$
energy, work, quantity of heat, electric energy	joule	J	$J = N.m = V.A.s = kg.m^2/s^2$
force	newton	N	$N = kg.m/s^2$
frequency	hertz	Hz	$Hz = s^{-1}$
magnetic field strength	ampere per metre	–	A/m
magnetic flux	weber	Wb	$Wb = V.s = kg.m^2/A.s^2$
magnetic flux density, magnetic induction, polarization	tesla	T	$T = Wb/m^2 = kg/A.s^2$
magnetic inductance	henry	H	$H = Wb/A = kg.m^2/A^2.s^2$
power	watt	W	$W = J/S = V.A = kg.m^2/s^3$
pressure, stress	pascal	Pa	$Pa = N/m^2$

Multiples of SI units

The multiples and sub-multiples are formed by orders of magnitude (e.g. 10^2, 10^3, etc.) which are the same irrespective of the units to which they are applied. Details are given below:

Prefix name	Prefix symbol	Factor by which the unit is multiplied	
exa	E	10^{18}	1 000 000 000 000 000 000
peta	P	10^{15}	1 000 000 000 000 000
tera	T	10^{12}	1 000 000 000 000
giga	G	10^{9}	1 000 000 000
mega	M	10^{6}	1 000 000
kilo	k	10^{3}	1 000
hecto	h	10^{2}	100
deca	da	10^{1}	10
deci	d	10^{-1}	0.1
centi	c	10^{-2}	0.01
milli	m	10^{-3}	0.001
micro	μ	10^{-6}	0.000 001
nano	n	10^{-9}	0.000 000 001
pico	p	10^{-12}	0.000 000 000 001
femto	f	10^{-15}	0.000 000 000 000 001
atto	a	10^{-18}	0.000 000 000 000 000 001

Examples

gigahertz (GHz), megavolt (MV), kilowatt (kW), milliampere (mA), microgram (μg), nanosecond (ns), picofarad (pF).

Notes

1) Because the name of the base unit for mass, 'kilogram', contains the name of the SI prefix, 'kilo', the names of the decimal multiples and sub-multiples of the unit of mass are formed by adding the prefixes to the word 'gram', e.g. milligram (mg), not microkilogram (μkg).

2) Compound prefixes shall not be used e.g. nanometre is nm and not mμm. Similarly, only one prefix should be used when forming a multiple of a compound SI unit.

3) The prefixes hecto (10^2) and deca (10^1), although frequently used, are not recommended.

4) A space rather than a comma should be used to divide large numbers into groups of three.

Rules for writing unit symbols

There are a number of rules for printing and using SI unit symbols, including:

1) Roman (upright) type, usually lower case is used for the unit symbols. If, however, the name of the unit is derived from a proper name, the first letter of the symbol is in upper case (capitals). For example, the unit of energy is written as joule but the symbol is J and the unit for frequency is hertz but the symbol is Hz. An exception to this rule is the case of Celsius, (C), the favoured replacement for Centigrade.

2) All symbols are used in the singular, e.g. 7.0 J.

3) Unit symbols are not followed by a full stop, e.g. 5.7 N (not 5.7 N.) except when they fall at the end of a sentence.

4) The product of two or more units may be indicated in any of the following ways,

$$N \cdot m, \quad N.m \quad \text{or} \quad N\,m$$

5) A solidus (oblique stroke, /), a horizontal line or negative exponents may be used to express a derived unit formed from two others by division, e.g.

$$m/s, \quad \frac{m}{s} \quad \text{or} \quad m.s^{-1}$$

6) The solidus must not be repeated on the same line unless ambiguity is avoided by parentheses (round brackets). In complicated cases negative exponents or parentheses should be used, e.g.

$$m/s^2 \quad \text{or} \quad m.s^{-2} \quad \text{but not} \quad m/s/s$$

$$m.kg/(s^3.A) \quad \text{or} \quad m.kg.s^{-1}.A^{-1}$$

$$\text{but not} \quad M.kg/s^3/A$$

7) For temperature measurement, kelvin is the unit of measurement not the method of calibration. Thus it is correct to write 273.15 K but incorrect to write 273.15°K. Degree Celsius (°C) may be used for practical representation of temperature values.

Non-SI units

There are certain units outside the SI system which are recognized as having to be retained because of their practical importance or for use in specialized fields. In a limited number of cases, compound units are formed with non-SI units and their multiples e.g. kg/h, km/h. However, it is intended that such units, e.g. the kilowatt hour (kWh), will eventually be discontinued.

Quantity	Name	Symbol	Definition
Area	are	a	1 å = 10^2 m^2
Length	ångström	Å	1 Å = 10^{-10} m
Mass	tonne	t	1 t = 10^3 kg
Plane angle	degree	°	1° = (n/180) rad
	minute	'	1' = (1/60)°
	second	"	1" = (1/60)'
Pressure	bar[1]	bar	1 bar = 10^5 Pa
	standard atmosphere	atm	1 atm = 101 325 Pa
Time[2]	minute	min	1 min = 60 s
	hour	h	1 h = 3 600 s
	day	d	1 d = 24 h
Velocity	knot[3]	kt	1 kt = 1 international nautical mile per hour = 0.514 444 m/s
Volume	litre[4]	l or L	1 l = 1 dm^3

Notes
1) The use of the bar should be restricted to the field of fluid pressure.
2) Units such as week, month and year are also in common use.
3) The UK knot is 0.514 773 m/s.
4) The litre is a special name for the cubic decimetre (dm^3) and should not be used for high precision measurements.

Some units particularly relevant to the rubber industry

Quantity	SI unit	Non SI unit
Adhesion (peel)	kN/m	lb/in^2
Adhesion (tensile)	MPa	lb/in^2
Flexural modulus	MPa	lb/in^2
Force	N	lbf
Impact	J/cm	ft.lb/in
Length	mm	mil (0.001 in)
	cm	in
	m	ft or yard
	km	mile
Mass	g	oz
	kg	lb
Pressure	kPa	lb/in^2
Tear strength	kN/m	lb/in
Temperature	°C	°F
Tensile modulus	MPa	lb/in^2
Tensile strength	MPa	lb/in^2
Velocity	km/h	mile/h

Conversion factors

The conversion factors are tabulated in alphabetical order of the quantity name, with adjectives being ignored. For convenience certain factors have been rounded off.

SI quantity	symbol	× factor	=	Non SI quantity	symbol	× factor	=	SI symbol
Acceleration								
metre per second^2	m/s^2	× 3.281	=	foot/sec^2	ft/s^2	× 0.3048	=	m/s^2
Acceleration (angular)								
radian per second^2	rad/s^2	× 0.159	=	rev per second^2	rev/s^2	× 6.289	=	rad/s^2
Adhesion (peel)								
kilonewton per metre	kN/m	× 5.714	=	pound force per inch	lbf/in	× 0.175	=	kN/m
Adhesion (tensile)								
megapascal	MPa	× 1.45 × 10^2	=	pound per square inch	lb/in^2	× 0.6899 × 10^{-3}	=	MPa
Amount of substance								
mole	Mol	× (see (a))	=	gram per litre	g/l	× (see (a))	=	mol

(a) The value of the mole is dependent upon the atomic or molecular weight of a substance. The elemental entities must be specified when using this unit.

SI quantity	symbol	× factor	=	Non SI quantity	symbol	× factor	=	SI symbol
Angle (plane)								
Radian	rad	× 63.6537	=	grade or gon	g	× 0.01571	=	rad
	rad	× 57.295	=	degree	°	× 0.017	=	rad
	rad	× 3437.7	=	minute	'	× 2.9 × 10^{-4}	=	rad
	rad	× 206 262	=	second	"	× 4.8 × 10^{-6}	=	rad
Angle (solid)								
Steradian	sr			no direct equivalent				
Area								
square metre	m^2	× 1 550	=	square inch	in^2	× 6.452 × 10^{-4}	=	m^2
	m^2	× 10.764	=	square foot	ft^2	× 0.0929	=	m^2
	m^2	× 1.196	=	square yard	yd^2	× 0.8361	=	m^2
	m^2	× 2.47 × 10^{-4}	=	acre	acre	× 4 047	=	m^2
hectare	ha	× 2.471	=	acre	acre	× 0.4047	=	ha
(10,000 m^2)	ha	× 3.86 × 10^{-3}	=	square mile	mile^2	× 259	=	ha
Density								
kilogram per cubic metre	kg/m^3	× 0.062	=	pound per cubic foot	lb/ft^3	× 16.02	=	kg/m^3
	kg/m^3	× 0.010	=	pound per Imperial gallon	lb/gal (imp)	× 99.776	=	kg/m^3
	kg/m^3	× 1 × 10^{-3}	=	gram per cubic centimetre	g/cm^3	× 1 × 10^3	=	kg/m^3
	kg/m^3	× 8.345 × 10^{-3}	=	pound per US gallon	lb/gal (US)	× 119.8264	=	kg/m^3

SI quantity	symbol	× factor	=	Non SI quantity	symbol	× factor	=	SI symbol
Density (linear)								
decitex	dtex	× 0.9	=	denier	den	× 1.1111	=	dtex
(dg/km)	dtex	× 10^4	=	newton metre	NM	× 10^{-4}	=	dtex
Energy, Work								
joule	J	× 10^7	=	erg	erg	× 10^{-7}	=	J
	J	× 0.238 × 10^{-3}	=	kilocalorie	kcal	× 4.19 × 10^3	=	J
	J	× 0.278 × 10^{-6}	=	kilowatt hour	kWh	× 3.6 × 10^6	=	J
	J	× 0.947 × 10^{-3}	=	British thermal unit	Btu	× 1.055 × 10^3	=	J
	J	× 0.102	=	kilogram force metre	kgf.m	× 9.810	=	J
Energy (specific)								
joule per kilogram	J/kg	× 0.238 × 10^{-3}	=	kilocalorie per kilogram	kcal/kg	× 4.19 × 10^3	=	J/kg
	J/kg	× 0.429 × 10^{-3}	=	British thermal unit per pound	Btu/lb	× 2.236 × 10^3	=	J/kg
	J/kg	× 0.334	=	foot pound force per pound	ft.lbf/lb	× 2.989	=	J/kg
Force								
newton	N	× 0.102	=	kilogram force	kgf	× 9.81	=	N
	N	× 0.225	=	pound force	lbf	× 4.45	=	N
Frequency								
hertz	Hz	× 1.0	=	cycle per second	c/s	× 1.00	=	H
Frequency (rotational)								
reciprocal second	s^{-1}	× 0.0167	=	revolution per minute	rev/min (rpm)	× 60	=	s^{-1}
Heat capacity (entropy)								
joule per kelvin	J/K	× 2.388 × 10^{-4}	=	kilocalorie per kelvin	kcal/K	× 4.19 × 10^3	=	J/K
Heat capacity (specific)								
joule per kilogram kelvin	J/kg.K	× 2.388 × 10^{-4}	=	kilocalorie per kilogram kelvin	kcal/kg.K	× 4.19 × 10^3	=	J/K
Heat flow rate								
watt	W	× 0.8598	=	kilocalorie per hour	kcal/h	× 1.163	=	W
Heat flow rate (density of)								
watt per metre	W/m	× 0.8598	=	kilocalorie per hour metre	kcal/h.m	1.163	=	W/m
Heat quantity (enthalpy)								
joule	J	× 2.388 × 10^{-4}	=	kilocalorie	kcal	× 4.19 × 10^3	=	J
Heat radiation								
watt per square metre kelvin^4	W/m^2.K^4	× 0.0598	=	kilocalorie per hour square metre kelvin^4	kcal/h.m.^{2}.K^4	× 1.163	=	w/m^2.K^4

Conversion factors

SI quantity	symbol	× factor	=	Non SI quantity	symbol	× factor	=	SI symbol
Heat transfer (coefficient of)								
watt per square metre kelvin	W/m².K	× 0.8598	=	kilocalorie per hour square metre kelvin	kcal/h.m².K	× 1.163	=	W/m².K
Impact strength								
joule per centimetre	J/cm	× 1.886	=	foot pound per inch	ft.lb/in	× 0.53	=	J/cm
Items per length unit								
items per metre	items/m	× 0.9144	=	items per yard	items/yd	× 1.094	=	items/m
	items/m	× 0.3048	=	items per foot	items/ft	× 3.281	=	items/m
	items/m	× 254.1	=	items per inch	items/in	× 39.37	=	items/m
Length								
metre	m	× 39.37	=	inch	in	× 0.0254	=	m
	m	× 3.28	=	foot	ft	× 0.3048	=	m
	m	× 1.0936	=	yard	yd	× 0.9164	=	m
	m	× 10^{10}	=	ångström	Å	× 10^{-10}	=	m
kilometre	km	× 0.6214	=	mile	mile	× 1.6093	=	km
	km	× 0.54	=	nautical mile	n.mile	× 1.852	=	km
Mass								
gram	g	× 0.0353	=	ounce (avoir)	oz	× 28.3495	=	g
	g	× 0.0322	=	ounce (troy)	oz tr	× 31.1035	=	g
kilogram	kg	× 2.204	=	pound	lb	× 0.45359	=	kg
tonne (1 000 kg)	t or Mg	× 2 204.62	=	pound	lb	× 4.536 × 10^{-4}	=	t or Mg
	t or Mg	× 0.9842	=	UK ton (long)	ton	× 1.0161	=	t or Mg
	t or Mg	× 1.1023	=	US ton (short)	sh ton	× 0.9072	=	t or Mg
Mass (volumetric)								
cubic metre per kilogram	m³/kg	× 99.7763	=	Imperial gallon per pound	gal/lb (imp)	× 1.002 × 10^{-2}	=	m³/kg
	m³/kg	× 119.8264	=	US gallon per pound	gal/lb (US)	× 8.345 × 10^{-3}	=	m³/kg
	m³/kg	× 16.02	=	cubic foot per pound	ft³/lb	× 0.062	=	m³/kg
	m³/kg	× 99.76	=	gallon per pound	gal/lb	× 0.01	=	m³/kg
	m³/kg	× 1 × 10^3	=	cubic centimtre per gram	cm³/g	× 1 × 10^{-3}	=	m³/kg
Momentum								
kilogram metre per second	kg.m/s	× 7.246	=	pound foot per second	lb.ft/s	× 0.138	=	kg.m/s
Momentum (angular)								
kilogram square metre per second	kg.m²/s	× 23.73	=	pound square foot per second	lb.ft²/s	× 4.214 × 10^{-2}	=	kg.m²/s
Moment of force								
newton metre	N.m	× 3.292 × 10^{-4}	=	ton force foot	tonf.ft	× 3.037 × 10^3	=	N.m
	N.m	× 0.737	=	pound force foot	lbf.ft	× 1.356	=	N.m
	N.m	× 8.85	=	pound force inch	lbf.in	× 0.113	=	N.m
	N.m	× 1.416 × 10^2	=	ounce force inch	ozf.in	× 7.061 × 10^{-3}	=	N.m

SI quantity	SI symbol	× factor	= Non SI quantity	Non SI symbol	× factor	= SI symbol
Moment of inertia						
Kilogram square metre	kg.m²	× 23.730	= pound square foot	lb.ft²	× 0.042	= kg.m²
	kg.m²	× 3.417 × 10³	= pound square inch	lb.in²	× 2.926 × 10⁻⁴	= kg.m²
	kg.m²	× 5.464 × 10³	= ounce square inch	oz.in²	× 0.183 × 10⁻⁴	= kg.m²
	kg.m²	× 0.737	= slug square foot	slug.ft²	× 1.356	= kg.m²
Power						
watt	W	× 0.102	= kilogram force metre per second	kgf.m/s	× 9.81	= W
	W	× 1.36 × 10⁻³	= metric horsepower	CV or PS	× 7.35 × 10²	= W
	W	× 1.34 × 10⁻³	= horsepower	hp	× 7.46 × 10²	= W
Pressure						
pascal	Pa	× 10⁻⁵	= bar	bar	× 10⁵	= Pa
	Pa	× 0.102	= kilogram force per square metre	kgf/m²	× 9.81	= Pa
	Pa	× 0.102	= millimetre of water	mm H₂O	× 9.81	= Pa
	Pa	× 0.75 × 10⁻⁴	= torr	Torr	× 1.333 × 10²	= Pa
	Pa	× 0.145 × 10⁻³	= pound force per square inch	lbf/in²	× 6.895 × 10³	= Pa
	Pa	× 0.987 × 10⁻⁵	= standard atmosphere	atm	× 1.013 × 10⁵	= Pa
	Pa	× 0.102 × 10⁻⁴	= technical atmosphere	at	× 9.807 × 10⁴	= Pa
Surface tension						
newton per metre	N/m	× 10³	= dyne per centimetre	dyne/cm	× 10⁻³	= N/m
Tear strength						
kilonewton per metre	kN/m	× 5.714	= pounds per inch	lb/in	× 0.175	= kN/m
Temperature						
Celsius	C	(°C + 273.15) × 9/5	= degree rankine	°R	(°R × 5/9) − 273.15	= C
	C	°C + 273.15	= kelvin	K	K − 273.15	= C
	C	(°C × 9/5) + 32	= degree fahrenheit	°F	(°F − 32) × 5/9	= C
Temperature (thermodynamic)						
kelvin	K	× 9/5	= rankine	R	× 5/9	= K
	K	K − 273.15	= degree Celsius	°C	°C + 273.15	= K
	K	1.8K − 459.67	= degree fahrenheit	°F	5/9 (°F − 32) + 273.15	= K
Thermal conductivity						
watt per metre kelvin	W/m.K	× 0.8598	= kilocalorie per hour metre kelvin	kcal/h.m.k	× 1.163	= W/m.K
Velocity						
metre per second	m/s	× 1.942	= knot	knot	× 0.515	= m/s
	m/s	× 2.237	= mile per hour	mile/h	× 0.447	= m/s
	m/s	× 3.289	= foot per second	ft/s	× 0.3048	= m/s
	m/s	× 2 × 10²	= foot per minute	ft/min	× 5.08 × 10⁻³	= m/s
	m/s	× 3.6	= kilometre per hour	km/h	× 0.278	= m/s
Velocity (angular)						
radian per second	rad/s	× 57.295	= degree per second	°/s	× 0.017	= rad/s

SI		× factor	=	Non SI		× factor	=	SI
quantity	symbol			quantity	symbol			symbol

Viscosity (dynamic)

pascal second	Pa.s	× 10	=	poise	p	× 0.1	=	Pa.s

Viscosity (kinematic)

square metre per second	m²/s	× 10⁴	=	stokes	S	× 10⁻⁴	=	m²/s

Volume and capacity

millilitre	ml	× 0.0352	=	UK fluid ounce	UK fl oz	× 28.412	=	ml
	ml	× 0.0338	=	US fluid ounce	US fl oz	× 29.573	=	ml
litre (10⁻³m³)	l	× 1.7598	=	UK pint	UK pt	× 0.5682	=	l
	l	× 2.1134	=	US liquid pint	US liq pt	× 0.4732	=	l
	l	× 0.22	=	UK gallon	UK gal	× 4.5461	=	l
	l	× 0.2642	=	US gallon	US gal	× 3.7854	=	l
cubic metre	m³	× 61 024	=	cubic inch	in³	× 1.639 × 10⁻⁵	=	m³
	m³	× 35.3147	=	cubic feet	ft³	× 0.0283	=	m³
	m³	× 1.308	=	cubic yard	yd³	× 0.7646	=	m³

Volume rate of flow

cubic metre per second	m³/s	× 35.314	=	cubic foot per second	ft³/s	× 0.028	=	m³/s
	m³/s	× 7.919 × 10⁵	=	Imperial gallon per hour	gal(imp)/h	× 1.263 × 10⁻⁶	=	m³/s
	m³/s	× 9.514 × 10⁵	=	U.S. gallon per hour	gal(US)/h	× 1.051 × 10⁻⁶	=	m³/s
	m³/s	× 1.319 × 10⁴	=	Imperial gallon per minute	gal(imp)/min	× 7.576 × 10⁻⁵	=	m³/s
	m³/s	× 1.585 × 10⁴	=	U.S. gallon per minute	gal(US)/min	× 6.309 × 10⁻⁵	=	m³/s
	m³/s	× 0.366	=	cubic inch per minute	in³/min	× 2.731 × 10⁻⁷	=	m³/s

Weight (area related)

gram per square metre	g/m²	× 2.95 × 10⁻²	=	ounce per yard	oz/yd²	× 33.91	=	g/m²

Index

Absorption, 104, 118
Abradants, 64, 66
Abrasion, 63, 64–67
  definition, 64
  resistance, 9, 17
  resistance index, 65, 66
  tests, 5, 64–67
Accelerated tests (*see also* weather resistance), 23, 93,
  120, 121, 127
Acrylate elastomer (*see also* rubber), 133
Adhesion (*see also* bonding), 97–103
  to cord, 98, 102
  to fabrics, 9, 97, 98, 100–102, 130
  to metals, 9, 97–100, 102
  strength, 101
Adhesives, 97
Ageing, 27, 29, 130
  accelerated tests, 93
  heat, 77, 93–96
ALCRYN, 4
  classification, 124
  compression set, 20
  effect of weathering, 111, 112
  electrical properties, 75
  hardness, 12
  general properties, 8, 9, 10
  storage, 121
  tensile strength, 16
Alternating current (a.c.), 69, 74
Aluminium oxide, 64, 65,
Akron abrasion tester, 64, 66
American National Standards Institute (ANSI), 6
American Society for Testing and Materials (ASTM), 3,
  6, 122
  ASTM C177, 78
  ASTM C509, 86
  ASTM C518, 78
  ASTM D149, 69, 75
  ASTM D150, 69, 75
  ASTM D257, 69, 75
  ASTM D394, 64, 66
  ASTM D395, 19
  ASTM D412, 14, 19, 24
  ASTM D413, 98
  ASTM D429, 98
  ASTM D430, 58, 102
  ASTM D454, 95
  ASTM D470, 113
  ASTM D471, 118

American Society for Testing and Materials – *continued*
  ASTM D518, 110
  ASTM D572, 95
  ASTM D573, 95
  ASTM D575, 35
  ASTM D623, 58
  ASTM D624, 35
  ASTM D746, 86
  ASTM D750, 110
  ASTM D797, 86, 88
  ASTM D813, 58
  ASTM D864, 78
  ASTM D865, 95
  ASTM D945, 40, 46
  ASTM D991, 69
  ASTM D1043, 86
  ASTM D1052, 58
  ASTM D1053, 86
  ASTM D1054, 40
  ASTM D1149, 113
  ASTM D1171, 110, 113
  ASTM D1229, 19, 86
  ASTM D1329, 86
  ASTM D1390, 27
  ASTM D1456, 34, 35
  ASTM D1460, 118
  ASTM D1630, 64, 67
  ASTM D1415, 12,
  ASTM D1871, 98
  ASTM D2000, 8, 122–124, 128
  ASTM D2137, 86
  ASTM D2138, 98
  ASTM D2228, 64, 67
  ASTM D2229, 98
  ASTM D2231, 40
  ASTM D2240, 12, 86
  ASTM D2630, 98, 102
  ASTM D2632, 40, 45
  ASTM D3137, 118
  ASTM D3389, 64, 67
  ASTM D3395, 113
  ASTM E96, 105
  ASTM No.3 oil, 8, 123, 124
Amplitude
  dynamic testing, 41–43, 53–55
  fatigue, 59, 61, 129
  forced vibration, 47–51
Antioxidants, 26, 56
Antiozonants, 63, 113, 115

Applications
  airships, 104
  balloons, 104, 105, 108
  belts, 17, 57, 58, 78
  bearings, 63
  boats, 97, 104
  bridge bearings, 26, 36, 95, 112
  bushings, 26
  conveyor belting, 64, 97, 102, 121
  damping, 35, 39
  diaphragms, 104
  flooring, 4, 72,
  footwear, 5, 57, 58, 64, 66, 72
  fuel tanks, 104
  gaskets, 4, 131
    permanent set, 22, 23
    permeability, 104
    stress relaxation, 27,
    window, 4, 111
  hose, 4, 5
    abrasion, 64
    adhesion, 97
    coolant, 129, 130
    electrical, 72
    permeability, 104, 105
    storage, 121
    tension, 17
  linings, 104
  military, 97, 121
  mouldings, 72
  mountings, 17, 23, 39, 52, 54, 55, 78
  O-rings, 17, 27, 32, 81, 82, 131
  packings, 17
  seals (*see also* elastomeric seals), 27, 32, 33, 78, 104, 105
  sheeting, 4, 72
  springs, 35, 81–83
  tubing, 4
  tyres, *see* tyres
  V-belts (*see also* V-belts), 52, 140–144
  wire and cables, 75, 76, 110
Arbor press, 27
Arizona dust, 132
Arrhenius plot, 93
Association Française de Normalisation (NF), 6
  NF C26–210, 69
  NF C26–215, 69
  NF C26–218, 69
  NF C26–220, 69
  NF C26–221, 69
  NF C26–225, 69
  NF C26–230, 69
  NF C41–101, 69
  NF ISO 813, 98
  NF ISO 814, 98
  NF ISO 2528, 105
  NF ISO 4647, 98
  NF T46–002, 14
  NF T46–003, 12
  NF T46–004, 95

Association Française de Normalisation – *continued*
  NF T46–005, 95
  NF T46–006, 95
  NF T46–007, 35
  NF T46–008, 98
  NF T46–009, 19
  NF T46–011, 19
  NF T46–013, 118
  NF T46–015, 58
  NF T46–016, 58
  NF T46–018, 86
  NF T46–019, 113
  NF T46–020, 98
  NF T46–021, 58
  NF T46–023, 35
  NF T46–025, 86
  NF T46–026, 40
  NF T46–027, 86
  NF T46–032, 86
  NF T46–033, 35
  NF T46–034, 105
  NF T46–036, 40
  NF T46–037, 105
  NF T46–038, 113
  NF T46–039, 113
  NF T46–040, 110
  NF T46–041, 110
  NF T46–042, 110
  NF T46–044, 27
  NF T46–052, 12
Autoclaves, 4,
Automotive
  bumpers, 100
  mountings, 12, 23, 26, 39, 54
  constant velocity bellows, 34, 134
  seals, 131–135
  suspension, 82, 83
  V-belts, 142–144

Barco Industries (*see also* Dynaliser™), 50
Blue dyed wool standards, *see* weather resistance
Bonding rubber (*see also* adhesion)
  to fabric, 58, 67, 97
  to metal, 26, 47, 97, 100
Bridge bearings, *see* applications
British Standards Institution (BSI), 6
  BS 148, 73
  BS 874, 78
  BS 903:Part A2, 14
  BS 903:Part A3, 35
  BS 903:Part A4, 35
  BS 903:Part A5, 19
  BS 903:Part A6, 19
  BS 903:Part A8, 40
  BS 903:Part A9, 64, 66, 67
  BS 903:Part A10, 58
  BS 903:Part A11, 58
  BS 903:Part A12, 98
  BS 903:Part A13, 86

British Standards Institution – *continued*
  BS 903:Part A14, 35
  BS 903:Part A15, 24
  BS 903:Part A16, 118
  BS 903:Part A17, 105
  BS 903:Part A18, 118
  BS 903:Part A19, 95
  BS 903:Part A21, 98
  BS 903:Part A24, 40
  BS 903:Part A25, 86
  BS 903:Part A26, 12
  BS 903:Part A27, 98
  BS 903:Part A29, 86
  BS 903:Part A30, 105
  BS 903:Part A31, 40
  BS 903:Part A39, 19, 86
  BS 903:Part A40, 98
  BS 903:Part A42, 27
  BS 903:Part A43, 113
  BS 903:Part A44, 113
  BS 903:Part A46, 105
  BS 903:Part A48, 98
  BS 903:Part A49, 58
  BS 903:Part A50, 58
  BS 903:Part C1 to C5/2782 Part 2, 68, 69, 73, 74
  BS 923, 69
  BS 2044, 69
  BS 2050, 69
  BS 3734, 78
  BS 4370, 78
  BS 5176, 122, 123
  BS 5294, 86
  BS 5555, 239
  BS 5775, 239
  BS 5730, 73
  BS 5901, 69
  BS 6233, 69
Brittle point, 84, 90
Brittleness temperature, 90, 91, 123
Bursting pressure, 130
Butyl rubber, *see* rubber

Cables, *see* applications
Carbon black, *see* fillers
cis-polyisoprene (*see also* rubber), 3
Chalking (*see also* weather resistance), 109, 112
Chemical resistance, 120, 130
  air, 113
  acids, 7, 9
  alkalis, 7
  fats, 7
  fuels, 33, 120
  gases (*see also* gases), 113–117
  greases, 7, 134
  inorganic salts, 7
  oils, 5, 7, 8, 9, 120, 127
  oxygen (*see also* oxygen), 26, 56, 63, 109, 113, 120
  ozone (*see also* ozone), 7, 9, 56, 63, 109, 113
  petroleum products, 7
  salt, 109

Chemical resistance – *continued*
  solvents, 9
  water (*see also* weather resistance), 9, 26, 56, 68, 109, 120
  waxes, 7
  weather, (*see also* weather resistance), 7, 109–112
Chromatography, gas, 106
Clash and Berg apparatus, 86, 88, 90
Classification system for rubbers, 122–124
  Class, 122–124
  Group, 122, 123
  Line call-out, 123, 124
  Type, 122–124
Coated fabrics, *see* fabrics, coated
Colour changes, 109, 111
Comparative tracking index (CTI), 73
Compounding (*see also* fillers and rubbers), 45, 109, 120
Compression set, *see* permanent set
Contact resistance, 69
Conversion factors, 243–247
Conveyor belting, *see* applications
Corrosion, 97
Corundum, *see* aluminium oxide
Crack growth, 56, 58, 60, 63
  length, 62
Cracking, surface, 56, 58, 90, 102, 109, 134
  ozone testing, 113, 115–117
Creep (*see also* strain relaxation), 24–26, 77, 78
  in compression, 24, 58
  rate, 24, 25
  in shear, 25
Crosslinking (*see also* vulcanisation), 95, 96, 122
Crystallization (*see also* low temperature properties), 53, 84–86, 88
  test, 91, 92
Cut growth, 57, 58, 60
Cutting resistance, 38, 64
Cyanoacrylate adhesive, 47, 101

Damping, 48, 52, 54, 55
Dead-weight tester, 142
Decibel, 54
Dehdrating agents (dessicants)
  anhydrous calcium chloride, 106
Delft test piece (*see also* tearing energy), 36, 37
De Mattia machine, 56–58, 63
Deutsches Institut für Normung (DIN), 6
  DIN 1301, 239
  DIN 3761, 131, 132
  DIN 52612, 78
  DIN 52616, 78
  DIN 53122, 105
  DIN 53386, 110
  DIN 53387, 110
  DIN 53388, 110
  DIN 53504, 14
  DIN 53505, 12
  DIN 53506, 35
  DIN 53507, 35
  DIN 53508, 95

Deutsches Institut für Normung – *continued*
  DIN 53509, 113
  DIN 53512, 40
  DIN 53513, 40
  DIN 53515, 35
  DIN 53516, 64
  DIN 53517, 19, 86
  DIN 53518, 19
  DIN 53519, 12
  DIN 53521, 118
  DIN 53530, 98
  DIN 53531, 98
  DIN 53532, 105
  DIN 53536, 105
  DIN 53537, 27
  DIN 53538, 118
  DIN 53541, 86
  DIN 53546, 86
  DIN 53548, 86
  DIN 7753, 143
  DIN ISO 1827, 98
  DIN ISO 8013, 24
  DIN VDE 0432:Parts 1 to 4, 69
  DIN VDE 0303:Parts 1 to 3, 69
  EDIN 53509, 113
Dielectric, 74
Dielectric constant, *see* electrical properties
Differential scanning calorimetry (DSC), 78
Diffusion, 104, 108
Dimension tolerances, 78
DIN abrader, 64–66
Dunlop
  abrasion tester, 64, 66
  pendulum, 44
  tripsometer, 44
Du Pont, 76, 93, 112, 120
  abrasion tester, 64, 66
  expansion joint test rig, 137–139
  flexer, 58, 102
Dynaliser™, 50, 51
Dynamic fatigue, *see* fatigue
Dynamic mechanical analyser (DMA), 48–51, 77, 78
Dynamic pressure pulsation test, 129
Dynamic properties
  strain, 117
  stress, 36
  stress-strain, 39–53,
Dynamic testing, 34, 39–53, 54, 55, 100–103, 134
  electromagnetic, 48
  forced oscillation, 41
  hydraulic, 48, 61
  mechanical, 48
  ozone, 113, 116
  vibration, 43
  Wallace-RAPRA, 48
Dynamometer, 14, 100, 141

Elastomer (*see also* rubber)
  conductive, 69
  definition, 3

Elastomeric seals, 131–135
  bellows (boots), 131, 134
  dynamic, 131
  gaiters, 131
  gaskets, 131
  exclusion, 131, 134
  shaft, 132
  static, 23, 131
  valve stem seals, 133
Electrical properties, 68–76
  antistatic, 68–72
  capacitance, 74
  conductance, 68
  conductivity, 68
  electric strength (dielectric strength), 9, 68, 69, 72, 73, 76
  electrodes, 68–74
  high voltage, 69
  insulation, 9
  insulation resistance, 68–72, 76
  permittivity (dielectric constant), 68, 69, 74–76
  power factor, 68, 69, 74–76
  resistance, 86
  resistivity, 68, 72, 75
    surface, 68–72
    volume (specific resistance), 68, 69, 70, 75, 76
  surface resistance, 70
  tracking resistance, 68, 69, 73, 74
Electrolyte, 73, 74
Electrometer, 72
Elongation, 17, 34, 68, 80, 134
  classification, 123
  definition, 14
  effect of weathering on, 111
  effect of liquids on, 119
  low temperature, 86
ELVAX (EVA), 7–9, 12, 16
Epichlorohydrin, 8
Ethylene-propylene, 8
Ethylene-propylene terpolymer (EPDM), (*see also* NORDEL), 8, 112
Extraction, 118, 119
Extrusions, 110, 121

Fabrics, coated, 106
Fatigue, 56–63, 129
  belt, 141
  cracking, 56, 117
  dynamic, 58, 102
  life, 56, 59, 63
  resistance, 61
  tension, 56, 61–63
Fillers, 25, 31, 34, 42, 56, 79, 122
  Carbon black, 3, 16, 40, 42, 55, 68,
    effect on weather resistance, 109
  fatty acids, 3
  oils, 3, 13, 64
  resins, 13
  silica, 3
  waxes, 3, 13, 64

Films, 108
Fire resistance, 76
Firestone flexometer, 58
Flex cracking, 57, 58
Flexing, 78
Flex life, 17
Flexometers, 58–60, 100
   compression, 58, 59
   rotary, 58
Flexometer testing, 58, 63
Florida, exposure in, *see* weather resistance
Flooring (rubber), *see* applications
Flue duct expansion joints, 136–139
   acid rain, 136
   ash, 136
   corrosion, 136
   reinforcement, 136, 137
   scrubbers, 136
   synthetic flue gas, 138
   test rig, 137–139
Fluoroelastomer, *see* VITON
Fluorosilicone, *see* rubber
Footwear, *see* applications
Fracture energy, 60
Fracture mechanics, 60–63
Frequency, 62, 129
   dynamic testing, 42, 43, 46, 47
   electrical, 74
   electromagnetic vibrators, 48–51
   transmissibility, 54, 55
Freudenberg, Carl, 131–135
Friction, 78, 81

Gaskets, *see* applications
Gases, resistance to, 113–117
Gates Corporation, The, 140
Gates Europe, 140–144
Gehman test, 86, 88–90
General Conference of Weights and Measures, 239
Glass transition temperature, 77, 78
Goodrich flexometer, 58
Goodyear-Healey pendulum, 44
GOST, *see* State Committee for Standards, former Soviet Union
Grey scale, *see* weather resistance
Guayule, 3
Gough-Joule effect, 80–83
Griffiths A A, 60

H-pull (block) test, 102
Hardness, 5, 11–13, 122, 123, 130
   dead-load testers, 11, 88
   Du Pont elastomers, 9, 12, 16, 17
   durometer, 11, 86
   effect of liquids, 118, 119
   International rubber hardness degrees (IRHD), 26, 60
      of Du Pont elastomers, 13
      for abrasion testing, 65

Hardness – *continued*
   International rubber hardness degrees – *continued*
      for low temperature testing, 91
      for pendulum testing, 44, 45
      scale, 11
      for typical products, 12,
   classification, 123
   low temperature, 88
   Monsanto Durolab™, 11
   Shore, 11, 13
   Wallace, 11
Heat ageing (*see also* ageing), 93–96, 122
Heat buildup, 56–59, 62
   in dynamic testing, 39
   low resilience, 52
Heat flow rate (flux), 78
Heat resistance, 8, 68
Hevea brasiliensis, 3
Hooke's Law, 39
Hose, *see* applications
Hydrocarbons, 9, 108
HYPALON (CSM), 7, 51
   compression set, 20
   electrical properties, 75, 76
   hardness, 12
   ozone resistant paint, 115
   general properties, 8, 9
   storage, 121
   tensile strength, 16, 17
Hysteresis, 39, 40, 47, 52, 56
HYTREL (YBPO), 4, 26
   bellows, 135
   compression set, 20
   classification, 124
   electrical properties, 75, 76
   general properties, 8, 9
   hardness, 12, 13
   tensile strength, 16
Humidity, 70

Impact test, 84, 100
Infra-red radiation, 111
International Electrical Commission (IEC), 68
   IEC 60, 69, 73
   IEC 93, 68, 69
   IEC 112, 69, 74
   IEC 167, 69
   IEC 227, 76
   IEC 243, 69, 73
   IEC 245, 76
   IEC 250, 69, 74
   IEC 502, 76
   IEC 885, 76
International Organization for Standardization (ISO), 6, 130
   ISO 31, 239
   ISO 34, 35, 37, 38
   ISO 36, 98, 100
   ISO 37, 14

International Organization for Standardization – *continued*
  ISO 48, 11, 12, 91
  ISO 105, 110
  ISO 132, 57, 58
  ISO 133, 57, 58
  ISO 188, 93–95
  ISO 471, 123
  ISO 813, 97, 98
  ISO 814, 98, 99
  ISO 815, 18, 19
  ISO 816, 35, 37, 38
  ISO 1000, 239
  ISO 1399, 104, 105
  ISO 1400, 91
  ISO 1431, 113, 115, 116
  ISO 1432, 86, 88
  ISO 1653, 19, 20, 86
  ISO 1747, 98, 100
  ISO 1817, 118–120, 123
  ISO 1818, 91
  ISO 1827, 35, 36, 100
  ISO 1853, 69, 71
  ISO 2285, 18, 19
  ISO 2528, 105, 106
  ISO 2781, 65
  ISO 2782, 104, 105
  ISO 2790, 143
  ISO 2856, 40, 43
  ISO 2878, 69, 72
  ISO 2882, 69, 72
  ISO 2883, 69, 72
  ISO 2921, 86
  ISO 2951, 69, 71
  ISO 3302, 78, 79
  ISO 3384, 27, 31
  ISO 3387, 86, 91
  ISO 3601, 131
  ISO 4547, 102
  ISO 4632, 122–124
  ISO 4637, 98, 101
  ISO 4647, 98
  ISO 4649, 64, 66
  ISO 4662, 40, 44, 45
  ISO 4663, 40, 45
  ISO 4664, 40, 47
  ISO 4665, 109–111
  ISO 4666, 58
  ISO 5287, 143, 144
  ISO 5470, 64, 67
  ISO 5600, 98, 99
  ISO 5603, 98, 102
  ISO 6133, 38, 101
  ISO 6179, 105, 106
  ISO 6194, 131
  ISO 6471, 86, 91
  ISO 6914, 95, 96
  ISO 6943, 58, 62
  ISO 7619, 12
  ISO 7743, 35

International Organization for Standardization – *continued*
  ISO 8013, 24
  ISO R812, 86, 90
International System of Units (SI), 239–242
  derived units, 240
  multiples of, 241
  non-SI units, 242
  rules for, 241
ISO abrader, 64–66
Izod impact test, 100

Joule, *see* Gough-Joule effect

KALREZ (FFKM), 8, 9, 12, 16, 20, 124
KEVLAR para aramid fibre, 130
Kinetic energy, 71
Kinetic theory, 95

Linear variable differential transformer, 49
Liquids, resistance to, 118–120
  change in volume, 118
  one surface tests, 119
Loughborough University, 30
Loss angle (δ), (*see also* phase angle) 40, 42, 47, 52, 75
  factor, 41, 75
  tangent of (tan δ), 40–42, 44, 47, 51, 54, 75
Low temperature properties, 77, 84–92, 122
  compression set, 88
  crystallization, 84, 85, 86, 88
  flexibility, 135
  rate of recovery, 86
  retraction, 87
  tests, 86–92
Lucas compression stress relaxometer, 28–30
  test jig, 28, 29
Lüpke pendulum, 44

Malaysian Rubber Producers' Research Association (MRPRA), 34, 36, 52, 53, 60,(0E)62, 63
Mechanical conditioning (*see also* scragging), 81
  for dynamic tests, 41, 43, 45, 47, 51, 53
  for fatigue tests, 59
  for stress relaxation, 31
  for stress-strain, 34–36
Materials Engineering Research Laboratory (MERL), 61
  MERL crack growth fatigue machine, 61
Modulus (*see also* tensile stress), 14, 34, 39, 77, 122
  chord (secant), 42, 53
  complex, 40, 41
  compression, 35, 40, 53
  dynamic, 42, 46, 52, 53, 55
    at low temperatures, 86
  effect of frequency, 42
  effect of temperature, 41
  effect of weathering, 111

Modulus – *continued*
  elastic, 40, 51, 80
  in-phase, 41
  out-of-phase (loss), 41
  shear, 35, 36, 39, 40, 53, 88, 90, 100
  static (tangent), 42, 43, 53
  relaxed, 34, 35
  tension, 40
  torsion, 88, 89
  viscous, 40, 51
Monsanto
  Durolab™, 11
  fatigue tester, 62
Mould shrinkage, 79

National Bureau of Standards (NBS) abrasion tester, 64, 66
Natural rubber, 3, 7
  classification, 124
  compression set, 20, 22
  crystallization, 84
  dynamic modulus, 52, 53
  fatigue, 61
  gas permeability of, 108
  general properties, 8, 9
  gum stock, 16, 104
  hardness, 12
  heat ageing, 93
  resilience, 45
  tearing energy, 60
  tensile strength, 16, 17
  transmissibility, 54, 55
  weather resistance of, 109
Neoprene (polychloroprene), 3, 7
  bridge bearings, 36
  bellows, 134, 135
  compression set, 20, 21
  crystallization, 84, 85
  dynamic modulus, 52
  electrical properties, 75, 76
  fatigue resistance, 61
  gas permeability of, 108
  general properties, 8, 9
  gum, 16
  hardness, 12
  heat ageing, 93, 94
  heat resistant, 93
  resilience, 45
  storage, 121
  stress-strain, 40
  tearing energy, 60
  tensile strength, 16, 17
  transmissibility, 55
  V-belts, 140–144
Nitrile rubber (NBR), *see* rubber
Noise control, 54
NORDEL (EPDM), 7
  compression set, 20
  electrical properties, 75, 76

NORDEL – *continued*
  general properties, 8, 9
  hardness, 12
  tensile strength, 16, 17

O-rings, *see* applications
Ovens, 93
Oxygen (*see also* chemical resistance)
  diffusion, 94
  pressure chamber, 93
Ozone (*see also* chemical resistance)
  exposure apparatus, 114
  tests, 114–117
Ozonised air, 116

Peel tests, 97–99
Pendulum, *see* rebound resilience
Perfluoroelastomer, *see* KALREZ
Permanent set, 17–23, 23, 130
  classification, 123
  compression, 9, 18, 19, 20, 21, 32, 59, 91
  low temperatures, 19, 86, 88, 130
  shear, 17,18
  tension set, 18, 19
  test pieces for, 18, 19
Permeability, 104–108
  gas, 104, 113
  tests, 104–108
    carrier gas, 105
    constant volume, 104
    constant pressure, 104, 105
    relative to gases, 108, 113
Permittivity, *see* electrical properties
Phase angle (*see also* loss angle), 41, 55, 74
Pico abrasion tester, 64, 67
Plasticisers, 13, 76, 85, 88, 93, 119
Polyacrylate, *see* rubber
Polychloroprene, *see* Neoprene
Polynorbornene, *see* rubber
Polysulphide, *see* rubber
Power factor, *see* electrical properties
Pre-conditioning, *see* mechanical conditioning
Primers, 97, 100

Quality (production) control, 5, 63, 94, 103, 117, 119, 120, 127, 139, 143,
QUV apparatus, *see* weather resistance

RAPRA Technology Ltd (ex Rubber and Plastics Research Association), 30
Rate of recovery, *see* low temperature properties
Rebound, 9
  falling weight apparatus, 45
  pendulum, 44, 45
  resilience, 40, 43–45
Relative humidity, 106, 121
Relative volume loss (abrasion), 65

Relaxometer, 28–30
Relaxation, 24
Relaxed modulus, 34
Resilience, 17, 22, 39, 52, 64
    definition, 40
    at low temperatures, 86
    of natural rubber, 45
    of Neoprene, 45
Resonance, 48–50, 129
    dynamic tests, 52
    forced vibration, 43
    transmissibility, 54, 55
Retraction, *see* low temperature properties
Rheovibron™, 48
Rivlin R S, 60
Roelig vibrator, 48
Ross flexer, 58
Rubber (*see also* elastomer)
    acrylate, 32, 133, 135
    butyl (IIR), 8, 55, 84
    chloroprene, *see* Neoprene
    compounding, 3, 5, 122
    crepe, 3
    definition, 3
    ECO, 135
    EPDM, 60, 61
    fluorosilicone (FMQ), 20, 32
    gum, 42, 45
    isoprene, 3, 63
    latex, 3
    methyl, 3
    moulding, 4,
    natural rubber, *see* natural rubber
    nitrile (NBR)
        compression set, 20
        heat and oil resistance, 8,
        seals, 135
        stress relaxation of, 32
        tearing energy, 60, 61
        transmissibility, 55
    norbene, 20
    olefinic TPE, 124
    processing, 4,
    reinforcement, 3
    sheeting, 4,
    silicone, 8, 20, 32
    smoked sheet, 3
    styrene butadiene (SBR), 58
        fatigue, 63
        heat and oil resistance, 8, 124
        permeability, 108
        transmissibility, 55
    sulphide, 8
    synthetic, 3, 17, 22
Rubber machinery
    calenders, 4
    extruders, 4,
    mills, 4,
    mixers, 4
    presses, 4

Safety factors, 127
Schering bridge, 74
Schiefer abrasion tester, 64
Schob pendulum, 44
Schopper rings, 62
Scott flexer, 58
Scragging (*see also* mechanical conditioning), 25, 34
Seals, *see* applications and elastomeric seals
Shape factor, 35, 36, 41, 99
Shawbury-Wallace compression stress relaxometer, 29, 30
Shear tests, 51, 100
Shore,
    durometer, 11
    hardness, 11
    scleroscope, 45
Silica, finely ground, 3
Silicone, *see* rubber
Silicone lubricant, 31
Simulated service tests, 5, 13, 23, 57, 64, 120, 125–144
    automobile coolant hose, 129, 130
    elastomeric seals, 131–135
    flue duct expansion joints, 136–139
    introduction, 127
    V-belts, 140–144
Society of Automotive Engineers (SAE), 122
    SAE J200, 8, 122–124
    SAE J636, 143
    SAE J637, 143, 144
Solubility, 108
Specific gravity, 9
Specific inductive capacity (permittivity), *see* electrical properties
Specifications, purchase, 127, 128
St Joe flexometer, 58
Standards, 6
Standard laboratory tests, 1–124, 127
Standard rubbers for abrasion tests, 65, 67
State Committee for Standards, former Soviet Union (GOST), 6
    GOST 262, 35
    GOST 265, 35
    GOST 12014, 35
Stiffness, 52, 64, 80, 134
    bridge bearings, 36
    definition, 34
    dynamic, 47, 55
    vs hardness, 13
    shape factor, 35
    at low temperatures, 84, 85, 88
Storage, 121
Stored energy density, 60
Strain, 39–41, 51, 71, 113
    creep dependent on, 25
    compression, 27, 35, 91
    energy, 44
    shear, 36, 37
    stress relaxation, 95
    tensile, 115
    threshold, 115, 117
    effect on weathering, 109, 110

Strain relaxation (creep), 22, 23, 50
Strain energy release rate, 60
Stress, 35, 39–41
   shear, 36, 37
   static, 56
   and stress relaxation, 23, 27, 95
Stress relaxation, 22, 23, 27–33, 50, 77, 116
   compression, 27, 31, 32
   fatigue, 56
   shear, 27, 36
   and strain, 95, 96
   tension, 27
Stress-strain, 53, 62, 81, 82, 140
   short-term, 34–38
   tensile, 16, 60, 119
Styrene butadiene rubber (SBR), *see* rubber
Swelling, 33, 108, 118, 122–124

T A Instruments Inc
   983 dynamic mechanical analyser, 48–50
   thermal analyzer, 77
Tan δ, *see* loss angle
Taber abrasion tester, 64, 67
Tack (green), 97
Tear resistance, 9, 64
Tear strength, 35, 37, 38, 61, 122
   test pieces, 37, 38
Tearing energy, 60–63
   in compression, 60
   in shear, 60
Temperature retraction test (TR), 86, 87
Tensile strength, 9, 14, 17, 68, 114
   classification, 123
   definition, 5
   at high temperatures, 93
   at low temperatures, 86
   effect of weathering, 111
Tensile stress-strain, 14–17, 34, 118
Tensile testing machine (universal), 14, 15
   for adhesion, 97–101
   for compression and shear modulus, 35, 36
   for stress relaxation, 31
   for tear strength, 38
Tension set, 86
Thermal analysis (TA), 77
Thermal conditioning, 31
Thermal energy, 84, 85
Thermal properties, 77–79
   coefficients of expansion, 77–80
   conductivity, 77, 78
   diffusivity, 77
   heat transfer coefficients, 77
   specific heat, 77 78
Thermal stability, 64
Thermogravimetric analysis (TGA), 77
Thermomechanical analysis (TMA), 77, 78
Thermoplastic elastomers (TPE), 4, 7, 31
Thomas A G, 60

Torsion
   modulus, 89
   pendulum, 45
   springs, *see* applications
   tests, 88, 89
Toughness, 122
Tracking, *see* electrical properties
Transition temperature, 21, 84, 88
Transmission
   rate of gas, 104, 106, 108
      cell for measurement of, 107
   volatile liquids, 106–108
Transmissibility, 54–55
Trouser test, 37
Tubing, *see* applications
Tyres, 4, 17, 52, 57, 64, 72, 97, 102
   inner tubes, 105

Ultraviolet light (UV), *see* weather resistance
Ultrasound, 43
'Use of Rubber in Engineering' conference, 54

V-belts, 140–144
   automotive, 142–144
   dead-weight tester, 142
   dynamometer tests, 141, 142
   fatigue of, 141
   industrial, 141, 142
   Micro-V, 140
   ribbed, 140
   toothed, 140
Vacuum, 133
   resistance to, 130
VAMAC (AEM), 10
   bellows, 135
   compression set, 20
   electrical properties, 75, 76
   general properties, 8, 9
   hardness, 12
   tensile strength, 16
Vibration, 52, 54, 55, 78, 129
   damping, 39, 52
   electromagnetic, 48
   forced, 40, 43, 46–48
   free, 40, 43, 45, 46, 49, 50
Viscoelasticity, 23, 39, 46, 49, 77
VITON (FKM), 7
   classification, 124
   compression set, 20
   electrical properties, 75, 76
   general properties, 8, 9
   hardness, 12
   expansion joints, 136–139
   seals, 133
   stress relaxation, 32
   tensile strength, 16, 17

Vocabulary of the rubber industry, 145–217
    French index, 218–223
    German index, 224–229
    Italian index, 230–235
Volvo Car Corporation, 129, 130
Volatile liquids, *see* permeability
Volume loss (abrasion), 66
Volume swell, 32
Vulcanisation, 4, 34, 91, 122
    bonding by, 47, 97, 99
    characteristics, 9
    degree of, 21, 38
    liguid curing medium (LCM), 4
    temperature of, 77
    ultra-high frequency (UHF), 4,

Wagner earth, 74
Wallace-MRPRA fatigue tester, 62, 63
Wallace-Shawbury age tester, 95
Water vapour, 106
Wax, 106, 113, 116
Wear, 64
Weather resistance, 5, 9, 109–112
    accelerated tests, 110–112
    effect of,
        artificial light, 109
        moisture, (dew, rain), 109–112
        oxygen, 109
        ozone, 109–112

Weather resistance – *continued*
    effect of - *continued*
        sunlight, 5, 109–112
        temperature, 109
        UV light, 109, 110, 112, 113
    artificial, 111
    assessment of deterioration, 109
        appearance, 110
        blue dyed wool, 110, 111
        chalking, 109, 112
        grey scale, 110, 111
        surface properties, 110
        visual, 109
    Exposure equipment, 110, 111
        carbon arc, 111
        fluorescent tubes, 111
        mercury arc, 111
        QUV apparatus, 101, 112
        Weather-Ometer™, 110
    Florida, exposure in, 110, 112
    natural weathering, 109, 110
Wires, *see* applications
Wykeham Farrance compression stress relaxometer, 30

Yerzley oscillograph, 46
Young's modulus, 12, 14, 39, 86, 88

Zerbini pendulum, 44